CONTEMPORARY MATHEMATICS

703

Higher Genus Curves in Mathematical Physics and Arithmetic Geometry

AMS Special Session
Higher Genus Curves and Fibrations in Mathematical Physics and Arithmetic Geometry
January 8, 2016
Seattle, Washington

Andreas Malmendier
Tony Shaska
Editors

AMERICAN MATHEMATICAL SOCIETY

2010 *Mathematics Subject Classification.* Primary 11G30, 11G50, 11G42, 14J27, 14J28, 14H40, 14H45, 14H52, 14H55.

Library of Congress Cataloging-in-Publication Data

Names: AMS Special Session on Higher Genus Curves and Fibrations in Mathematical Physics and Arithmetic Geometry (2016: Seattle, Wash.) | Malmendier, Andreas, 1976– editor. | Shaska, Tony, 1967– editor.

Title: Higher genus curves in mathematical physics and arithmetic geometry: AMS Special Session on Higher Genus Curves and Fibrations in Mathematical Physics and Arithmetic Geometry, January 8, 2016, Seattle, Washington / Andreas Malmendier, Tony Shaska, editors.

Description: Providence, Rhode Island: American Mathematical Society, [2018] | Series: Contemporary mathematics; volume 703 | Includes bibliographical references.

Identifiers: LCCN 2017042709 | ISBN 9781470428563 (alk. paper)

Subjects: LCSH: Arithmetical algebraic geometry–Congresses. | Mathematical physics–Congresses. | AMS: Number theory – Arithmetic algebraic geometry (Diophantine geometry) – Curves of arbitrary genus or genus $\neq 1$ over global fields. msc | Number theory – Arithmetic algebraic geometry (Diophantine geometry) – Heights. msc | Number theory – Arithmetic algebraic geometry (Diophantine geometry) – Arithmetic mirror symmetry. msc | Algebraic geometry – Surfaces and higher-dimensional varieties – Elliptic surfaces. msc | Algebraic geometry – Surfaces and higher-dimensional varieties – $K3$ surfaces and Enriques surfaces. msc | Algebraic geometry – Curves – Jacobians, Prym varieties. msc | Algebraic geometry – Curves – Special curves and curves of low genus. msc | Algebraic geometry – Curves – Elliptic curves. msc | Algebraic geometry – Curves – Riemann surfaces; Weierstrass points; gap sequences. msc

Classification: LCC QA242.5 .A4827 2018 | DDC 516.3/5–dc23

LC record available at https://lccn.loc.gov/2017042709

DOI: `http://dx.doi.org/10.1090/conm/703`

∞ The paper used in this book is acid-free and falls within the guidelines
established to ensure permanence and durability.
Visit the AMS home page at `http://www.ams.org/`

10 9 8 7 6 5 4 3 2 1 23 22 21 20 19 18

Contents

Preface

Algebraic curves and their fibrations have played a major role in both mathematical physics and arithmetic geometry. The goal of this book is to focus on the role of higher genus curves; in particular, hyperelliptic and superelliptic curves in algebraic geometry and mathematical physics.

These proceedings are based on the Special Session entitled Higher Genus Curves and Fibrations in Mathematical Physics and Arithmetic Geometry that was held at the AMS Joint Mathematics Meetings on January 8, 2016, in Seattle, Washington.

The first three papers of this book are focused on automorphism groups of curves. In the first paper, Russell and Wootton study the action of a cyclic group C_p on a compact, oriented surface S of genus $\sigma \geq 2$. Such action is said to be finitely maximal if there is no finite supergroup of homeomorphisms $G > C_p$. The authors prove that for sufficiently large genus σ, the number of topologically distinct finitely maximal C_p-actions on a surface of genus σ is at least linear in σ.

Broughton studies the quasi-platonic action of the group G on the Riemann surface S. This is a conformal action of G on S such that S/G is a sphere and the projection $\pi_G : S \to S/G$ is branched over $\{0, 1, \infty\}$. The projection π_G is a regular Belyi function and induces a regular dessin d'enfant on S, and so S is defined over a number field. The absolute Galois group $\mathrm{Gal}(\overline{\mathbb{Q}})$ acts on regular dessins, hence quasi-platonic actions, by acting on the coefficients of a defining equation of S. The author reconstructs the Galois action from the branch cycle description of the action and the structure of the group G.

Swirnaski presents an algorithm for computing equations of canonically embedded Riemann surfaces with automorphisms. This is used to produce equations of Riemann surfaces with large automorphism groups for genus 7. The main tools are the Eichler trace formula for the character of the action of the automorphism group on holomorphic differentials, algorithms for producing matrix generators of a representation of a finite group with a specified irreducible character, and Gröbner basis techniques for computing flattening stratifications.

A superelliptic curve $\mathcal{X}$ of genus $g \geq 2$ is not necessarily defined over its field of moduli, but it can be defined over a quadratic extension of it. While a lot of work has been done by many authors to determine which hyperelliptic curves are defined over their field of moduli, less is known for superelliptic curves. Hidalgo and Shaska observe that if the reduced group of a genus $g \geq 2$ superelliptic curve $\mathcal{X}$ is different from the trivial or cyclic group, then $\mathcal{X}$ can be defined over its field of moduli; in the cyclic situation we provide a sufficient condition for this to happen. We also determine those families of superelliptic curves of genus at most 10 which might not be definable over their field of moduli.

Beshaj studies the Weierstrass equations for genus 2 curves defined over a ring of integers $\mathcal{O}_\mathbb{F}$ which correspond to reduced binary sextics. This is done via reduction theory and Julia quadratic of binary sextics. The author shows that when a binary sextic has extra automorphisms, then it is usually easier to compute its Julia quadratic. Moreover, she shows that when the curve is given in the standard form $y^2z^4 = f(x^2, z^2)$ and defined over $\mathcal{O}_\mathbb{F}$, then the binary form f is reduced. Such curves have minimal height among integral models defined by sextics in x^2, z^2, even up to twist.

Continuing the study of genus 2 curves, Beshaj, Hidalgo, Kruk, Malmendier, Quispe, and Shaska describe how to build a database of genus 2 curves defined over $\mathbb{Q}$ which contains all curves with minimal absolute height $\mathfrak{h} \leq 5$, all curves with moduli height $\mathfrak{H} \leq 20$, and all curves with extra automorphisms in standard form $y^2 = f(x^2)$ defined over $\mathbb{Q}$ with height $\mathfrak{h} \leq 101$. Each isomorphism class of genus 2 curves in the database is characterized by its automorphism group and Clebsch and Igusa invariants. Moreover, an equation over its minimal field of definition is provided. The distribution of rational points in the moduli space $\mathcal{M}_2$ for which the field of moduli is a field of definition is discussed and some open problems are presented.

Magyar and Whitcher study strong arithmetic mirror symmetry and toric isogenies. A mirror pair of Calabi-Yau varieties exhibits strong arithmetic mirror symmetry if the number of points on each variety over a finite field is equivalent, modulo the order of that field. The authors search for strong mirror symmetry in pencils of toric hypersurfaces generated using polar dual pairs of reflexive polytopes. They characterize the pencils of elliptic curves where strong arithmetic mirror symmetry arises and provide experimental evidence that the phenomenon generalizes to higher dimensions and that pencils of $K3$ surfaces with the same Picard-Fuchs equation have related point counts.

Kumar and Kuwata describe two constructions of elliptic $K3$ surfaces starting from the Kummer surface of the Jacobian of a genus 2 curve. These parallel the base-change constructions of Kuwata for the Kummer surface of a product of two elliptic curves. One of these also involves the analogue of an Inose fibration. The authors use these methods to provide explicit examples of elliptic $K3$ surfaces over the rationals of geometric Mordell-Weil rank 15.

Shor considers the problem of calculating the higher-order Weierstrass weight of the branch points of a superelliptic curve C. For any $q > 1$, he gives an exact formula for the q-weight of an affine branch point and also finds a formula for the q-weight of a point at infinity in the case where n and d are relatively prime.

Previato studies Poncelet's porism and projective fibrations. Poncelet's porism theorem is used to produce a natural compactification of several moduli spaces. The monodromy of the polygons, viewed as torsion points on a fibration by elliptic curves, can be tested computationally for an action of the full symmetric group. Analogous constructions can be implemented for hyperelliptic fibrations corresponding to higher-dimensional versions of Poncelet's porism.

Levin formulates and proves a general version of Runge's method, suited to combination with other methods for integral points on varieties. He gives some applications of the main theorem, including results for integral points on curves which recover as a special case some known results on rational points on elliptic curves with prime power denominators.

Joyner and Shaska study connections between self-inversive and self-reciprocal polynomials, reduction theory of binary forms, and minimal models of curves. The authors prove that if $\mathcal{X}$ is a superelliptic curve defined over $\mathbb{C}$ and its reduced automorphism group is nontrivial or not isomorphic to a cyclic group, then we can write its equation as $y^n = f(x)$ or $y^n = xf(x)$, where $f(x)$ is a self-inversive or self-reciprocal polynomial.

Deopurkar and Patel describe a sequence of effective divisors on the Hurwitz space $H_{d,g}$ for d dividing $g-1$ and compute their cycle classes on a partial compactification. These divisors arise from vector bundles of syzygies canonically associated to a branched cover. They find that the cycle classes are all proportional to each other. These computations are motivated by the question of determining the effective cone and ultimately the birational type of $H_{d,g}$.

There is considerable activity in the area of algebraic curves of higher genus due to their importance in pure mathematics and applications. We hope that this volume will help further our understanding of algebraic curves and their connections to other areas of mathematics.

Both editors want to thank all authors for their contributions to this volume. We would especially like to thank the referees for their tireless work that they put toward this volume.

Andreas Malmendier

Tony Shasksa

Contemporary Mathematics
Volume **703**, 2018
http://dx.doi.org/10.1090/conm/703/14127

A lower bound for the number of finitely maximal C_p-actions on a compact oriented surface

Jacob Russell and Aaron Wootton

ABSTRACT. Let C_p denote a cyclic group of prime order $p \geq 7$. A topological action of C_p on a compact, oriented surface S of genus $\sigma \geq 2$ is said to be finitely maximal if there is no finite supergroup of homeomorphisms $G > C_p$. In the following, for sufficiently large genus σ, when non-zero, we show that the number of topologically distinct finitely maximal C_p-actions on a surface of genus σ is at least linear in σ.

1. Introduction

A finite subgroup H of the full group of orientation preserving homeomorphisms $\text{Homeo}^+(S)$ of a compact oriented surface S of genus $\sigma \geq 2$ is said to be *finitely maximal* if there does not exist a finite $G < \text{Homeo}^+(S)$ with $G > H$. Two subgroups $H_1, H_2 \leq \text{Homeo}^+(S)$ are said to define topologically equivalent actions if they are conjugate in $\text{Homeo}^+(S)$. For a given finite group G and genus $\sigma \geq 2$, let $N_{G,\sigma}$ denote the number of distinct topological G-actions on a surface of genus σ which are finitely maximal. In the following, we provide a lower bound for the number $N_{C_p,\sigma}$ for C_p a cyclic group of prime order $p \geq 7$. Specifically, for sufficiently large genus σ, when non-zero, we show that this number is at least linear in σ.

Though interesting in its own right, motivation for this work comes from a number of different places. For example, finitely maximal C_p-actions are in one-to-one correspondence with the isolated strata in the branch locus, $\mathcal{B}_\sigma$ of the moduli space $\mathcal{M}_\sigma$ of compact Riemann surfaces of genus σ. Providing bounds for $N_{C_p,\sigma}$ in turn provides bounds for the number of distinct isolated strata in $\mathcal{B}_\sigma$, and hence provides a bound for the total number of disconnected components in the branch locus. For further reading on the branch locus of moduli space, see also [**1**], [**2**], [**3**], [**4**], [**10**], [**11**], [**18**].

This work also has implications for counting conjugacy classes of finite subgroups of the mapping class group. Specifically, if $\mathfrak{M}_\sigma$ denotes the mapping class group in genus σ, then there is a natural one-to-one correspondence between conjugacy classes of finite subgroups of $\mathfrak{M}_\sigma$ and equivalence classes of finite topological group actions on a smooth oriented surface of genus σ. Moreover, if $H < G$ both act on a surface of genus σ, then we have the corresponding containment in $\mathfrak{M}_\sigma$. As such, our results provide a lower bound for the number of distinct conjugacy

2010 *Mathematics Subject Classification.* Primary 14H37, 14H15; Secondary 30F20, 32G15.

classes in $\mathfrak{M}_\sigma$ of subgroups isomorphic to C_p that are finitely maximal in $\mathfrak{M}_\sigma$. See [**6**], [**14**], and [**26**] for work in this area.

Classification and enumeration of group actions on compact oriented surfaces has spanned the literature for over a century dating back to Hurwitz, see for example [**15**], [**16**] and [**17**], and is still of current significant interest today. Due to the simplicity of their structure, cyclic group actions have been extensively studied. In [**13**], necessary and sufficient conditions for the existence of the action of a cyclic group are given. The problem of enumeration of classes of cyclic group actions was considered in [**20**] using the theory of generating functions, with an explicit generating function provided in the special case of a cyclic group of prime order. Other work, also using generating functions, to count classes of cyclic group actions appear in [**14**], and more recently, [**22**]. Similar results exist for other classes of groups. For example, conditions for the existence of an Abelian group action are considered in [**21**], and enumeration of such actions in [**22**] using generating functions, and [**6**] using more direct methods. There have also been significant contributions in the classification and enumeration of "large" automorphism groups and quasiplatonic groups. For example, in [**25**], it is shown that the number of classes of quasiplatonic Riemann surfaces of genus at most σ has growth type $\sigma^{\log(\sigma)}$. Using computers, other more direct methods have found explicit counts of classes of large automorphism groups for small (≤ 300) genus, see [**9**] for a recent survey of results in this area.

Our approach to the problem of enumeration is direct. A starting point for our work is the paper [**24**] in which it is shown that for sufficiently large genus, $N_{C_p,\sigma} = 0$ if and only if $\sigma \equiv (p-3)/2 \mod (p-1)/2$. The method used in [**24**] to prove this result is explicit – when $\sigma \equiv (p-3)/2 \mod (p-1)/2$, it is shown that any such action always has to extend (and in fact extend to the cyclic group of order $2p$) and outside of this sequence, to show $N_{C_p,\sigma} \neq 0$, an explicit action is constructed which cannot possibly extend to a larger finite group. Our general approach is to adapt and extend this method to construct additional finitely maximal actions, the number of which depend upon the number of fixed points of C_p. We shall then use this to construct a linear (in σ) lower bound.

Our work is outlined as follows. In Section 2, we provide all the necessary terminology and background results. Our approach to the problem is fairly standard, using the theory of Fuchsian groups and generating vectors. In Section 3, we develop the main results required to prove the result. The proof we offer is direct, providing explicit descriptions of these actions via generating vectors. We finish in Section 4 by proving the main result – that outside of an infinite sequence of genera, the number $N_{C_p,\sigma}$ is at least linear in σ.

2. Preliminaries

We approach the study of topological group actions via the theory of surface kernel epimorphisms and generating vectors as introduced in [**12**]. Since we are only considering actions of cyclic groups of prime order p, we simplify the notation, terminology and preliminary results to this case. For a more general approach see, for example [**5**].

A surface S of genus $\sigma \geq 2$ is topologically equivalent to a quotient of the upper half plane $\mathbb{H}/\Lambda$ where Λ is any torsion free Fuchsian group isomorphic to the fundamental group of S, also called a *surface group* for S. A cyclic group C_p

of prime order p acts on S if and only if $C_p = \Gamma/\Lambda$ for some Fuchsian group Γ containing such a Λ as a normal subgroup of index p. We call the map $\rho\colon \Gamma \to C_p$ a *surface kernel epimorphism.*

A presentation of Γ is completely determined by the genus h of the quotient surface S/C_p and the number r of fixed points of C_p on S and is given by

$$\Gamma = \left\langle a_1, b_1, \ldots, a_h, b_h, c_1, \ldots, c_r \mid c_i^p, \prod_{j=1}^{h}[a_j, b_j] \prod_{i=1}^{r} c_i \right\rangle \tag{1}$$

where

$$\sigma = 1 + p(h-1) + r\left(\frac{p-1}{2}\right).$$

Note that the map ρ is completely determined by the images of the generators of Γ so a convenient way of representing a surface kernel epimorphism is through a so-called generating vector, defined as follows:

DEFINITION 2.1. A vector of group elements $(\alpha_1, \beta_1, \ldots, \alpha_h, \beta_h, \eta_1, \ldots, \eta_r)$ belonging to C_p is called a *(h,r)-generating vector* for C_p with genus σ if all of the following hold:

(1) $C_p = \left\langle \alpha_1, \beta_1, \ldots, \alpha_h, \beta_h, \eta_1, \ldots, \eta_r \right\rangle$.
(2) $\prod_{j=1}^{r} \eta_j = 1$.
(3) Each η_i is non-trivial.
(4) The Riemann-Hurwitz formula holds:

$$\sigma = 1 + p(h-1) + r\left(\frac{p-1}{2}\right)$$

For conciseness, in a generating vector for C_p, we adopt the notation $(\alpha)^k$ to mean k copies of α and α^k to mean a single α raised to the kth power. Since it will be important later, we call the vector of group elements $(\eta_1, \ldots, \eta_r)$ containing the last r elements of a generating vector the the *tail* of the generating vector.

A topological group action gives rise to a generating vector via the corresponding surface kernel epimorphism. Likewise, a generating vector gives rise to a topological group action by defining a surface kernel epimorphism. Therefore, we shall often state that a generating vector defines a topological group action of C_p, and by this, we mean the group action determined by the corresponding surface kernel epimorphism.

Distinguishing between topological equivalence classes of group actions was first considered in [**23**] (see also [**12**] for cyclic prime group actions, [**8**, Theorem 7] for all cyclic groups and [**20**] for all groups.) When applied to the special case of a cyclic group of prime order p, this classification implies when $r = 0$, all generating vectors define topologically equivalent actions, and when $r > 0$, we have the following criteria to distinguish between topologically distinct actions in terms of their generating vectors:

THEOREM 2.2. *Fix a prime p. For $r > 0$ two (h,r)-generating vectors*

$$(\alpha_1, \beta_1, \ldots, \alpha_h, \beta_h, \eta_1, \ldots, \eta_r) \text{ and } (\alpha_1', \beta_1', \ldots, \alpha_h', \beta_h', \eta_1', \ldots, \eta_r')$$

for C_p define topologically equivalent group actions if and only if there exists a permutation $\chi \in S_r$ and $\tau \in \mathrm{Aut}(C_p)$ such that

$$(\tau(\eta_{\chi(1)}), \ldots, \tau(\eta_{\chi(r)})) = (\eta_1', \ldots, \eta_r')$$

i.e. *the tails differ by permutation and/or automorphism of C_p.*

For brevity, given the tail $\mathcal{T} = (\eta_1, \ldots, \eta_r)$ of a generating vector, a permutation $\chi \in S_r$ and an automorphism $\tau \in \mathrm{Aut}(C_p)$, we let $\tau(\mathcal{T}(\chi))$ denote the composition $(\tau(\eta_{\chi(1)}), \ldots, \tau(\eta_{\chi(r)}))$.

Since our primary goal is to determine when a given C_p-action is finitely maximal we adopt this term for generating vectors themselves. That is, when we say a generating vector is (or is not) finitely maximal, it is understood that the corresponding topological group action is (or is not) finitely maximal.

In [**24**], it is shown the genus of the quotient surface S/C_p for a finitely maximal C_p-action satisfies $h < (p-3)/2$. The following result, also from [**24**], proved using the techniques employed in [**7**], provides necessary and sufficient conditions for when a given generating vector defines a finitely maximal C_p-action in terms of its tail:

THEOREM 2.3. *Let $\mathcal{V} = (\eta_1, \ldots \eta_r)$ be the tail of a (h, r)-generating vector for C_p where $h < (p-3)/2$. If $\mathcal{V}$ defines an action which is not finitely maximal, then C_p is a subgroup of either $C_p \times C_p$, C_{pq} or $C_p \rtimes C_q$ for some prime q. Moreover:*

1. *$C_p < C_{pq}$ if and only if there exist integers $g, k, m \geq 0$ such that $h = gq + (k+m-2)\left(\frac{q-1}{2}\right)$ and after a reordering of the η_i's, $\mathcal{V}$ has the form $((c_1)^q, \ldots, (c_n)^q, f_1, \ldots, f_k)$ where $r = qn + k$.*
2. *$C_p < C_p \rtimes C_q$ if and only if there exist integers $g, m \geq 0$ such that $h = gq + (m-2)\left(\frac{q-1}{2}\right)$ and after a reordering of the η_i's, there exists an integer α with $\alpha^q \equiv 1 \mod (p)$ so $\mathcal{V}$ has the form $\left(c_1, c_1^{\alpha}, \ldots, c_1^{\alpha^{q-1}}, c_2, c_2^{\alpha}, \ldots c_n^{\alpha^{q-1}}\right)$ where $r = qn$.*
3. *$C_p < C_p \times C_p$ if and only if there exist integers $g, m \geq 0$ such that $h = gp + (m-2)\left(\frac{p-1}{2}\right)$ and after a reordering of the η_i's, $\mathcal{V}$ has the form $((c_1)^p, \ldots (c_n)^p))$ where $r = np$.*

If none of these conditions is satisfied then $\mathcal{V}$ defines a finitely maximal C_p-action

The following Corollary is immediate.

COROLLARY 2.4. *Let $\mathcal{V} = (\eta_1, \ldots \eta_r)$ be the tail of a (h, r)-generating vector for C_p and suppose $\eta \in C_p$ appears exactly $n > 1$ times in $\mathcal{V}$ i.e. there are exactly n η_i's equal to η. If no other element of C_p appears exactly n times, then C_p does not extend to $C_p \rtimes C_q$ or $C_p \times C_p$.*

Note that if $\mathcal{V}$ is the tail of a generating vector for $C_p = \langle x \rangle$, then by Theorem 2.2, it is equivalent to a tail of the form $((x)^{\alpha_1}, (x^2)^{\alpha_2}, \ldots, (x^{p-1})^{\alpha_{p-1}})$ i.e. we simply permute all like powers to be consecutive in the generating vector. With this in mind, we have the following useful consequence of Theorem 2.3

COROLLARY 2.5. *Let $\mathcal{V} = ((x)^{\alpha_1}, (x^2)^{\alpha_2}, \ldots, (x^{p-1})^{\alpha_{p-1}})$, be the tail of a (h, r)-generating vector for C_p which extends to C_{pq} for some prime q. Then for k defined in Theorem 2.3, we have $k \geq r_1 + r_2 + \cdots + r_{p-1}$ where r_i is the remainder of α_i after division by q.*

PROOF. Let $\alpha_i = q\beta_i + r_i$ where r_i is the remainder of α_i after division by q. By Theorem 2.3, the tail of a C_p-generating vector which extends to the action of C_{pq} will be equivalent to a tail of the form $((c_1)^q, \ldots, (c_n)^q, f_1, \ldots, f_k)$. We rewrite the vector $\mathcal{V} = ((x)^{\alpha_1}, (x^2)^{\alpha_2}, \ldots, (x^{p-1})^{\alpha_{p-1}})$ to be of this form. First, since there are $\beta_1 q + r_1$ total copies of x and $r_1 < q$, then there are at most β_1 repetitions of q-copies at the start of the tail. This leaves a minimum of r_1 copies of x which must appear as individual terms at the end of the tail. Using a similar argument with x^i for each i we see that there are minimally r_i copies of x^i that must appear individually at the end of tail. Thus $k \geq r_1 + \cdots + r_{p-1}$. □

3. Bounding Actions by the Length of the Tail

In order to determine a lower bound for the number of C_p actions, we shall first describe a method to create tails of generating vectors that do not satisfy any of the conditions of Theorem 2.3 necessary for extension to a larger group. We emphasis that our work in this section only considers tails of generating vectors, so in particular, the bound we develop will not be in terms of the genus of the corresponding surface on which it acts, but rather the length of the tail r. We shall consider how this relates to the genus in the next section.

Henceforth, let x denote a generator of C_p.

LEMMA 3.1. *Let* $\mathcal{T} = (x, x^2, \ldots, x^{p-1}, (x)^i, (x^a)^j, (x^b)^l)$ *and* $r = i+j+l+(p-1)$ *for positive integers* i, j, l, a, b *satisfying:*

(1) $l \in \{4, 6\}$ *if* r *is even, and* $l \in \{1, 3\}$ *if* r *is odd*
(2) $ja \not\equiv -i \mod (p)$
(3) $lb \equiv -(i + ja) \mod (p)$
(4) $1 < a \leq p - 1$
(5) i *is even*
(6) $i > l$
(7) $j > i + l + 1$

Then provided $h < (p-3)/2$, $\mathcal{T}$ *defines the tail of a finitely maximal* (h, r)*-generating vector for* C_p.

PROOF. First we check $\mathcal{T}$ defines the tail of a generating vector. In order to do this, we need to show it satisfies the second condition of Definition 2.1 – that is, we need to show the product of the elements in $\mathcal{T}$ is the identity. Now we know $1 + 2 + \cdots + (p-1) \equiv 0 \mod (p)$, so the product of the first $p-1$ elements is the identity. Therefore, we need to show the product of the remaining elements is also the identity, or equivalently that $i + aj + bl \equiv 0 \mod (p)$. This is ensured by the assumption $lb \equiv -(i + ja) \mod (p)$ provided we can solve for b. However, we are also assuming $ja \not\equiv -i \mod (p)$ and thus there is always a solution for b. Therefore, $\mathcal{T}$ defines the tail of a generating vector for C_p.

Next we check maximality. Since $j > i + l + 1$ and $a \neq 1$, x^a appears at least $j+1$ times (possibly more depending upon the value of b). No other element of C_p appears in $\mathcal{T}$ this many times and therefore containment in $C_p \times C_p$ or $C_p \rtimes C_q$ is not possible by Corollary 2.4. This leaves containment in C_{pq} as the only possibility.

Suppose that $\mathcal{T}$ does extend to C_{pq} for some q. Then there exists k and m such that $\mathcal{T}$ can be written in the form given in Theorem 2.3. Since x, x^a and x^b are the only elements to appear more than one time, by Corollary 2.5, we must have $k \geq (p-4)$. This means when $q \geq 5$ we have

$$h = gq + (k + m - 2)\left(\frac{q-1}{2}\right) \geq (p-6)\left(\frac{q-1}{2}\right) > \frac{p-5}{2}$$

contradicting our assumption that $h < (p-3)/2$. Therefore, we only need consider $q = 3$ and $q = 2$.

For $q = 3$, if $b = 1$ or $b = a$, then there are at most two distinct elements which appear more than once in the tail, and therefore by Corollary 2.5, we must have

$k \geq (p-3)$. This means we have

$$h = 3g + (k+m-2)\left(\frac{3-1}{2}\right) \geq (p-5) > \frac{p-5}{2}$$

contradicting our assumption that $h < (p-3)/2$. If b is distinct from 1 and a, then x^b appears precisely $l+1$ times. For each given value of l, $l+1$ is never a multiple of 3. Therefore, since there are $p-4$ elements which each appear precisely once, and x^b appears $l+1$ times which has a positive remainder after division by 3, we must again have $k \geq p-3$ and we obtain the same contradiction.

For $q=2$, a similar argument holds. In this case, we first observe that since i is even, for each given choice of l, since $r = i+j+l+(p-1)$, j is also even. It follows that if 1, a and b are all distinct, then x appears $i+1$ times and x^a appears $j+1$ times. In particular, since $i+1$ and $j+1$ are odd, and there are at least $p-4$ elements which appear exactly once, by Corollary 2.5, we must have $k \geq (p-2)$. This means we have

$$h = 2g + (k+m-2)\left(\frac{2-1}{2}\right) \geq \frac{(p-4)}{2} > \frac{p-5}{2}$$

again contradicting our assumption that $h < (p-3)/2$. If $b = a$ or $b = 1$, similar reasoning also implies $k \geq p-2$ and the result follows. □

Next we shall provide conditions for when the tails defined in Lemma 3.1 define distinct C_p-actions on a given surface for a fixed h, the genus of the quotient surface, and r the number of fixed points of the action. In order to do this, we first need the following simple Lemma which provides bounds on j and i.

LEMMA 3.2. *For any tail $\mathcal{T}$ satisfying Lemma 3.1, we have:*

(1) $j > \dfrac{r-p}{2} + 1$
(2) $i < \dfrac{r-p}{2} - l$

PROOF. We know $r = i+j+l+(p-1)$, so we have $j = r-(i+l+(p-1))$. Therefore, the last condition of Lemma 3.1 implies

$$j = r-(i+l+(p-1)) = r-(i+l+1+(p-2)) > r-(j+(p-2)).$$

Therefore we get

$$2j > r-(p+2) \text{ and so } j > \frac{r-(p-2)}{2} = \frac{r-p}{2} + 1.$$

The inequality for i then follows from the last condition of Lemma 3.1. □

LEMMA 3.3. *For the tail defined in Lemma 3.1, for a given l and r, any choice of the ordered pair (a,i) completely determines the topological equivalence class of a C_p-action. Moreover, if two pairs (a_1,i_1) and (a_2,i_2) define the same action, then either $(a_1,i_1)=(a_2,i_2)$ or $|i_1-i_2| = l$.*

PROOF. For a given l and r, given a pair (a,i), the remaining integers are found using the conditions of Lemma 3.1. Specifically, $j = r-i-l-(p-1)$ and b is the unique solution to $lb \equiv -(i+ja) \mod (p)$. Therefore, the generating vector and hence the topological equivalence class of a C_p-action is determined by the pair (a,i).

Next we show that when $(a_1, i_1) \neq (a_2, i_2)$, the tails

$$\mathcal{T}_1 = \left(x, x^2, \ldots, x^{p-1}, (x)^{i_1}, (x^{a_1})^{j_1}, (x^{b_1})^l\right)$$

and

$$\mathcal{T}_2 = \left(x, x^2, \ldots, x^{p-1}, (x)^{i_2}, (x^{a_2})^{j_2}, (x^{b_2})^l\right)$$

define topologically distinct actions with the possible exception of when $|i_1 - i_2| = l$. By Theorem 2.2, in order to do this, we need to show that there is no automorphism τ of C_p and permutation χ of r such that $\tau(\mathcal{T}_1(\chi)) = \mathcal{T}_2$, so we assume there is.

First note that any automorphism of the first $p-1$ elements of $\mathcal{T}_1$ is simply a permutation of those elements, and therefore, after an appropriate permutation can be put back in the same order. Secondly, observe that the number of elements repeated more than once and the occurrences of those repeated elements will be the same after application of an automorphism of C_p and a permutation. Therefore, since we are assuming $\tau(\mathcal{T}_1(\chi)) = \mathcal{T}_2$, we can actually assume

$$(x, x^2, \ldots, x^{p-1}, (\tau(x))^{i_1}, (\tau(x)^{a_1})^{j_1}, (\tau(x)^{b_1})^l) = (x, x^2, \ldots, x^{p-1}, (x)^{i_2}, (x^{a_2})^{j_2}, (x^{b_2})^l).$$

Now, if x, x^{a_1} and x^{b_1} are all distinct, then so are $\tau(x)$, $\tau(x)^{a_1}$ and $\tau(x)^{b_1}$. Therefore, since $l < i_1 < j_1$ and $l < i_2 < j_2$, it follows that $\tau(x) = x$, $i_1 = i_2$, $j_1 = j_2$, $a_1 = a_2$ and $b_1 = b_2$. In particular, $(a_1, i_1) = (a_2, i_2)$, a contradiction.

If x, x^{a_1} and x^{b_1} are not distinct, then neither are x, x^{a_2} and x^{b_2}, and we either have $b_1 = a_1$ or $b_1 = 1$ and similarly $b_2 = a_2$ or $b_2 = 1$. This means we have one of the four following possibilities:

(1) $(x, x^2, \ldots, x^{p-1}, (\tau(x))^{i_1+l}, (\tau(x)^{a_1})^{j_1}) = (x, x^2, \ldots, x^{p-1}, (x)^{i_2+l}, (x^{a_2})^{j_2})$
(2) $(x, x^2, \ldots, x^{p-1}, (\tau(x))^{i_1+l}, (\tau(x)^{a_1})^{j_1}) = (x, x^2, \ldots, x^{p-1}, (x)^{i_2}, (x^{a_2})^{j_2+l})$
(3) $(x, x^2, \ldots, x^{p-1}, (\tau(x))^{i_1}, (\tau(x)^{a_1})^{j_1+l}) = (x, x^2, \ldots, x^{p-1}, (x)^{i_2+l}, (x^{a_2})^{j_2})$
(4) $(x, x^2, \ldots, x^{p-1}, (\tau(x))^{i_1}, (\tau(x)^{a_1})^{j_1+l}) = (x, x^2, \ldots, x^{p-1}, (x)^{i_2}, (x^{a_2})^{j_2+l})$

By Lemma 3.2,

$$i_1 < i_1 + l < (r-p)/2 < (r-p)/2 + 1 < j_2 < j_2 + l,$$

so in particular, it follows that

(1) $i_1 + l < j_2$
(2) $i_1 + l < j_2 + l$
(3) $i_1 < j_2$
(4) $i_1 < j_2 + l$.

Therefore, for all four cases, we must have $\tau(x) = x$ and $i_1 + l = i_2 + l$ in case (1), $i_1 + l = i_2$ in case (2), $i_1 = i_2 + l$ in case (3) and $i_1 = i_2$ in case (4). In particular, $a_1 = a_2$ in all cases, and in cases (1) and (4) we have $i_1 = i_2$, a contradiction, and in cases (2) and (3) we have $|i_1 - i_2| = l$. □

Though ideally we would like the topological equivalence class of an action to be uniquely determined by a pair (a, i), we note that, as Lemma 3.2 indicates, this is not always the case. For example, if l is even, there always exists equivalent actions given by different pairs provided there exists a tail in which $b = a$ or $b = 1$. Specifically, if $b = 1$, then

$$(x, x^2, \ldots, x^{p-1}, (x)^i, (x^a)^j, (x)^l)$$

with pair (a, i) is equivalent to

$$(x, x^2, \ldots, x^{p-1}, (x)^{i+l}, (x^a)^{j-l}, (x^a)^l)$$

with pair $(a, i+l)$. On the other hand, when $b = a$,

$$(x, x^2, \ldots, x^{p-1}, (x)^i, (x^a)^j, (x^a)^l)$$

with pair (a, i) is equivalent to

$$(x, x^2, \ldots, x^{p-1}, (x)^{i-l}, (x^a)^{j+l}, (x)^l)$$

with pair $(a, i-l)$.

We are now ready to provide a lower bound for the number of distinct actions using the tails defined in Lemma 3.1.

THEOREM 3.4. *When non-zero, the number $N_{C_p.\sigma}$ of the topologically distinct finitely maximal C_p-actions with tail of length $r > p+20$ on a surface S of genus $\sigma \geq 2$ satisfies*

$$N_{C_p,\sigma} \geq \frac{(p-3)(r-p-20)}{8}.$$

PROOF. In order to find a lower bound, we shall filter the set of tails of the form given in Lemma 3.1 to a subset where each define a topologically distinct action. In order to do this, we must first determine which tails of this form satisfy the seven conditions of Lemma 3.1.

Suppose $\mathcal{T} = (x, x^2, \ldots, x^{p-1}, (x)^i, (x^a)^j, (x^b)^l)$ for positive integers i, j, l, a, b. Since we are assuming $r > p+20$, all conditions of Lemma 3.1 can be imposed on $\mathcal{T}$ except possibly conditions (2) and (3). Also note that satisfaction of condition (2) guarantees that there is a choice of b that satisfies condition (3). Therefore, we need to filter our set of tails to a subset for which condition (2) is guaranteed to hold.

If condition (2) fails, then $i + aj \equiv 0 \mod (p)$. Since $\gcd(a, p) = 1$, it follows that either p divides both i and j, or p divides neither i nor j. In the latter case, we get $a \equiv j^{-1}(-i) \mod (p)$. In particular, a is completely determined by i and j, so for a fixed i and j, there is only one possibility for a. Therefore, for each i and j, we can simply exclude this value of a to ensure condition (2) holds.

If p divides both i and j then condition (2) never holds, so we shall impose conditions on the tail to avoid this happening. Now, if p divides both i and j, then it must divide $i + j$, so it suffices to provide conditions ensuring that p does not divide $i + j$. For a fixed r and l, since $r = (p-1) + i + j + l$, we have $i + j = r - (p-1) - l$ and so $i + j \equiv r - l + 1 \mod (p)$. Therefore, if r is even, we choose $l = 4$ when $r \not\equiv 3 \mod (p)$ and $l = 6$ else, and when r is odd, we choose $l = 1$ when $r \not\equiv 0 \mod (p)$ and $l = 3$ else. Through these choices of l, we never have $i+j \equiv 0 \mod (p)$ and so condition (2) will always be satisfied (except for the choice of a previously excluded).

Next we need to filter this set of tails further so that each one defines a unique topological equivalence class. By Lemma 3.3, we know that if two tails with pairs (a_1, i_1) and (a_2, i_2) define the same action, then either they are the same pair, or $|i_1 - i_2| = l$. Therefore, in order to ensure every tail defines a distinct action, we simply restrict the values of i to a maximal subset of integers so that the difference of two members is never l. Specifically, for l odd, since we have already restricted i to the even numbers, there are no values for which $|i_1 - i_2| = l$, so we use the same set. For l even, we use the set $\alpha l + \beta$ where α is odd and β runs over the even integers $\{0, 2, \ldots, l-2\}$. Note that in each case, this will always be at least half of the possible values of i.

We are now ready to count, and for this, we need to count the maximum number of pairs (a, i) with the restrictions we have imposed. Since $a \neq 1$ and for a given i and j, $a \not\equiv j^{-1}(-i) \mod (p)$, there are precisely $p-3$ possible choices for a. Now we are assuming $i > l$ and by Lemma 3.2 we have $i < (r-p)/2 - 2$. We are also assuming i is even so $i = 2t$ for some t. Since $l \leq 6$, this means $8 \leq 2t < (r-p)/2 - 2$, and so $4 \leq t < (r-p)/4 - 1$. Therefore, there are at most

$$\frac{r-p}{4} - 1 - 4 = \frac{r-p-20}{4}$$

choices for t, and hence for i. Filtering this set further to ensure that no two values differ by l leaves at least half of these. Thus the total number of possibilities for i is at least $\frac{r-p-20}{8}$, and therefore, the total number of actions is at least

$$\frac{(p-3)(r-p-20)}{8}.$$

□

4. A Lower Bound for $N_{C_p,\sigma}$

When non-zero, Theorem 3.4 provides a lower bound for $N_{C_p,\sigma}$ in terms of r, the length of the tail of a generating vector. We shall now use this result to show that the number $N_{C_p,\sigma}$, when non-zero, is always bounded below by a linear function in σ for sufficiently high σ.

THEOREM 4.1. *If $\sigma \equiv (p-3)/2 \mod (p-1)/2$, then $N_{C_p,\sigma} = 0$. Else, for $\sigma > p^2 + 7p - 9$, the number $N_{C_p,\sigma}$ of finitely maximal C_p actions on a surface of genus σ is bounded below by a linear function of σ.*

PROOF. By [**24**, Corollary 1], if $\sigma \equiv (p-3)/2 \mod (p-1)/2$, then $N_{C_p,\sigma} = 0$. Therefore, we shall henceforth assume $\sigma \not\equiv (p-3)/2 \mod (p-1)/2$.

Now, if C_p acts on a surface S with h the genus of the quotient surface S/C_p and r the number of fixed points then the Riemann-Hurwitz formula holds:

$$\sigma = 1 + p(h-1) + r\left(\frac{p-1}{2}\right).$$

In order to apply Theorem 3.4, we need $r > p + 20$. Therefore, since for a finitely maximal action we know $0 \leq h \leq \frac{p-5}{2}$ and since the Riemann-Hurwitz formula is an increasing function in both r and h, if we restrict to

$$\sigma > 1 + p\left(\frac{p-3}{2} - 1\right) + (p+20)\left(\frac{p-1}{2}\right) = p^2 + 7p - 9,$$

then the condition $r > p + 20$ is guaranteed to hold. Therefore, we henceforth assume σ satisfies this bound.

Now, since $\sigma > p^2 + 7p - 9 > \frac{1}{2}((p(p-4)+1)$, there always exists a C_p-action on a surface S of genus σ, see [**19**, Corollary 5.4]. In addition, by [**24**, Theorem 4], provided $\sigma \not\equiv (p-3)/2 \mod (p-1)/2$, then there always exists a C_p-action with S/C_p of genus h for some unique $h < (p-3)/2$ and r, the number of fixed points, which can be found using the Riemann-Hurwitz formula:

$$r = \frac{2(\sigma - 1 - p(h-1))}{p-1}.$$

Given the assumptions made on σ, we know $r > p + 20$ and so we can use the bound developed in Theorem 3.4 to bound $N_{C_p,\sigma}$ in terms of the length of the tail r. Specifically, we have:

$$N_{C_p,\sigma} \geq \frac{(p-3)(r-p-20)}{8} = \frac{(p-3)(\frac{2(\sigma-1-p(h-1))}{p-1} - p - 20)}{8}$$
$$= \frac{p-3}{4(p-1)}\sigma - \frac{((p-3)(p(2h+p+17)-18))}{8(p-1)}$$

Since $0 \leq h \leq \frac{p-5}{2}$, this gives

$$N_{C_p,\sigma} \geq \frac{p-3}{4(p-1)}\sigma - \frac{((p-3)(p(2h+p+17)-18))}{8(p-1)}$$
$$\geq \frac{p-3}{4(p-1)}\sigma - \frac{(p-3)(p^2+6p-9)}{4(p-1)} = A\sigma + B$$

for constants A and B dependent only on p. The result follows. □

References

[1] G. Bartolini, A. F. Costa, and M. Izquierdo, *On the connectivity of branch loci of moduli spaces*, Ann. Acad. Sci. Fenn. Math. **38** (2013), no. 1, 245–258, DOI 10.5186/aasfm.2013.3820. MR3076808

[2] G. Bartolini, A. F. Costa, and M. Izquierdo, *On isolated strata of p-gonal Riemann surfaces in the branch locus of moduli spaces*, Albanian J. Math. **6** (2012), no. 1, 11–19. MR2965666

[3] G. Bartolini and M. Izquierdo, *On the connectedness of the branch locus of the moduli space of Riemann surfaces of low genus*, Proc. Amer. Math. Soc. **140** (2012), no. 1, 35–45, DOI 10.1090/S0002-9939-2011-10881-5. MR2833515

[4] S. A. Broughton, *The equisymmetric stratification of the moduli space and the Krull dimension of mapping class groups*, Topology Appl. **37** (1990), no. 2, 101–113, DOI 10.1016/0166-8641(90)90055-7. MR1080344

[5] S. A. Broughton, *Classifying finite group actions on surfaces of low genus*, J. Pure Appl. Algebra **69** (1991), no. 3, 233–270, DOI 10.1016/0022-4049(91)90021-S. MR1090743

[6] S. A. Broughton and A. Wootton, *Finite abelian subgroups of the mapping class group*, Algebr. Geom. Topol. **7** (2007), 1651–1697, DOI 10.2140/agt.2007.7.1651. MR2366175

[7] E. Bujalance, F. J. Cirre, and M. Conder, *On extendability of group actions on compact Riemann surfaces*, Trans. Amer. Math. Soc. **355** (2003), no. 4, 1537–1557, DOI 10.1090/S0002-9947-02-03184-7. MR1946404

[8] M. Carvacho B., *Nonequivalent families of group actions on Riemann surfaces*, J. Pure Appl. Algebra **217** (2013), no. 12, 2345–2355, DOI 10.1016/j.jpaa.2013.03.011. MR3057315

[9] M. D. E. Conder, *Large group actions on surfaces*, Riemann and Klein surfaces, automorphisms, symmetries and moduli spaces, Contemp. Math., vol. 629, Amer. Math. Soc., Providence, RI, 2014, pp. 77–97, DOI 10.1090/conm/629/12550. MR3289634

[10] A. F. Costa and M. Izquierdo, *On the existence of connected components of dimension one in the branch locus of moduli spaces of Riemann surfaces*, Math. Scand. **111** (2012), no. 1, 53–64, DOI 10.7146/math.scand.a-15213. MR3001358

[11] A. F. Costa and M. Izquierdo, *Equisymmetric strata of the singular locus of the moduli space of Riemann surfaces of genus 4*, Geometry of Riemann surfaces, London Math. Soc. Lecture Note Ser., vol. 368, Cambridge Univ. Press, Cambridge, 2010, pp. 120–138. MR2665007

[12] J. Gilman, *On conjugacy classes in the Teichmüller modular group*, Michigan Math. J. **23** (1976), no. 1, 53–63. MR0430320

[13] W. J. Harvey, *Cyclic groups of automorphisms of a compact Riemann surface*, Quart. J. Math. Oxford Ser. (2) **17** (1966), 86–97, DOI 10.1093/qmath/17.1.86. MR0201629

[14] W. J. Harvey, *On branch loci in Teichmüller space*, Trans. Amer. Math. Soc. **153** (1971), 387–399, DOI 10.2307/1995564. MR0297994

[15] A. Hurwitz, *Ueber Riemann'sche Flächen mit gegebenen Verzweigungspunkten* (German), Math. Ann. **39** (1891), no. 1, 1–60, DOI 10.1007/BF01199469. MR1510692

[16] A. Hurwitz, *Ueber algebraische Gebilde mit eindeutigen Transformationen in sich* (German), Math. Ann. **41** (1892), no. 3, 403–442, DOI 10.1007/BF01443420. MR1510753

[17] A. Hurwitz, *Ueber die Anzahl der Riemann'schen Flächen mit gegebenen Verzweigungspunkten* (German), Math. Ann. **55** (1901), no. 1, 53–66, DOI 10.1007/BF01448116. MR1511135

[18] M. Izquierdo and D. Ying, *Equisymmetric strata of the moduli space of cyclic trigonal Riemann surfaces of genus 4*, Glasg. Math. J. **51** (2009), no. 1, 19–29, DOI 10.1017/S0017089508004497. MR2471672

[19] R. S. Kulkarni and C. Maclachlan, *Cyclic p-groups of symmetries of surfaces*, Glasgow Math. J. **33** (1991), no. 2, 213–221, DOI 10.1017/S0017089500008247. MR1108745

[20] E. K. Lloyd, *Riemann surface transformation groups*, J. Combinatorial Theory Ser. A **13** (1972), 17–27. MR0302893

[21] C. Maclachlan, *Abelian groups of automorphisms of compact Riemann surfaces*, Proc. London Math. Soc. (3) **15** (1965), 699–712, DOI 10.1112/plms/s3-15.1.699. MR0179348

[22] C. Maclachlan and A. Miller, *Generating functions for finite group actions on surfaces*, Math. Proc. Cambridge Philos. Soc. **124** (1998), no. 1, 21–49, DOI 10.1017/S0305004197002417. MR1620568

[23] J. Nielsen, *Die Struktur periodischer Transformationen von Flächen*, Danske Vid Selsk. Mat-Fys. Medd., **1** (1937), 1–77.

[24] V. Peterson, J. Russell, and A. Wootton, *Maximal group actions on compact oriented surfaces*, J. Algebra **472** (2017), 1–14, DOI 10.1016/j.jalgebra.2016.10.004. MR3584868

[25] J.-C. Schlage-Puchta and J. Wolfart, *How many quasiplatonic surfaces?*, Arch. Math. (Basel) **86** (2006), no. 2, 129–132, DOI 10.1007/s00013-005-1464-1. MR2205227

[26] M. Stukow, *Conjugacy classes of finite subgroups of certain mapping class groups*, Turkish J. Math. **28** (2004), no. 2, 101–110. MR2062555

Department of Mathematics, CUNY Graduate Center, The City College of New York, New York 10016
Email address: jrussellmadonia@gradcenter.cuny.edu

Department of Mathematics, The University of Portland, Portland, Oregon 97203
Email address: wootton@up.edu

Contemporary Mathematics
Volume **703**, 2018
http://dx.doi.org/10.1090/conm/703/14128

Galois action on regular dessins d'enfant with simple group action

S. Allen Broughton

ABSTRACT. A quasi-platonic action of the group G on the Riemann surface S is a conformal action of G on S such that S/G is a sphere and the projection $\pi_G : S \to S/G$ is branched over $\{0, 1, \infty\}$. The action is induced by a triple of $(a, b, c) \in G^3$, generating G, with $abc = 1$. The projection π_G is a regular Belyi function and induces a regular dessin d'enfant on S, and so S is defined over a number field. The absolute Galois group $\mathrm{Gal}(\overline{\mathbb{Q}})$ acts on regular dessins, hence quasi-platonic actions, by acting on the coefficients of a defining equation of S. The action of $\psi \in \mathrm{Gal}(\overline{\mathbb{Q}})$ on triples is $(a, b, c) \to (ua^t u^{-1}, vb^t v^{-1}, wc^t w^{-1})$ for some $(u, v, w) \in G^3$, according to the branch cycle argument. The integer t is characterized by the action of ψ on cyclotomic subfields of $\overline{\mathbb{Q}}$, and (u, v, w) is determined by the action of ψ away from cyclotomic subfields. In this paper we try to reconstruct the Galois action from the branch cycle description of the action and the structure of the group G. To this end, we focus on groups which are simple or covers of simple groups.

1. Introduction

1.1. Quasi-platonic actions and dessins d'enfant. Let S be a closed Riemann surface, of genus $\sigma \geq 2$; we denote the group of conformal automorphisms by $\mathrm{Aut}(S)$. We say that a group G acts conformally on S if there is a monomorphism

$$\epsilon : G \hookrightarrow \mathrm{Aut}(S). \tag{1.1}$$

Subgroups of $\mathrm{Aut}(S)$ act conformally in an obvious, natural way. Extending the notion of conformal action to any (allowable) finite group allows us to abstractly work with the structure of G without getting bogged down with concrete representations of G as transformations of S.

Quasi-platonic G-actions extend the notion of automorphism groups of platonic solids. A G-action is called *quasi-platonic* if the quotient S/G is the Riemann sphere $\widehat{\mathbb{C}} = P^1(\mathbb{C})$, and the quotient map $\pi_G : S \to S/G$ is branched over three points, typically $0, 1$, and ∞. A surface is called quasi-platonic if the action of $\mathrm{Aut}(S)$ is quasi-platonic. It turns out that if the action of $G \leq \mathrm{Aut}(S)$ is quasi-platonic then all intermediate groups H, $G \leq H \leq \mathrm{Aut}(S)$ have quasi-platonic actions. See Proposition 2.8 for more details. There is great interest in quasi-platonic actions and surfaces for the following reasons:

2010 *Mathematics Subject Classification.* Primary 17B20, 20H15, 51F15.

(1) They are rigid, i.e., the conformal structure of the surfaces cannot be infinitesimally deformed without losing symmetry. They form special points in the moduli space of curves.
(2) The surface S has a defining equation with coefficients in a number field k_S.
(3) The G-action defines a *regular dessin d'enfant* $\mathcal{D}$, a bipartite graph on S. To construct the dessin, assume that $\pi_G : S \to S/G$ is branched over $\{0, 1, \infty\}$. Let $[0, 1] \subseteq \widehat{\mathbb{C}}$ be the standard unit interval. Then $\mathcal{D} = \pi_G^{-1}([0, 1])$ is a bipartite graph in S called a *dessin d'enfant*. The group G acts simply transitively on the edges of $\mathcal{D}$; hence, $\mathcal{D}$ is called a regular dessin d'enfant. The complement $S - \mathcal{D}$ is a disjoint union of open, congruent convex hyperbolic polygons, permuted transitively by G. Each polygon is the lift to S of $\widehat{\mathbb{C}} - I$. This geometric structure on the surface, and its invariance under G, generalizes the notion of a platonic solid and the tetrahedral, cubic, octahedral, dodecahedral and icosahedral tilings and automorphism groups of the sphere.
(4) There is a rich interplay between dessins and the action of the absolute Galois group on surfaces defined over number fields. We discuss this in detail in Section 3.

Dessins d'enfant and Belyi functions may be defined more generally. Belyi's theorem [**1**] allows for an identification and interplay among these objects:

- surfaces defined over number fields;
- surfaces S with a meromorphic function:

$$\beta : S \to \widehat{\mathbb{C}} \tag{1.2}$$

  called a Belyi function, branched over $\{0, 1, \infty\}$;
- embedded bipartite graphs $\mathcal{D} = \beta^{-1}([0, 1]) \subseteq S$ such that $S - \mathcal{D}$ is a disjoint union of open topological disks; and,
- triples (π_1, π_2, π_3) of permutations generating a transitive permutation group, with $\pi_1\pi_2\pi_3 = 1$.

We do not describe the interplay among the objects above, but see [**6**] for a comprehensive overview of this material. We restrict our attention to the regular case defined by a quasi-platonic G-action:

- The Belyi function

$$\beta = \pi_G : S \to S/G \tag{1.3}$$

  is a regular branched cover.
- The group of automorphisms $\epsilon(G) \subseteq \mathrm{Aut}(S)$ permutes the edges of $\mathcal{D}$ simply transitively and the disks in $S - \mathcal{D}$ transitively.
- The dessin is associated to a triple $(a, b, c) \in G^3$, generating G, with $abc = 1$.

If we start off with a regular Belyi function $\beta : S \to \widehat{\mathbb{C}}$, then we denote the group of covering transformations of β by $\mathrm{Cov}(\beta) \subseteq \mathrm{Aut}(S)$. In case $\beta = \pi_G$, then $\mathrm{Cov}(\beta) = \epsilon(G)$.

1.2. Action of the absolute Galois group $Gal(\overline{\mathbb{Q}})$**.** Given a quasi-platonic surface S defined over a number field K and a field automorphism $\psi \in \mathcal{G} = \mathrm{Gal}(\overline{\mathbb{Q}})$ of the algebraic closure of $\mathbb{Q}$, we may construct another quasi-platonic surface S^ψ

over K^ψ by applying ψ to the coefficients of defining equations of S. Correspondingly, we get a transformed dessin $\mathcal{D}^\psi \subseteq S^\psi$, Belyi function β^ψ, and generating triple $(a,b,c)^\psi$ described in the abstract. Once the appropriate equivalence relations are made, we get an action on all three sets. The goal of this paper is to describe, as explicitly as possible, the Galois action on the generating triples by using the group structure of G. This line of attack was inspired by the success of this approach on actions of $PSL_2(q)$ in a previous article [**3**]. Our example calculations for the alternating groups in Section 4 show that the problem is complex, as might be expected.

One approach to studying the Galois action is to study the Belyi function 1.2 and the dessin $\mathcal{D}$ under the action of under the action of $\mathcal{G}$. Unfortunately, for the surfaces we wish to study, the number field k_S may be unknown, the Belyi function may have a very large degree, and may be difficult to construct. An alternate method would be to factor $\pi_G : S \to S/H \to S/G$. The last map will be a Belyi function that may be easier to analyze. See the paper [**7**] for examples. We shall adopt a different approach in which we translate the Galois action to generating triple (a,b,c) and an associated tiling $\mathcal{T}$ of the surface. The tilings are constructed from the vertex set $\pi_G^{-1}(0) \cup \pi_G^{-1}(1) \cup \pi_G^{-1}(\infty)$, the dessin $\mathcal{D} = \mathcal{D}_1$ and two companion dessins $\mathcal{D}_2$, $\mathcal{D}_3$, which we introduce shortly.

Because of the limited scope of this paper, we only develop the first steps of this analysis, setting up a framework for some calculations using Magma. We recall some of the exposition of an earlier paper [**3**] to give a smoother exposition in this paper.

1.3. Overview of the paper. The remainder of this paper is organized as follows. In Section 2, we describe the construction of surfaces, with a quasi-platonic action for a given group G via Fuchsian triangle groups. In Section 3, we discuss the action of the absolute Galois group on the quasi-platonic actions. Finally, in Section 4, we give some examples of the Galois action on quasi-platonic actions. The main result is a precise characterization of the Galois action on quasi-platonic $PSL_2(q)$ actions. Our result is given in terms of the "cyclotomic splitting" of the absolute Galois group alluded to in the abstract.

2. Quasi-platonic group actions and regular Belyi functions

Our approach in this section is similar to that in the paper of Jones and Singerman [**9**].

2.1. *G*-actions and covering groups. The universal cover of S is the hyperbolic plane $\mathbb{H}$ with covering map $\pi_S : \mathbb{H} \to S$. We denote the group of covering transformations of π_S by $\Pi \simeq \pi_1(S)$. The conformal group action of G on S has a covering action by a Fuchsian group Γ defined by an exact sequence

$$\Pi \hookrightarrow \Gamma \overset{\eta}{\twoheadrightarrow} G. \tag{2.1}$$

The induced isomorphism $\overline{\eta} : \Gamma/\Pi \leftrightarrow G$ defines an action $\epsilon = \overline{\eta}^{-1}$ of G on S through the natural action of Γ/Π on $S = \mathbb{H}/\Pi$.

We can construct our surfaces, groups, and dessins through tilings of the hyperbolic plane by triangles as in Figure 1. Though it is not typical to use the triangular tiling in studying dessins d'enfant, we will use the idea in our analysis of generating triples.

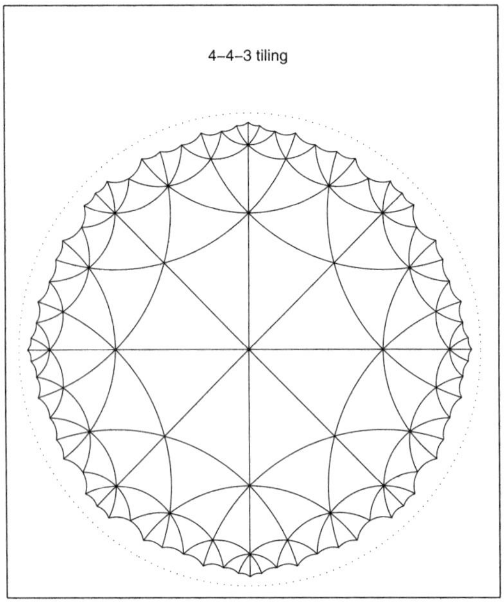

FIGURE 1

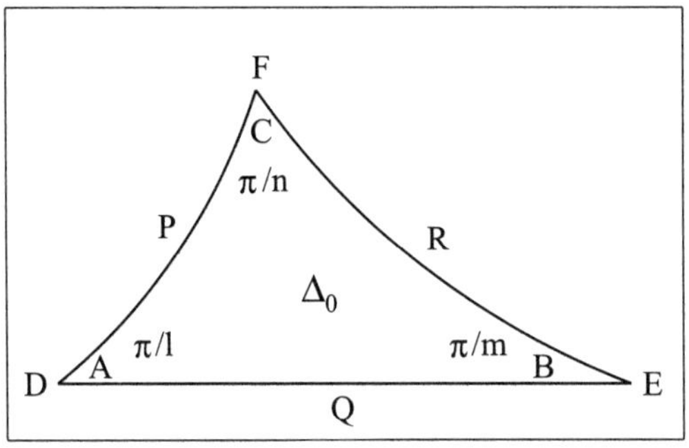

FIGURE 2

In Figure 2 we picture a (counterclockwise oriented) (l,m,n) triangle $\triangle DEF$ in the hyperbolic plane $\mathbb{H}$ (or Poincaré disc). The line segments $\overline{FD}$, $\overline{DE}$, and $\overline{EF}$ meet in the angles $\frac{\pi}{l}$, $\frac{\pi}{m}$, and $\frac{\pi}{n}$, respectively, where l, m, and n are integers ≥ 2. An (l,m,n)-triangle exists if and only if $\frac{1}{l}+\frac{1}{m}+\frac{1}{n}<1$. All such triangles are congruent. The triangles in Figure 1 are $(4,4,3)$ triangles.

Let $P,Q,R \in \mathrm{Aut}^*(\mathbb{H})$ be the hyperbolic reflections in the lines $\overline{FD}$, $\overline{DE}$, and $\overline{EF}$, respectively, and define the rotations:

$$A = PQ,\ B = QR,\ C = RP.$$

The mappings A,B,C are counterclockwise rotations, centered at D,E,F, respectively, through the angles $\frac{2\pi}{l}$, $\frac{2\pi}{m}$, $\frac{2\pi}{n}$, respectively. It is well known that $T_{l,m,n} = \langle A,B,C\rangle \subset PSL_2(\mathbb{R})$ is a discrete group of conformal isometries of the hyperbolic plane with the following presentation,

$$T_{l,m,n} = \langle A, B, C | A^l = B^m = C^n = ABC = 1 \rangle. \tag{2.2}$$

We call (l, m, n) the *signature* of $T_{l,m,n}$ and also call (l, m, n) the *signature* or *branching data* of the G-action on S, and the *signature* $\text{sig}(\pi_G)$ of the Belyi function $\pi_G : S \to S/G$. Next we construct actions directly from the group G.

CONSTRUCTION 2.1. Suppose that (a, b, c) is a triple of elements of G such that

$$\begin{aligned} G &= \langle a, b, c \rangle \\ o(a) &= l, o(b) = m, o(c) = n, \text{ and} \\ abc &= 1. \end{aligned} \tag{2.3}$$

The triple (a, b, c) is a called a *generating* (l, m, n)*-triple*. If $\langle a, b, c \rangle$ is a proper subgroup of G, we just call (a, b, c) an (l, m, n)*-triple*. For some selection of a generating triple (a, b, c), the epimorphism of equation 2.1 is given by

$$\eta : T_{l,m,n} \to G, \; A \to a, \; B \to b, \; C \to c. \tag{2.4}$$

The kernel $\Pi = \ker(\eta)$ is torsion free and defines a closed Riemann surface $S = \mathbb{H}/\Pi$ whose genus σ satisfies the Riemann-Hurwitz equation,

$$\frac{2\sigma - 2}{|G|} = 1 - \frac{1}{l} - \frac{1}{m} - \frac{1}{n}. \tag{2.5}$$

Tilings and the three dessins. The (l, m, n) tiling on $\mathbb{H}$ is invariant under the action of Π so there is an induced tiling $\mathcal{T}$ on S. The tiling $\mathcal{T}$ on S is invariant under the action of G. On $\widehat{\mathbb{C}} = S/G$ the image of the tiling has a very simple structure. The images of vertices of type D, E, and F are $0, 1$, and ∞, respectively. The images of edges of type $\overline{DE}$, $\overline{EF}$, and $\overline{FD}$ are $I_1 = [0, 1]$, $I_2 = [1, \infty]$, and $I_3 = [\infty, 0]$, respectively. The images of clockwise oriented and counterclockwise oriented triangles are mapped to the upper half and the lower half of the Riemann sphere, respectively. Now define $\mathcal{D}_i = \pi_G^{-1}(I_i)$. The polygons of the complements of $\mathcal{D}_1$, $\mathcal{D}_2$, and $\mathcal{D}_3$ in S are dihedrally symmetric $2n$-gons, $2l$-gons and $2m$-gons surrounding the vertices of type F, D and E, respectively. The inverse images of $\mathcal{D}_1$, $\mathcal{D}_2$, and $\mathcal{D}_3$ in S are the unions of all edges of type $\overline{DE}$, $\overline{EF}$, and $\overline{FD}$, respectively. The tiling $\mathcal{T}$ encodes the information of all three dessins simultaneously. Also note that the 1-skeleton of S is $\mathcal{D}_1 \cup \mathcal{D}_2 \cup \mathcal{D}_3$ and that $S - (\mathcal{D}_1 \cup \mathcal{D}_2 \cup \mathcal{D}_3)$ is a union of congruent hyperbolic (l, m, n) triangles.

CONSTRUCTION 2.2. Observe that given a triple (a, b, c), the covering $\beta : S \to \widehat{\mathbb{C}}$ and the tiling T on S can be constructed directly from the upper and lower hemisphere tiling structure on $\widehat{\mathbb{C}}$. Construct a covering $\beta : S^\circ \to \widehat{\mathbb{C}} - \{0, 1, \infty\}$ as follows. We know that $\pi_1(\widehat{\mathbb{C}} - \{0, 1, \infty\}) = \langle \gamma_1, \gamma_2, \gamma_3 \mid \gamma_1\gamma_2\gamma_3 = 1 \rangle$ and, hence, we may define a surjective map $\pi_1(\widehat{\mathbb{C}} - \{0, 1, \infty\}) \to G$ by $\gamma_1 \to a$, $\gamma_2 \to b$, $\gamma_3 \to c$. The kernel of this map determines a covering space $\beta : S^\circ \to \widehat{\mathbb{C}} - \{0, 1, \infty\}$. Pull up everything up by β and compactify.

Epimorphisms and equivalence. We can use the tiling on S to construct an epimorphism for the G-action.

CONSTRUCTION 2.3. Pick a clockwise oriented triangle $\overline{\Delta}$ on S, which we call the *master tile*. The point $\overline{D}$ on S corresponding to D in $\triangle DEF$ is $\pi_G^{-1}(0) \cap \overline{\Delta}$. The stabilizer $G_{\overline{D}}$ of $\overline{D}$ is cyclic of order l. The rotation number map rot:$G_{\overline{D}} \to \mathbb{C}$, given by the map $g \to dg$ on the tangent plane $T_{\overline{D}}(S)$, is an isomorphism of $G_{\overline{D}}$ onto the lth roots of unity. Pick a in $G_{\overline{D}}$ so that $\mathrm{rot}(a) = \mathrm{rot}(A, \overline{D}) = \exp(\frac{2\pi i}{l})$. Do the same to get b and c such that $\mathrm{rot}(b) = \exp(\frac{2\pi i}{m})$ and $\mathrm{rot}(c) = \exp(\frac{2\pi i}{n})$. Using homotopy arguments with lifts of curves, it can be shown that $abc = 1$, and that $A \to a$, $B \to b$, $C \to c$ is a uniformizing epimorphism. The selection of a different counterclockwise oriented triangle gives the triple $\mathrm{Ad}_g \cdot (a,b,c) = (gag^{-1}, gbg^{-1}, gcg^{-1})$, for some $g \in G$.

The enumeration of quasi-platonic actions is the same as the determination of $\mathrm{Aut}(G)$ equivalence classes of generating (l,m,n)-triples of G. First, we define our notions of equivalence of actions.

DEFINITION 2.4. We say that two conformal actions $\epsilon_1, \epsilon_2 : G \hookrightarrow \mathrm{Aut}(S)$ are *algebraically equivalent* if $\epsilon_2 = \epsilon_1 \circ \omega$ for some $\omega \in \mathrm{Aut}(G)$ or, equivalently, if $\epsilon_1(G)$ and $\epsilon_2(G)$ are the same subgroup of $\mathrm{Aut}(S)$. Two actions $\epsilon_1 : G \hookrightarrow \mathrm{Aut}(S_1)$ and $\epsilon_2 : G \hookrightarrow \mathrm{Aut}(S_2)$ on possibly different surfaces are *conformally equivalent* if there is a conformal equivalence $h : S_1 \leftrightarrow S_2$ such that,

$$\epsilon_2(g) = h \circ \epsilon_1(\omega(g)) \circ h^{-1}, \ g \in G. \tag{2.6}$$

Specifically, two actions of G on the same surface are conformally equivalent if and only if they determine conjugate subgroups of $\mathrm{Aut}(S)$. If we may take the automorphism ω to be trivial, then we say the two actions are *geometrically equivalent*. In terms of Belyi functions, two actions are algebraically equivalent if and only if

$$\pi_{\epsilon_1(G)} = \pi_{\epsilon_2(G)},$$

and they are conformally equivalent if and only if

$$\pi_{\epsilon_1(G)} = \pi_{\epsilon_2(G)} \circ h, \tag{2.7}$$

for an automorphism h of S.

Let (a,b,c) be a generating (l,m,n)-triple of G, $\Gamma = T_{l,m,n}$, and $\omega \in \mathrm{Aut}(G)$. Then, the equation 2.1 can be expanded to a commutative diagram,

$$\begin{array}{ccccc} \Pi & \hookrightarrow & \Gamma & \overset{\eta}{\twoheadrightarrow} & G \\ \downarrow id & & \downarrow id & & \downarrow \omega \\ \Pi & \hookrightarrow & \Gamma & \overset{\omega\circ\eta}{\twoheadrightarrow} & G \end{array} \tag{2.8}$$

Both epimorphisms determine the same group of automorphisms of $\Gamma/\Pi \subseteq \mathrm{Aut}(S)$ acting on $S = \mathbb{H}/\Pi$. The transformed generating triple (a', b', c'), determined by $\omega \circ \eta$, is $(\omega(a), \omega(a), \omega(a))$. Thus, each equivalence class determined by the action $\omega \cdot (a,b,c) = (\omega(a), \omega(a), \omega(a))$, $\omega \in \mathrm{Aut}(G)$, determines a unique surface $S = \mathbb{H}/\Pi$ and unique subgroup of $\mathrm{Aut}(S)$. Correspondingly, given two epimorphisms η_1, η_2 with the same kernel as in the left half of the diagram, we have $\eta_2 = \omega \circ \eta_1$ for an $\omega \in \mathrm{Aut}(G)$ and, so, η_1 and η_2 determine the equivalent triples. Equivalent epimorphisms determine equivalent conformal actions.

REMARK 2.5. There is a braid action on triples generated by these transformations:

$$(a,b,c) \to (b, b^{-1}ab, c), \ (a,b,c) \to (a, c, c^{-1}bc), (a,b,c) \to (a^{-1}ca, b, a) \tag{2.9}$$

and their inverses. The action commutes with the $\mathrm{Aut}(G)$-action on triples and defines $\mathrm{Aut}(G)$-invariant bijections of (l, m, n) triples to (m, l, n) and (n, m, l) triples. The other permutations of indices are obtained by composition. The permutation of signatures does not produce any new actions. Consider, for instance, the permutation $(l, m, n) \to (m, l, n)$. Reflect the triangle $\triangle DEF$ in the side $\overline{EF}$ to obtain the (clockwise oriented) $\triangle ED'F$ triangle, an (m, l, n)-triangle. The rotations, in order, at the corners are $B, B^{-1}AB$, and C. The very same map $\eta : \Gamma \twoheadrightarrow G$ given in equation 2.4 takes the triple $(B, B^{-1}AB, C)$ to $(b, b^{-1}ab, c)$ and, so, the same surface $S = \mathbb{H}/\Pi$, $\ker(\eta) = \Pi$ is determined. As $\Gamma = \langle B, B^{-1}AB, C\rangle$, then the same subgroup of automorphisms of S is determined, and the image of G in $\mathrm{Aut}(S)$ is the same. Therefore no new actions are determined. There is a similar argument for all other permutations. Therefore, we may assume the signature has standard lexicographic form $l \leq m \leq n$.

If the any two of l, m, n are equal, then the braid action imposes further equivalence relations on the generating triples. In terms of the Belyi functions, this amounts to a post composition action $\beta \to T \circ \beta$ where T is a fractional linear transformation preserving $\{0, 1, \infty\}$. For instance, if $l = m$ then the two Belyi functions are β and $1 - \beta$. To take the braid action into account, we need consider conformal equivalence of actions.

We have two approaches to classifying quasi-platonic actions, equivalence classes of actions and Belyi functions. The following is easily proven.

PROPOSITION 2.6. *For a surface S, the set of regular Belyi functions with group of covering transformations isomorphic to G is in 1-1 correspondence with the algebraic equivalence classes of quasi-platonic G actions on S.*

2.2. Uniqueness of Belyi functions and actions. We next show that regular Belyi functions on surfaces are determined, up to a finite indeterminacy, by their signatures. The finite determinacy can be related to algebraic relations between their covering groups. The best result would be conformal equivalence as in equation 2.7. An analogous problem for non-regular dessins is the case of uniform dessins, where conformal equivalence fails. See [**15**] for a representative work on uniform dessins. Uniqueness is trivially true is when $\epsilon(G) = \mathrm{Aut}(S)$. To put this trivial case and the next proposition into the context of the previous research [**14**], we make the following definition.

DEFINITION 2.7. A quasi-platonic action $\epsilon : G \hookrightarrow \mathrm{Aut}(S)$ of G on S is called maximal if and only if $\epsilon(G) = \mathrm{Aut}(S)$.

PROPOSITION 2.8. *Suppose we are given regular Belyi functions $\beta_1, \beta_2 : S \to \widehat{\mathbb{C}}$, with the same signature whose covering groups $G_1 = \mathrm{Cov}(\beta_1), G_2 = \mathrm{Cov}(\beta_2)$ are non-maximal. Then there is a normal subgroup $N \lhd \mathrm{Aut}(S)$ satisfying:*

(1) $N \lhd G_1, G_2$,
(2) $S \to S/N = S'$ *is a Galois cover,*
(3) $\mathrm{Aut}(S)/N \leq \mathrm{Aut}(S')$, *and*
(4) *there is an automorphism θ of $\mathrm{Aut}(S)/N$ such that $G_2/N = \theta(G_1/N)$.*

PROOF. Let Π uniformize S so that there are exact sequences uniformizing the actions of G_1 and G_2:

$$\Pi \hookrightarrow \Gamma_i \overset{\eta_i}{\twoheadrightarrow} G_i, \ i = 1, 2.$$

TABLE 1. Pairs of triangle groups

Case	sig(Γ_1)	sig(Δ)	$\lvert\Delta/\Gamma_1\rvert$
N1	(m, m, n), $m \geq 3, m + n \geq 7$	$(2, m, 2n)$	2
N2	(n, n, n), $n \geq 4$	$(3, 3, n)$	3
N3	(n, n, n), $n \geq 4$	$(2, 3, 2n)$	6
T1	$(2, n, 2n)$, $n \geq 4$	$(2, 3, 2n)$	3
T2	$(3, n, 3n)$, $n \geq 4$	$(2, 3, 2n)$	4
T3	$(n, 2n, 2n)$, $n \geq 3$	$(2, 4, 2n)$	4
T4	$(n, 4n, 4n)$, $n \geq 2$	$(2, 3, 4n)$	6
T5	$(4, 4, 5)$	$(2, 4, 5)$	6
T6	$(9, 9, 9)$	$(2, 3, 9)$	12
T7	$(3, 8, 8)$	$(2, 3, 8)$	10
T8	$(4, 8, 8)$	$(2, 3, 8)$	12
T9	$(3, 3, 7)$	$(2, 3, 7)$	8
T10	$(2, 7, 7)$	$(2, 3, 7)$	9
T11	$(7, 7, 7)$	$(2, 3, 7)$	24

The normalizer $\Delta = \mathrm{Nor}_{\mathrm{Aut}(\mathbb{H})}(\Pi)$ is a Fuchsian group, and according to Singerman's paper [**14**] the pairs $\Gamma_1 < \Delta$ and $\Gamma_2 < \Delta$ consist of two triangle groups which must both be in the same row of Table 1.

The cases in the table with an N in the notation have $\Gamma_i \triangleleft \Delta$ the others are not normal pairs. Since Δ is a triangle group, the permutation representation of Δ on the cosets of Γ_1 yields a triple (π_1, π_2, π_3) in a symmetric group whose degree is given in the last column of the table. In Singerman's paper [**14**], it is shown that the triple is unique up to conjugacy in the symmetric group. The permutation representations ρ_i of the pairs $\Gamma_i < \Delta$ satisfy $\rho_2 = \alpha\rho_1\alpha^{-1}$ for an appropriate permutation α. It follows that the kernels of the representations equal a common subgroup Π' which satisfies

$$\Pi \triangleleft \Pi' \triangleleft \Gamma_1, \Gamma_2.$$

We define $N = \Pi'/\Pi$. The group $\mathrm{Aut}(S)/N = \Delta/\Pi'$ and isomorphic the "monodromy" group $\langle \pi_1, \pi_2, \pi_3 \rangle$ of the pair $\Gamma_1 < \Delta$. Since the monodromy groups determined by both pairs are conjugate in the symmetric group, the conjugating permutation induces an automorphism θ of $\mathrm{Aut}(S)/N$ with the required properties. □

REMARK 2.9. It is beyond the scope of this paper to determine when the automorphism θ is inner and when $\mathrm{Aut}(S)/N = \mathrm{Aut}(S')$. For a quick example, in the Hurwitz cases $T9$, $T10$ and $T11$, θ is inner and $\mathrm{Aut}(S)/N = \mathrm{Aut}(S')$.

From the above we obtain the following finiteness result in Corollary 2.10. Corollary 2.11 is straight forward There are other ways to prove finiteness but the Proposition shows that we actually have some hope of finding nice forms for the Belyi functions.

COROLLARY 2.10. *For a given surface S there are only finitely many regular Belyi functions.*

COROLLARY 2.11. *Suppose two regular Belyi functions on a surface S have signatures which are permutations of each other. Then $\beta_2 = T \circ \beta_1$ where T is a fractional linear transformation preserving $\{0, 1, \infty\}$.*

2.3. Triple sets and G-signatures. We saw above that the automorphism classes of generating triples form a classifier of actions. Later, when considering the Galois action, we will see that conjugacy classes of generating triples will also be important. We are going to define a refinement of the signature that takes this into account. To this end, and to help enumerate triples, we define an "approximate automorphism group" to be a group of automorphisms of G satisfying $\mathrm{Inn}(G) \subseteq L \subseteq \mathrm{Aut}(G)$. The extreme cases $\mathrm{Inn}(G)$ and $\mathrm{Aut}(G)$ are denoted by K and A respectively. This notion is most appropriate when the index $|\mathrm{Aut}(G) : \mathrm{Inn}(G)| = |\mathrm{Out}(G)|$ is small, say, when the center of G is small. A specific case with interesting intermediate L is $G = PGL_2(q), L = PGL_2(q)$, when q is a prime power. In this case $A/L \simeq \mathrm{Gal}(\mathbb{F}_q)$.

For $g \in G$ and L as above we define the L-conjugacy class of g, $g^L = \{\omega(g) : \omega \in L\}$. When $L = K$ we get standard conjugacy classes $g^K = g^G$. We define the G-signature of a G-action with a generating triple (a, b, c) to be the triple (a^A, b^A, c^A). In [**13**] the triple, (a^G, b^G, c^G) is called *ramification type* and is also used as a classifier of actions.

For $(a, b, c) \in G^3$, we define *L-triple sets and (generating) L-triple sets,*

$$L_G(a, b, c) = \{(x, y, z) : x \in a^L, y \in b^L, z \in c^L, xyz = 1\}, \text{ and} \tag{2.10}$$

$$L^\circ_G(a, b, c) = \{(x, y, z) : (x, y, z) \in L_G(a, b, c), \langle x, y, z\rangle = G\}. \tag{2.11}$$

If $L = \mathrm{Inn}(G)$ or $\mathrm{Aut}(G)$, we use the notation $K_G(a, b, c)$ and $K^\circ_G(a, b, c)$ or $A_G(a, b, c)$ and $A^\circ_G(a, b, c)$ respectively. Observe that $K_G(a, b, c) \subseteq L_G(a, b, c) \subseteq A_G(a, b, c)$ and that $A_G(a, b, c)$ is a union of $\mathrm{Aut}(G)$ classes of triples.

The sets $A^\circ_G(a, b, c)$ provide a natural classifier of actions of a similar type and an action space for the action of the absolute Galois group and braid groups in Hurwitz spaces. Likewise, the set $K^\circ_G(a, b, c)$ is a suitable classifier for the action of the absolute Galois group and braid groups in Hurwitz spaces. The two types of classifiers are closely linked when $|\mathrm{Out}(G)|$ is small. The set $A^\circ_G(a, b, c)$ consists of one or more $\mathrm{Aut}(G)$ classes representing distinct actions with the same G-signature. We call these *companion actions* or *companion orbits.* Likewise $K^\circ_G(a, b, c)$ may contain several different K orbits, which we shall call companion orbits. Companion orbits in $K^\circ_G(a, b, c)$ may or may not determine distinct actions.

REMARK 2.12. If the character theory of G is tractable, then the following formula (see [**2**], [**8**]) may be used for computing $|K_G(a, b, c)|$:

$$|K_G(a, b, c)| = \frac{|G|^2}{|\mathrm{Cent}(a)| \cdot |\mathrm{Cent}(b)| \cdot |\mathrm{Cent}(c)|} \sum_\chi \frac{\chi(a)\chi(b)\chi(c)}{\chi(1)}. \tag{2.12}$$

3. The Galois action on quasi-platonic actions

In this section we first recall the action of the Galois group on surfaces and generating triples. Though this is well known, we recall in section 3.1 some of the exposition in [**3**] for a smooth discussion in section 3.2. First we make a remark about where the field automorphisms live.

REMARK 3.1. Every element of $\mathrm{Gal}(\overline{\mathbb{Q}})$ is the restriction of an element of $\mathrm{Gal}(\mathbb{C})$. It is often convenient to apply field automorphisms to any element of $\mathbb{C}$. Thus, we need our automorphism to be in $\mathrm{Gal}(\mathbb{C})$, though the effect of the automorphism is not changed on elements of $\overline{\mathbb{Q}}$. So, we often formulate definitions and results with field automorphisms fully defined on $\mathbb{C}$, though we are really only interested in $\mathrm{Gal}(\overline{\mathbb{Q}})$.

3.1. The Galois action. We assume we are given a quasi-platonic action of G with $\pi_G : S \to S/G = \widehat{\mathbb{C}}$, branched over $\{0, 1, \infty\}$. According to Belyi's theorem [**1**], S can be defined over a number field. The intersection of all such defining fields for all surfaces conformally equivalent to S is called the *moduli field* of S. Though we shall not need the following fact, it can be shown that S has a defining equation over its moduli field (see [**5**]). From now on we are going to assume that S is defined as a smooth, closed subvariety of some $\mathbb{P}^d$ over a number field which we denote by k_S.

If $\psi \in \mathrm{Gal}(\mathbb{C})$ then we define S^ψ, the ψ-Galois transform of S, to be the Riemann surface obtained by applying ψ to the coefficients of the defining equation(s) of S. Since ψ is extended to all of $\mathbb{C}$, there is an induced map, still denoted ψ, $\psi : S \to S^\psi$ by applying ψ coordinatewise. The map is a bijection, but definitely not a morphism. Since S is defined over a number field, the surface S^ψ only depends on the action of ψ on the algebraic closure of $\mathbb{Q}$. For any two affine or projective varieties X, Y and map $f : X \to Y$, we define X^ψ, Y^ψ, $\psi : X \to X^\psi$, and $\psi : Y \to Y^\psi$ in a similar fashion. The map $f^\psi : X^\psi \to Y^\psi$ is defined by $f^\psi(\psi(x)) = \psi(f(x))$, $x \in X$ or $f^\psi = \psi f \psi^{-1}$. When f is expressed in local coordinates we simply apply ψ to the coefficients. Consequently, for every automorphism g of S, g^ψ is an automorphism of S^ψ, and $g \to \psi g \psi^{-1}$ is an isomorphism of $\mathrm{Aut}(S)$ to $\mathrm{Aut}(S^\psi)$. If $\epsilon : G \to \mathrm{Aut}(S)$ defines the G-action, then $\epsilon^\psi : g \to \epsilon(g)^\psi$ is a G-action on S^ψ, the quotient map $\pi_G^\psi : S^\psi \to S^\psi/G$, is a Belyi function, branched over $\{0, 1, \infty\}$, and the following diagram commutes,

$$\begin{array}{ccc} S & \xrightarrow{\psi} & S^\psi \\ \downarrow \pi_G & & \downarrow \pi_G^\psi \\ \widehat{\mathbb{C}} & \xrightarrow{\psi} & \widehat{\mathbb{C}} \end{array} \tag{3.1}$$

We will call S^ψ, ϵ^ψ, and π_G^ψ the ψ Galois transforms of S, ϵ, and π_G respectively.

REMARK 3.2. Observe that ψ fixes $0, 1, \infty \in \widehat{\mathbb{C}}$ so that ψ maps the G ramification points on S to those on S^ψ. Specifically,

$$\begin{aligned} \pi_G^{-1}(0) &\xrightarrow{\psi} \left(\pi_G^\psi\right)^{-1}(0), \\ \pi_G^{-1}(1) &\xrightarrow{\psi} \left(\pi_G^\psi\right)^{-1}(1), \\ \pi_G^{-1}(\infty) &\xrightarrow{\psi} \left(\pi_G^\psi\right)^{-1}(\infty). \end{aligned} \tag{3.2}$$

Rotation Numbers. Without actually knowing the equations of S and S^ψ, we can determine the action of ψ on rotation numbers, information we shall use shortly. Let a non-trivial automorphism $g \in \mathrm{Aut}(S)$ fix the point $x_0 \in S$ and let f be any function that vanishes at x_0 with order 1. Then, $f \circ g = \mathrm{rot}(g, x_0) f + k$ where k

vanishes at x_0 with order 2 or greater. Now apply ψ to get

$$g^{\psi}(\psi(x_0)) = \psi g \psi^{-1}(\psi(x_0)) = \psi(x_0),$$

and

$$f^{\psi} \circ g^{\psi} = \psi\left(\mathrm{rot}(g, x_0)\right) f^{\psi} + k^{\psi},$$

where k^{ψ} vanishes at $\psi(x_0)$ with order 2 or greater. We see that g^{ψ} fixes $\psi(x_0)$ and

$$\mathrm{rot}(g^{\psi}, \psi(x_0)) = \psi(\mathrm{rot}(g, x_0)). \tag{3.3}$$

Now let $N > 1$ be any integer and $\zeta = \exp(2\pi i/N)$. The cyclotomic field $\mathbb{Q}[\zeta]$ is a normal subfield of $\mathbb{C}$, and the action of ψ on $\mathbb{Q}[\zeta]$ is given by $\zeta \to \zeta^s$ for some number s relatively prime to N. Applying equation 3.3 to $h = \epsilon(g)$ we get

$$\mathrm{rot}(\epsilon^{\psi}(g), \psi(x_0)) = (\mathrm{rot}(\epsilon(g), x_0))^s. \tag{3.4}$$

Now we work out the Galois action on epimorphisms. There are epimorphisms

$$\eta_1 : T_{l,m,n} \to G,\ A \to a_1,\ B \to b_1,\ C \to c_1$$
$$\eta_2 : T_{l,m,n} \to G,\ A \to a_2,\ B \to b_2,\ C \to c_2,$$

such that η_1 uniformizes the ϵ action on S, with generating triple (a_1, b_1, c_1); and η_2 uniformizes the ϵ^{ψ} action on S^{ψ} with generating triple (a_2, b_2, c_2). Following the discussion in Section 2; let $\triangle\overline{DEF}$ be a triangle in S determining the triple (a_1, b_1, c_1) and let $\triangle\overline{D'E'F'}$ be a triangle in S^{ψ} determining (a_2, b_2, c_2). Since $G \cdot \overline{D} = \pi_G^{-1}(0)$ and $G \cdot \overline{D}' = \left(\pi_G^{\psi}\right)^{-1}(0)$, (equation 3.2) then there is a $u \in G$, so that $\psi(u\overline{D}) = \overline{D'}$, and the stabilizer of $\overline{D}'$ is $u\langle a_1\rangle u^{-1}$. To determine the element $a_2 \in u\langle a_1\rangle u^{-1}$ we work with rotation numbers. To this end, let $N = \mathrm{lcm}(l, m, n)$ and $\zeta = \exp(2\pi i/N)$ as above. Then the cyclotomic field $\mathbb{Q}[\zeta]$ contains the rotation numbers of a_1, b_1, c_1. Let t be an integer such that $st = 1 \bmod N$. Then

$$\begin{aligned} \mathrm{rot}(\epsilon^{\psi}(ua_1^t u^{-1}), \overline{D'}) &= (\mathrm{rot}(\epsilon(ua_1^t u^{-1}), u\overline{D}))^s \\ &= (\mathrm{rot}(\epsilon(a_1), \overline{D}))^{st} \\ &= \exp\left(\frac{2\pi i}{l}\right); \end{aligned}$$

and, similarly $\mathrm{rot}(\epsilon^{\psi}(vb_1^t v^{-1}), \overline{E'}) = \exp\left(\frac{2\pi i}{m}\right)$, $\mathrm{rot}(\epsilon^{\psi}(wc_1^t w^{-1}), \overline{E'}) = \exp\left(\frac{2\pi i}{m}\right)$. It follows that $(a_2, b_2, c_2) = (ua_1^t u^{-1}, vb_1^t v^{-1}, wc_1^t w^{-1})$. We summarize the preceding discussion by the following theorem. It is known as the branch cycle argument and a proof is given as Lemma 2.8 in [**16**].

THEOREM 3.3. *Let $\eta_1 : T_{l,m,n} \to G,\ A \to a_1,\ B \to b_1,\ C \to c_1$ be an epimorphism defining an (l, m, n) G-action on the Riemann surface S. Let $\psi \in \mathrm{Gal}(\mathbb{C})$, $N = \mathrm{lcm}(l, m, n)$, $\zeta = \exp(2\pi i/N)$, and suppose that $\psi(\zeta) = \zeta^s$. Select t so that $st = 1 \bmod N$. Then ϵ^{ψ}, the ψ Galois transform action on S^{ψ}, is induced by $\eta_2 : T_{l,m,n} \to G,\ A \to a_2,\ B \to b_2,\ C \to c_2$, such that*

$$(a_2, b_2, c_2) = (ua_1^t u^{-1}, vb_1^t v^{-1}, wc_1^t w^{-1}) \tag{3.5}$$

for some $u, v, w \in G$. Moreover, $a_2 b_2 c_2 = 1$ and $G = \langle a_2, b_2, c_2\rangle$.

As discussed in Section 2, we have $(a_1, b_1, c_1) \in K_G^{\circ}(a_1, b_1, c_1)$ and $(a_2, b_2, c_2) \in K_G^{\circ}(a_1^t, b_1^t, c_1^t)$. There is no simple formula for (a_2, b_2, c_2) known to the author, and the triple needs to be found computationally. Indeed, it is not immediately apparent that the $K_G^\circ(a_1^t, b_1^t, c_1^t)$ is non-empty from simple group theoretic considerations. In [**3**], using equation 2.12 and the action of ψ on characters

$$\chi(a^t) = \psi^{-1}(\chi(a)),\ \chi(b^t) = \psi^{-1}(\chi(b)),\ \chi(c^t) = \psi^{-1}(\chi(c)),$$

it was shown

$$\left|K_G^\circ(a^t, b^t, c^t)\right| = \left|K_G^\circ(a, b, c)\right|,$$

as expected, since we would expect ψ to be a bijection on generating triples.

Because of the possible presence of companion orbits in $K_G^\circ(a_1^t, b_1^t, c_1^t)$ the action determined by equation 3.5 may not be uniquely identifiable in $K_G^\circ(a_1^t, b_1^t, c_1^t)$. So we shall call $K_G^\circ(a_1^t, b_1^t, c_1^t)$ a *Galois $t-$target* for the generating triple (a_1, b_1, c_1). It is easily shown that equation 3.5 determines a bijection between the K orbits in $\left|K_G^\circ(a, b, c)\right|$ and $\left|K_G^\circ(a^t, b^t, c^t)\right|$.

3.2. Facts and Problems about Galois Actions. We start this section with two quick, illustrative examples where everything is computable.

EXAMPLE 3.4. The following example is not quite legitimate since the genus of S is zero. However, it can be used to illustrate some ideas later. Let G be the dihedral group D_n and $S = P^1(\mathbb{C})$. The group G has the presentation

$$G = D_n = \langle a, b : a^2 = b^n = 1, aba = b^{-1}\rangle.$$

The G action is given by $a : z \to 1/z$ and $b : z \to \omega z$, $\omega = \exp(\frac{2\pi i}{n})$. The fixed point sets of $\langle b\rangle$ and $\langle ab^{-j}\rangle$ are $\{0, \infty\}$ and $\{\pm\omega^{j/2}\}$, respectively. It is easily shown that a Belyi function for the action is

$$\beta(z) = \frac{z^n + z^{-n} + 2}{4}.$$

EXAMPLE 3.5. Suppose that S is any surface. Let $\overline{S}$ denote the surface obtained by taking the complex conjugates of the coefficients of a defining equation for S. If the defining equation has real coefficients then complex conjugation maps S to itself. We then say that S has a symmetry. In Figure 2 we can take the reflection of triangle ΔDEF in the side DE as the uniformizing triangle for $\overline{S}$. The induced map on generating triples is.

$$(a, b, c) \to (a^{-1}, b^{-1}, (ab)^{-1}c^{-1}ab)). \tag{3.6}$$

It follows then that S admits a symmetry (normalizing the G-action) if and only if equation 3.6 defines an automorphism of G. In this example, $t = 1$ and $u = 1, v = 1, w = (ab)^{-1}$ in equation 3.5.

3.2.1. *Setup for Galois action analysis.* One of the goals of studying the Galois action on surfaces is to gain some information about $\mathcal{G}$ itself. In this sub-section we set up some underlying structures for analysis of the action on $\mathcal{G}$ on Belyi functions and generating triples.

From Corollaries 2.10 and 2.11 it makes sense to consider the finite set of Belyi functions on a surface with a given group

$$\mathcal{B}(S, G) = \left\{\beta : S \to \widehat{\mathbb{C}} : \beta \text{ a regular Belyi function, } \mathrm{Cov}(\beta) \simeq G\right\}.$$

To use the equations 3.2 with a simplified notation we define the vertex set of a Belyi function.

DEFINITION 3.6. Let $\beta : S \to \widehat{\mathbb{C}}$ be a regular Belyi function branched over $\{0, 1, \infty\}$. Then, the vertex set of β is defined as:

$$\mathcal{V}_\beta = \beta^{-1}(0) \cup \beta^{-1}(1) \cup \beta^{-1}(\infty).$$

REMARK 3.7. The vertex set $\mathcal{V}_\beta$ is the set of vertices of the tiling $\mathcal{T}$ on S. Galois transformations do not mix the type of the vertices, namely $0, 1, \infty$, according to equation 3.2.

PROPOSITION 3.8. *Let $\beta : S \to \widehat{\mathbb{C}}$ be a regular Belyi function, then the points of the vertex set $\mathcal{V}_\beta$ have coordinates in a number field.*

PROOF. Let β be a regular Belyi function on S and suppose that $\psi \in \mathrm{Gal}(\mathbb{C})$ is trivial on k_S. Then, the orbit $\{\beta^{\psi^k}, k \in \mathbb{Z}\}$ is contained in the set of Belyi functions on S, which is finite by Corollary 2.10. Thus, $\beta = \beta^{\psi^e}$ for some positive integer e. It follows from equation 3.2 that ψ^e fixes the vertex set as a set. Suppose now that some coordinate $x \in \mathbb{C}$ of one of the points of the vertex set is transcendental. Let ψ be the automorphism on $k_S(x)$ that is trivial on k_s and $\psi(x) = x + 1$. Then for some e the automorphism ψ^e, when extended to $\mathbb{C}$ and applied repeatedly to the vertex set, produces infinitely many points, a contradiction. □

Let $S = S_1, \ldots, S_e$ denote the set of Galois transforms of S. Each is an embedded smooth curve in $\mathbb{P}^d$ defined over a conjugate of k_S. We note that some of the surfaces may be conformally equivalent, especially if the defining equations are chosen poorly. Here are some collections upon which the Galois group $\mathcal{G}$ acts:

(1) the set $\{S_1, \ldots, S_e\}$;
(2) the set of conformal equivalence classes of surfaces obtained by imposing conformal equivalence on $\{S_1, \ldots, S_e\}$;
(3) the set of all Belyi functions $\mathcal{B}(l, m, n)$ on any of the S_i with signature (l, m, n) and group of covering transformations G;
(4) algebraic equivalence classes of G actions with the same G-signature or classes of generating triples in $A^\circ(a, b, c)$ and $K^\circ(a, b, c)$;
(5) $Aut(G)$ classes of generating triples;
(6) tilings on the S_i; and
(7) the vertex cloud

$$\mathcal{VC}(l, m, n, G) = \bigcup_\beta \mathcal{V}_\beta$$

where the union is taken over all *regular* Belyi functions with signature (l, m, n) and $\mathrm{Cov}(\beta) \simeq G$.

In principle everything can be recovered from the action of $\mathcal{G}$ on $\mathcal{VC}(l, m, n, G)$. Without much knowledge of k_S and the defining equations we would like to propose these sample problems.

(1) Describe the permutation representation of $\mathcal{G}$ on $S_1, \ldots, S_e$.
(2) If $\mathcal{G}_S \subseteq \mathcal{G}$ is the stabilizer of S, describe the following actions:
   - the permutation representation of $\mathcal{G}_S$ on the set $\bigcup_{\beta \in \mathcal{B}(S,G)} \mathcal{V}_\beta$, and
   - the action of $\mathcal{G}_S$ on G-actions on S.
(3) Describe the action of $\mathcal{G}$ on generating triples of G.

3.2.2. *Restrictions on the Galois action.* The Galois action cannot be willy-nilly on the above sets, there are some restrictions. The vertex cloud has a decomposition

$$\mathcal{VC}(l,m,n,G) = \bigcup_{\beta} \mathcal{V}_\beta = \bigcup_{S_i} \bigcup_{\beta \in \mathcal{B}(S_i,G)} \mathcal{V}_\beta$$

Thus the permutation representation on $\mathcal{VC}(l,m,n,G)$ has several levels of quotient representations. First it acts on the set of $\{S_1,\ldots,S_e\}$. In turn this set may have non-trivial equivalence classes of conformally equivalent surfaces. In the other direction, the vertex clouds has a refinement by Belyi functions. $\bigcup_{\beta \in \mathcal{B}(S_i,G)} \mathcal{V}_\beta$. Finally, the stabilizer of a Belyi function maps into the automorphism group of G. Piecing all these actions together imposes some restrictions on $\mathcal{G}$. We give a few examples. The following proposition essentially rephrases Proposition 2.8.

PROPOSITION 3.9. *Suppose that G acts on S via $\epsilon : G \hookrightarrow \mathrm{Aut}(S)$ and that $\psi \in \mathcal{G}_S$. Then, ϵ and ϵ^ψ are conformally equivalent, as in equation* 2.6.

Next we look at a result that shows that a non-trivial Galois action really moves the vertex cloud around, tearing the dessin apart then putting them back together in a non-isomorphic way. See, for instance [**11**], in which the Wilson operations on a graph are related to cyclotomic fields of definition and the cyclotomic action of $\mathcal{G}$. In that paper the following concept of adjacency is used.

DEFINITION 3.10. We say that a $\psi \in \mathcal{G}$ preserves adjacencies in the dessin $\mathcal{D}$ if two vertices connected in $\mathcal{D}$ remain connected in $\mathcal{D}^\psi$.

PROPOSITION 3.11. *Suppose that a $\psi \in \mathcal{G}$ preserves adjacencies in each of the three dessins defining the tiling S. Then S and S^ψ are either conformally equivalent or anti-conformally equivalent.*

PROOF. By the definition of preserving adjacencies, the map of vertex sets $\psi : \mathcal{V}_\beta \to \mathcal{V}_{\beta^\psi}$ extends to a mapping of the 1-skeleton $\mathcal{D}_1 \cup \mathcal{D}_2 \cup \mathcal{D}_3$ of S to the 1-skeleton $\mathcal{D}_1^\psi \cup \mathcal{D}_2^\psi \cup \mathcal{D}_3^\psi$of S^ψ. The map either preserves of reverses the orientation of the triangles. In the case that ψ preserves orientation we get an extended map $S \to S^\psi$ which is a conformal equivalence. Otherwise we get a conformal equivalence $S \to \overline{S^\psi}$. □

The following result is trivial to prove but is worth recording. The action of $\mathrm{Inn}(G)$ on generating triples is easy to understand geometrically. We simply chose a different tile on S to construct the uniformizing $\Pi \hookrightarrow \Gamma \xrightarrow{\eta} G$. However, there is no such easy geometrical interpretation of $\mathrm{Out}(G)$ except as noted in the symmetry case in Example 3.5.

PROPOSITION 3.12. *Suppose that $\psi \in \mathcal{G}$ maps S to itself. Then $g \to g^\psi$ is an automorphism of* $\mathrm{Aut}(S)$.

Thus we are able to recover some of $\mathrm{Out}(G)$ as point transformations on $S(\overline{\mathbb{Q}})$ the algebraic points on S. However, other than complex conjugation they are not continuous and don't really qualify as geometric transformations. An obvious question is: Do we get all of $\mathrm{Out}(G)$? The following example shows that we can get everything.

EXAMPLE 3.13. Consider again the dihedral example given in Example 3.4. There is only one surface, namely the Riemann sphere. The Belyi function has rational coefficients and has no other transforms. The vertex cloud consists of 0, ∞, and the $2n$th roots of 1. Therefore, the only effective action on the vertex cloud is induced by cyclotomic automorphisms. We check easily that the induced action of $\mathcal{G}$ on $\mathrm{Aut}(G)$ is

$$(a, b) \to \left(a, b^t\right)$$

where t and n are relatively prime.

3.2.3. *Action on triples.* First, we are going to split the cyclotomic action from the non-cyclotomic action. Construct a "cyclotomic" splitting of $\mathcal{G}$ as follows: Let

$$\begin{aligned}
&\mathbb{Q}_r = \mathbb{Q}[\exp(2\pi i/r)], \ \mathbb{Q}_\infty = \bigcup\nolimits_r \mathbb{Q}_r \\
&\mathcal{K}_r = \mathcal{K}_r(\mathbb{Q}) = \left\{\theta \in \mathrm{Gal}(\overline{\mathbb{Q}}) : \theta|\mathbb{Q}_r = id\right\} \lhd \mathrm{Gal}(\overline{\mathbb{Q}}) \\
&\mathcal{Q}_r = \mathcal{Q}_r(\mathbb{Q}) = \mathrm{Gal}(\overline{\mathbb{Q}})/\mathcal{K}_r(\mathbb{Q})
\end{aligned}$$

and

$$\mathcal{K} = \mathcal{K}_\infty(\mathbb{Q}), \ \mathcal{Q} = \mathcal{Q}_\infty(\mathbb{Q}).$$

We have exact sequences:

$$\begin{aligned}
\mathcal{K}_r &\hookrightarrow \mathrm{Gal}(\overline{\mathbb{Q}}) \twoheadrightarrow \mathcal{Q}_r \\
\mathcal{K} &\hookrightarrow \mathrm{Gal}(\overline{\mathbb{Q}}) \twoheadrightarrow \mathcal{Q}
\end{aligned}$$

Since $\mathcal{Q}_r \backsimeq \mathrm{Gal}(\mathbb{Q}_r)$ is cyclic, the first sequence is split. The "splitting" descends to a splitting of the action on generating triples. The elements of $\mathcal{K}_\infty$ act trivially on rotation numbers at the fixed points of G. Therefore, in the branch cycle action, elements of $\mathcal{K}_\infty$ yield

$$(a, b, c) \to (uau^{-1}, vbv^{-1}, wcw^{-1}). \tag{3.7}$$

REMARK 3.14. We make the distinction because the action of $\mathcal{Q}_\infty$ is easier to determine, whereas the action of $\mathcal{K}_\infty$ might require a defining field of the surface to get anywhere. Presumably, elements of type $\mathcal{K}_\infty$ might have a nontrivial action if the field of definition is not cyclotomic. Surfaces with cyclotomic fields typically have nice actions on the dessins, see [**11**]. Given a quasi-platonic surface there will be a finite quotient $\overline{\mathcal{K}}$ of $\mathcal{K}_\infty$ such that semi-direct product $\mathcal{Q}_r \ltimes \overline{\mathcal{K}}$ induces all the Galois action on the generating triples. The orbits of $\mathcal{Q}_r$ and $\overline{\mathcal{K}}$ form two partitions of the generating triples. Any two orbits from the two partitions must intersect in a K class (conjugacy) since $\mathcal{Q}_r \cap \overline{\mathcal{K}}$ is trivial. Since the size of the $\mathcal{Q}_r$ orbits has an (easily determined) bound in terms of (l, m, n), we either have numerous Galois orbits or we have non-trivial $\mathcal{K}_\infty$ action.

We give a brief analysis of each type of transformations separately.
Action of $\mathcal{Q}_\infty$. The cyclotomic action is completely implemented by some finite $\mathcal{Q}_r$. This group is cyclic and so that the Galois orbit will be a cycle. The number of elements in the each of the distinct $K_G^\circ(a^t, b^t, c^t)$ will be the same. Thus, the critical question is to determine the number of transforms $(ua^tu^{-1}, vb^tv^{-1}, wcw^{-1})$ that lie in $K_G^\circ(a, b, c)$. This can be computed using the conjugacy class structure of G. We can relate the condition $K_G^\circ(a^t, b^t, c^t) = K_G^\circ(a, b, c)$ to rotation numbers of a, b, c.

PROPOSITION 3.15. *Suppose that for a generating triple (a, b, c) we have*

$$K_G^\circ(a^t, b^t, c^t) = K_G^\circ(a, b, c)$$

then if w is a rotation number of a then so is w^t. Likewise for b and c.

PROOF. Let D be a fixed point of a on S and let N_a be the normalizer of a in G. Then fixed points of a on S is the orbit $N_a \cdot D$ and contains $\frac{|N_a|}{|\langle a\rangle|}$ points. If $a^t = xax^{-1}$ then a fixes $x \cdot D$ and the rotation number of a at $x \cdot D$ is $(rot(a, D))^t$. □

Action of $\mathcal{K}_\infty$. Let us restrict the possible (u, v, w) By conjugating by w^{-1} and a change of variables $u \to w^{-1}u$, and $u \to w^{-1}v$ we may assume the form.

$$(a, b, c) \to (uau^{-1}, vbv^{-1}, c).$$

We can multiply both u and v on the left the same element of $\mathrm{Cent}_G(c)$ and maintain the same action class. Since we must have

$$abc = 1 \Longrightarrow bca = 1$$

and

$$uau^{-1}vbv^{-1}c = 1$$

then

$$\begin{aligned} uau^{-1}vbv^{-1}c &= 1 \\ a\left(a^{-1}uau^{-1}\right)vbv^{-1}b^{-1}bc &= 1 \\ \left(a^{-1}uau^{-1}\right)\left(vbv^{-1}b^{-1}\right)bca &= 1 \\ \left(a^{-1}uau^{-1}\right)\left(vbv^{-1}b^{-1}\right) &= 1 \\ a^{-1}uau^{-1} &= bvb^{-1}v^{-1} \end{aligned}$$

or

$$a^{-1}uau^{-1} = bvb^{-1}v^{-1}. \tag{3.8}$$

This gives us a necessary condition on u, v. We note further that we can act on both u and v on the right by elements of $\mathrm{Cent}_G(a)$ and $\mathrm{Cent}_G(b)$, respectively and still have the same action class, so we have a better chance of enumerating them. The last equation "defines" v in terms of u and we have one degree of freedom (modulo action by centralizers on the left and right). The mapping idea is a bit awkward in large simple groups but could work well in the following case: a tower $V \hookrightarrow G \twoheadrightarrow H$ where H is simple and V is a vector space over a prime and the natural conjugation action of G on V is given by a linear representation. Then pick (a, b, c) that projects to a known generating triple $(\overline{a}, \overline{b}, \overline{c})$ for H. Then the equation $a^{-1}uau^{-1} = bvb^{-1}v^{-1}$ becomes a linear algebra question for $u, v \in V$. This would give us a rich field to produce potential $\mathcal{K}_\infty$ actions.

4. Examples of Galois actions on quasi-platonic actions

4.1. $\mathbf{PSL}_2(q)$ examples. A detailed analysis of $PSL_2(q)$ actions is given in [**3**]. We recall some of the highlights. The following Proposition is proven as Proposition 34 in [**3**].

PROPOSITION 4.1. *For the Galois action of the Galois group $\mathcal{G}$ on the (l, m, n) actions of $PSL_2(q)$ all $\mathcal{G}$ orbits have the same size.*

We may also use the results from [**3**] to prove the following.

THEOREM 4.2. *For $G = PSL_2(q)$ the action of $\mathcal{K}_\infty$ is trivial.*

PROOF. We just sketch the proof. We set $G = PSL_2(q)$ and $L = PGL_2(q)$. If q is even then $G = L$. If q is an odd prime then $|L/K| = 2$ and $A = L$, Otherwise $K < L < A$. Here are the main steps in point form:

- Using Macbeath's results on generating triples we show that $L_G^\circ(a, b, c)$ has one or two L orbits.
- If $L_G^\circ(a, b, c)$ is a single L orbit, then $\mathcal{K}_\infty$ acts trivially since the $\mathcal{K}_\infty$ orbit must be contained in $L_G^\circ(a, b, c)$.
- Otherwise we can lift triples $(a, b, c) \in L_G^\circ(a, b, c)$ to covering triples $(\widetilde{a}, \widetilde{b}, \widetilde{c})$ in the Schur cover $\widetilde{G} = SL_2(q)$. Lifting to the Schur cover separates the two L orbits and it can be shown that the $\mathcal{K}_\infty$ action fixes both orbits rather than exchanging them.
- If q is even or a prime we are done. Otherwise, we have to work further with $A/L \simeq \mathrm{Gal}(\mathbb{F}_q)$.

□

We should discuss briefly the appearance of the Schur cover above. Let $\pi : \widetilde{G} \to G$ be the Schur cover. A lift $(\widetilde{a}, \widetilde{b}, \widetilde{c})$ of a triple (a, b, c) is a triple (usually with a different signature) such that $(\pi(\widetilde{a}), \pi(\widetilde{b}), \pi(\widetilde{c})) = (a, b, c)$. Two inequivalent companion actions in $K_G^\circ(a, b, c)$ and $A_G^\circ(a, b, c)$ can arise as images of lifts with different signatures. In the $PSL_2(q)$ case, lifting can resolve the indeterminacy of the Galois action. For this to work the number of companion actions should be comparable to the kernel of π. This works perfectly for $PSL_2(q)$ and fails miserably for the alternating group examples in the next section.

Tables 2 and 3 list examples of Galois orbits that appear in [**3**] using the above theorem and lots of Magma computation. The triple (α, β, γ) is given by

$$(\alpha, \beta, \gamma) = (\mathrm{Trace}(\widetilde{a}), \mathrm{Trace}(\widetilde{b}), \mathrm{Trace}(\widetilde{c})) \tag{4.1}$$

in the second table w is a primitive element in $\mathbb{F}_8$.

4.2. Alternating group examples. In Tables 4 and 5 below, using Magma, we have calculated the number of G actions, for selected signatures and alternating groups, up to algebraic equivalence. Since the Schur cover $\widetilde{G}$ is not much bigger than G is, it cannot be used to separate orbits when there are large numbers of actions. Let us look at the $(7, 7, 7)$ actions of Alt_8. There are two conjugacy classes of order 7, namely K_{11} and K_{12}, which are exchanged by a transposition in Sym_8. The generating $(7, 7, 7)$ triples break up into eight sets of K classes as given in Table 5. There are

$$(2 \times 36 + 6 \times 47)\,|G| = 354\,|G|$$

generating $(7, 7, 7)$ triples. This yields

$$\frac{354\,|G|}{2\,|G|} = 177$$

$\mathrm{Aut}(G)$ classes. Thus we either have a large number of Galois orbits or there is some nontrivial $\mathcal{K}_\infty$ action. Finally, we see that there are some classes on which $\mathcal{Q}_r$ acts trivially.

TABLE 2. Galois action on hyperbolic $PSL_2(7)$ actions

Signature (l,m,n)	Covering trace triple (α,β,γ)	Genus	Galois Orbit Sizes
$(2,3,7)$	$(0,1,2)$	3	$\{1\}$
$(2,4,7)$	$(0,3,2)$	10	$\{1\}$
$(2,7,7)$	$(0,2,2)$	19	$\{1\}$
$(3,3,4)$	$(1,1,3),(1,1,4)$	8	$\{2\}$
$(3,3,7)$	$(1,1,-2)$	17	$\{1\}$
$(3,4,4)$	$(-1,3,3)$	15	$\{1\}$
$(3,4,7)$	$(1,3,2),(1,3,-2)$	24	$\{1\}$
$(3,7,7)$	$(1,2,2),(-1,2,2)$	33	$\{1,1\}$
$(4,4,4)$	$(3,3,3),(3,3,4)$	22	$\{2\}$
$(4,4,7)$	$(3,3,-2)$	31	$\{1\}$
$(4,7,7)$	$(3,2,2),(-3,2,2)$	40	$\{2\}$
$(7,7,7)$	$(2,2,-2)$	49	$\{1\}$

TABLE 3. Galois action on hyperbolic $PSL_2(8)$ actions

Signature (l,m,n)	Covering trace triple (α,β,γ)	Genus	Galois Orbit Sizes
$(2,3,7)$	$(0,1,w^3)$	7	$\{1\}$
$(2,3,9)$	$(0,1,w^2)$	15	$\{1\}$
$(2,7,7)$	$(0,w^3,w^5),(0,w^3,w^6)$	55	$\{1,1\}$
$(2,7,9)$	$(0,w^3,w),(0,w^3,w^2),(0,w^3,w^4)$	63	$\{3\}$
$(2,9,9)$	$(0,w,w^2),(0,w,w^4)$	71	$\{1,1\}$
$(3,3,7)$	$(1,1,w^3)$	41	$\{1\}$
$(3,3,9)$	$(1,1,w)$	57	$\{1\}$
$(3,7,7)$	$(1,w^3,w^3),(1,w^3,w^5),(1,w^3,w^6)$	97	$\{3\}$
$(3,7,9)$	$(1,w^3,w),(1,w^3,w^2),(1,w^3,w^4)$	105	$\{3\}$
$(3,9,9)$	$(1,w,w),(1,w^2,w^2),(1,w^4,w^4)$	113	$\{1\}$
$(7,7,7)$	$(w^3,w^3,w^5),(w^3,w^3,w^6),$ $(w^3,w^5,w^3),(w^3,w^6,w^3)$	145	$\{1,1,1,1\}$
$(7,7,9)$	$(w^a,w^b,w),a,b=3,5,6$	153	$\{3,3,3\}$
$(7,9,9)$	$(w^3,w^a,w^b),a,b=1,2,4$	161	$\{3,3,3\}$
$(9,9,9)$	$(w,w,w),(w,w,w^4),(w,w^2,w^2)$ $(w,w^2,w^4),(w,w^4,w),(w,w^4,w^2)$	169	$\{1,1,1,1,1,1\}$

TABLE 4. Galois action on hyperbolic alternating group actions

Group Pair (G, A)	$\lvert Z(\widetilde{G})\rvert$	Signature (l, m, n)	K_G° tag (a^G, b^G, c^G)	Genus	Actions
(Alt_7, Sym_7)	6	$(3,3,5)$	(K_4, K_4, K_6)	169	1
(Alt_7, Sym_7)	6	$(3,3,7)$	(K_3, K_4, K_8)	241	1
(Alt_7, Sym_7)	6	$(4,4,4)$	(K_5, K_5, K_5)	316	12
(Alt_7, Sym_7)	6	$(4,4,5)$	(K_5, K_5, K_6)	379	11
(Alt_7, Sym_7)	6	$(4,5,6)$	(K_5, K_6, K_7)	484	4
(Alt_7, Sym_7)	6	$(5,5,5)$	(K_6, K_6, K_6)	505	1
(Alt_7, Sym_7)	6	$(5,5,7)$	(K_6, K_6, K_7)	577	12
(Alt_7, Sym_7)	6	$(5,6,7)$	(K_6, K_7, K_8)	619	7
(Alt_7, Sym_7)	6	$(6,6,6)$	(K_7, K_7, K_7)	631	1
(Alt_7, Sym_7)	6	$(7,7,7)$	(K_8, K_8, K_8)	721	23
(Alt_8, Sym_8)	2	$(4,4,15)$	(K_6, K_6, K_{13})	4369	3
(Alt_8, Sym_8)	2	$(4,4,15)$	(K_6, K_7, K_{13})	4369	12
(Alt_8, Sym_8)	2	$(4,4,15)$	(K_6, K_6, K_{13})	4369	15
(Alt_8, Sym_8)	2	$(7,7,7)$	(K_{11}, K_{11}, K_{11})	5761	177
(Alt_8, Sym_8)	2	$(15,15,15)$	(K_{13}, K_{13}, K_{13})	8065	20

TABLE 5. Sizes of K_G° sets Braid automorphisms not considered

Type of (a^G, b^G, c^G)	$\frac{\lvert K_G^\circ(a,b,c)\rvert}{\lvert G\rvert}$
(K_{11}, K_{11}, K_{11})	36
(K_{12}, K_{11}, K_{11})	47
(K_{11}, K_{12}, K_{11})	47
(K_{11}, K_{11}, K_{12})	47
(K_{11}, K_{12}, K_{12})	47
(K_{12}, K_{11}, K_{12})	47
(K_{12}, K_{12}, K_{11})	47
(K_{12}, K_{12}, K_{12})	36

References

[1] G. V. Belyĭ, *Galois extensions of a maximal cyclotomic field* (Russian), Izv. Akad. Nauk SSSR Ser. Mat. **43** (1979), no. 2, 267–276, 479. MR534593

[2] Thomas Breuer, *Characters and automorphism groups of compact Riemann surfaces*, London Mathematical Society Lecture Note Series, vol. 280, Cambridge University Press, Cambridge, 2000. MR1796706

[3] S. Allen Broughton, *Quasi-platonic $PSL_2(q)$-actions on closed Riemann surfaces*, Albanian J. Math. **9** (2015), no. 1, 31–61. MR3438844

[4] Marston D. E. Conder, *The symmetric genus of alternating and symmetric groups*, J. Combin. Theory Ser. B **39** (1985), no. 2, 179–186, DOI 10.1016/0095-8956(85)90047-4. MR811121

[5] Kevin Coombes and David Harbater, *Hurwitz families and arithmetic Galois groups*, Duke Math. J. **52** (1985), no. 4, 821–839, DOI 10.1215/S0012-7094-85-05243-3. MR816387

[6] Pierre Guillot, *An elementary approach to dessins d'enfants and the Grothendieck-Teichmüller group*, Enseign. Math. **60** (2014), no. 3-4, 293–375, DOI 10.4171/LEM/60-3/4-5. MR3342648

[7] Kenji Hoshino, *The Belyi functions and dessin d'enfants corresponding to the non-normal inclusions of triangle groups*, Math. J. Okayama Univ. **52** (2010), 45–60. MR2589845

[8] G. A. Jones, *Enumerating Regular maps and Hypermaps*, Disertaciones del Semiario de Matematics Fundematales num. 3, UNED, (1989).

[9] Gareth Jones and David Singerman, *Belyĭ functions, hypermaps and Galois groups*, Bull. London Math. Soc. **28** (1996), no. 6, 561–590, DOI 10.1112/blms/28.6.561. MR1405488

[10] Gareth Jones, David Singerman, and Paul Watson, *Symmetries of quasiplatonic Riemann surfaces*, Rev. Mat. Iberoam. **31** (2015), no. 4, 1403–1414, DOI 10.4171/RMI/873. MR3438394

[11] G. A. Jones, M. Streit, and J. Wolfart, *Wilson's map operations on regular dessins and cyclotomic fields of definition*, Proc. Lond. Math. Soc. (3) **100** (2010), no. 2, 510–532, DOI 10.1112/plms/pdp033. MR2595748

[12] MAGMA. Magma Computational Algebra System, Computational Algebra Group, University of Sydney.

[13] K. Magaard, T. Shaska, S. Shpectorov, and H. Völklein, *The locus of curves with prescribed automorphism group*, Sūrikaisekikenkyūsho Kōkyūroku **1267** (2002), 112–141. Communications in arithmetic fundamental groups (Kyoto, 1999/2001). MR1954371

[14] David Singerman, *Finitely maximal Fuchsian groups*, J. London Math. Soc. (2) **6** (1972), 29–38, DOI 10.1112/jlms/s2-6.1.29. MR0322165

[15] David Singerman and Robert I. Syddall, *The Riemann surface of a uniform dessin*, Beiträge Algebra Geom. **44** (2003), no. 2, 413–430. MR2017042

[16] Helmut Völklein, *Groups as Galois groups*, Cambridge Studies in Advanced Mathematics, vol. 53, Cambridge University Press, Cambridge, 1996. An introduction. MR1405612

Department of Mathematics, Rose-Hulman Institute of Technology, Terre Haute, Indiana 47803

Email address: `brought@rose-hulman.edu`

Contemporary Mathematics
Volume **703**, 2018
http://dx.doi.org/10.1090/conm/703/14129

Equations of Riemann surfaces with automorphisms

David Swinarski

ABSTRACT. We present an algorithm for computing equations of canonically embedded Riemann surfaces with automorphisms. This is used to produce equations of Riemann surfaces with large automorphism groups for genus 7. The main tools are the Eichler trace formula for the character of the action of the automorphism group on holomorphic differentials, algorithms for producing matrix generators of a representation of a finite group with a specified irreducible character, and Gröbner basis techniques for computing flattening stratifications.

Riemann surfaces (or algebraic curves) with automorphisms have been important objects of study in complex analysis, algebraic geometry, number theory, and mathematical physics for more than a century, as their symmetries often permit us to do calculations that would otherwise be intractable.

Such Riemann surfaces are special in the sense that a general Riemann surface of genus $g \geq 3$ has no nontrivial automorphisms. Moreover, the group of automorphisms of a Riemann surface of genus $g \geq 2$ is finite.

A great deal of progress has been made on classifying and studying these Riemann surfaces. Notably, for the important special cases of cyclic curves or superelliptic curves, the automorphism groups that may occur and affine plane equations of these curves have been published. See [**1**, **20**] for recent work by Sanjeewa, Shaska, Beshaj, and Zhupa on cyclic and superelliptic curves, and [**3**, **4**, **22**] for some prior work on hyperelliptic and cyclic trigonal curves.

In the general case, Breuer and Conder performed computer searches that for each genus g list the Riemann surfaces of genus g with large automorphism groups (that is, $|\operatorname{Aut}(X)| > 4(g_X - 1)$). Specifically, they list sets of surface kernel generators (see Definition 1.2 below), which describe these Riemann surfaces as branched covers of $\mathbb{P}^1$. Breuer's list extends to genus $g = 48$, and Conder's list extends to genus $g = 101$ [**2**, **5**]. Even for small values of g, these lists are extremely large, as a surface X may appear several times for various subgroups of its full automorphism group. In [**15**], Magaard, Shaska, Shpectorov, and Völklein refined Breuer's list by determining which surface kernel generators correspond to the full automorphism group of the Riemann surface. This reveals that many, but not all, of the Riemann surfaces with large automorphism groups are cyclic or superelliptic.

To our knowledge, at this time there is no algorithm published in the literature for producing equations of the noncyclic Riemann surfaces. Here, we present

2010 *Mathematics Subject Classification.* Primary 14H37, 14H45.

an algorithm to compute canonical equations of an arbitrary nonhyperelliptic Riemann surfacewith automorphisms. The main tools are the Eichler trace formula for the character of the action of $\mathrm{Aut}(X)$ on holomorphic differentials, algorithms for producing matrix generators of a representation of a finite group with a specified irreducible character, and Gröbner basis techniques for computing flattening stratifications. This algorithm has been used to produce equations of the nonhyperelliptic Riemann surfaces with genus $4 \le g_X \le 7$ satisfying $|\mathrm{Aut}(X)| > 4(g_X - 1)$.

A few more remarks may help distinguish Algorithm 1.1 from previous work. For cyclic or superelliptic curves, the equations in [**1**,**20**] are more concise than those obtained by Algorithm 1.1, so Algorithm 1.1 may be considered complementary to these papers. The input to Algorithm 1.1 need not be the full automorphism group, so it can be run from Breuer or Conder's data (though we recommend using the full automorphism group from [**15**] whenever possible). Furthermore, Algorithm 1.1 can be applied to arbitrary families of Riemann surfaces with automorphisms (not just those with 0- or 1-dimensional Hurwitz loci). Finally, the canonical ideal of a Riemann surface contains some information that the author does not know how to obtain using techniques of Fuchsian group theory alone. For example, the Betti table can be computed from the canonical ideal, and for low genus curves, this yields information about the linear series on the curve (see [**21**]).

Here is an outline of the paper. In Section 1, we describe the main algorithm. In Section 2, we describe one example in detail, a genus 7 Riemann surface with 54 automorphisms. In Section 3, we give equations of selected canonically embedded Riemann surfaces of genus 7 along with matrix surface kernel generators. The calculations were performed in `Magma`, `GAP`, and `Macaulay2` [**11**,**13**,**16**].

Online material. The webpage [**24**] contains links to the latest version of the author's `Magma` code, files detailing the calculations for specific examples, and many equations that are not printed in Section 3.

1. The main algorithm

We begin by stating the main algorithm. Then, in the following subsections, we discuss the steps in more detail, including precise definitions and references for terms and facts that are not commonly known.

ALGORITHM 1.1.
INPUTS:

(1) *A finite group G;*
(2) *an integer $g \ge 3$;*
(3) *a set of surface kernel generators $(a_1, \ldots, a_{g_0}; b_1, \ldots, b_{g_0}; g_1, \ldots, g_r)$ determining a family of nonhyperelliptic Riemann surfaces X of genus g with $G \subset \mathrm{Aut}(X)$*

OUTPUT: *A locally closed set $B \subset \mathbb{A}^n$ and a family of smooth curves $\mathcal{X} \subset \mathbb{P}^{g-1} \times B$ such that for each closed point $b \in B$, the fiber $\mathcal{X}_b$ is a smooth genus g canonically embedded curve with $G \subset \mathrm{Aut}(\mathcal{X}_b)$.*

Step 1. *Compute the conjugacy classes and character table of G.*
Step 2. *Use the Eichler trace formula to compute the character of the action on differentials and on quadrics and cubics in the canonical ideal.*
Step 3. *Obtain matrix generators for the action on holomorphic differentials.*
Step 4. *Use the projection formula to obtain candidate quadrics and cubics.*
Step 5. *Compute a flattening stratification and select the locus yielding smooth algebraic curves with degree $2g-2$ and genus g.*

We comment on Steps 2-5 below.

1.1. Step 2: Counting fixed points and the Eichler trace formula.

We begin by defining surface kernel generators.

DEFINITION 1.2 (cf. [**2**] Theorem 3.2, Theorem 3.14). A *signature* is a list of integers $(g_0; e_1, \ldots, e_r)$ with $g_0 \geq 0$, $r \geq 0$, and $e_i \geq 2$.

A set of *surface kernel generators* for a finite group G and signature $(g_0; e_1, \ldots, e_r)$ is a sequence of elements $a_1, \ldots, a_{g_0}, b_1, \ldots, b_{g_0}, g_1, \ldots, g_r \in G$ such that

(1) $\langle a_1, \ldots, a_{g_0}, b_1, \ldots, b_{g_0}, g_1, \ldots, g_r \rangle = G$;
(2) $\operatorname{Order}(g_i) = e_i$; and
(3) $\prod_{j=1}^{g_0} [a_j, b_j] \prod_{i=1}^{r} g_i = \operatorname{Id}_G$.

Surface kernel generators are called *ramification types* in [**15**] and *generating vectors* in [**19**].

As explained in [**2**, Section 3.11], surface kernel generators describe the quotient morphism $X \to X/G$ as a branched cover. Here X is a Riemann surface of genus g, G is a subgroup of $\operatorname{Aut}(X)$, the quotient X/G has genus g_0, the quotient morphism branches over r points, and the integers e_i describe the ramification over the branch points.

Surface kernel generators are used in the following formula for the number of fixed points of an automorphism:

THEOREM 1.3 ([**2**, Lemma 11.5]). *Let σ be an automorphism of order $h > 1$ of a Riemann surface X of genus $g \geq 2$. Let $(g_1, \ldots, g_r)$ be part of a set of surface kernel generators for X, and let $(m_1, \ldots, m_r)$ be the orders of these elements. Let $Fix_{X,u}(\sigma)$ be the set of fixed points of σ, where σ acts on a neighborhood of the fixed point by $z \mapsto \exp(2\pi i u/h) z$. Then*

$$|\operatorname{Fix}_{X,u}(\sigma)| = |C_G(\sigma)| \sum_{\substack{g_i \text{ s.t.} \\ h | m_i \\ \sigma \sim g_i^{m_i u/h}}} \frac{1}{m_i}$$

Here $C_G(\sigma)$ is the centralizer of σ in G, and $\sim$ denotes conjugacy.

Next we recall the Eichler Trace Formula. For a Riemann surface X, let Ω_X be the holomorphic cotangent bundle, and let $\omega_X = \bigwedge \Omega_X$ be the sheaf of holomorphic differentials. The Eichler Trace Formula gives the character of the action of $\operatorname{Aut}(X)$ on $\Gamma(\omega_X^{\otimes d})$.

THEOREM 1.4 (Eichler Trace Formula [9, Theorem V.2.9]). *Suppose $g_X \geq 2$, and let σ be a nontrivial automorphism of X of order h. Write χ_d for the character of the representation of* Aut(X) *on* $\Gamma(\omega_X^{\otimes d})$. *Then*

$$\chi_d(\sigma) = \begin{cases} 1 + \displaystyle\sum_{\substack{1 \leq u < h \\ (u,h)=1}} |\operatorname{Fix}_{X,u}(\sigma)| \frac{\zeta_h^u}{1-\zeta_h^u} & \text{if } d = 1 \\ \displaystyle\sum_{\substack{1 \leq u < h \\ (u,h)=1}} |\operatorname{Fix}_{X,u}(\sigma)| \frac{\zeta_h^{u(d\%h)}}{1-\zeta_h^u} & \text{if } d \geq 2 \end{cases}$$

Here $d\%h$ denotes the unique integer x between 0 and $h-1$ such that $d \equiv x \pmod{h}$.

Together, the previous two results give a group-theoretic method for computing the character of the Aut(X) action on $\Gamma(\omega_X^{\otimes d})$ starting from a set of surface kernel generators.

We can use the character of Aut(X) on $\Gamma(\omega_X^{\otimes d})$ to obtain the character of Aut(X) on quadrics and cubics in the canonical ideal as follows. Let S be the coordinate ring of $\mathbb{P}^{g-1}$, let $I \subset S$ be the canonical ideal, and let S_d and I_d denote the degree d subspaces of S and I.

By Noether's Theorem, the sequence

$$0 \to I_d \to S_d \to \Gamma(\omega_X^{\otimes d}) \to 0$$

is exact for each $d \geq 2$, and by Petri's Theorem, the canonical ideal is generated either by quadrics or by quadrics and cubics. Thus, beginning with the character of the action on $\Gamma(\omega_X) \cong S_1$, we may compute the characters of the actions on $S_2 = \operatorname{Sym}^2 S_1$ and $S_3 = \operatorname{Sym}^3 S_1$ and $\Gamma(\omega_X^{\otimes 2})$ and $\Gamma(\omega_X^{\otimes 3})$, and then obtain the characters of the actions on I_2 and I_3.

1.2. Step 3: matrix generators for a specified irreducible character. From Step 2 we have the character of the action on $\Gamma(\omega_X)$. We seek matrix generators for this action.

Finding efficient algorithms to produce matrix generators of a representation of a finite group G with a specified character is a subject of ongoing research [6, 7]. For lack of a suitably general reference, in [24], we outline an algorithm that was suggested by Valery Alexeev and James McKernan. This algorithm is not expected to perform efficiently; it is given merely to establish that Step 3 in Algorithm 1.1 can be performed algorithmically.

Finally, we note that in [23], Streit describes a method for producing matrix generators for the action of Aut(X) on $\Gamma(\omega_X)$ for some Belyĭ curves.

1.3. Step 4: the projection formula. Recall the projection formula for representations of finite groups (see for instance [10] formula (2.31)).

THEOREM 1.5 (Projection formula). *Let V be a finite-dimensional representation of a finite group G over $\mathbb{C}$. Let $V_1, \ldots, V_k$ be the irreducible representations of G, let χ_i be their characters, and let $V \cong \bigoplus_{i=1}^k V_i^{\oplus m_i}$. Let $\pi_i : V \to V_i^{\oplus m_i}$ be the projection onto the ith isotypical component of V. Then*

$$\pi_i = \frac{\dim(V_i)}{|G|} \sum_{g \in G} \overline{\chi_i(g)} g.$$

From Step 3, we have matrix generators for the G action on $\Gamma(\omega_X) = S_1$. Thus, we can compute matrix generators for the actions on S_2 and S_3, and use the projection formula to compute the isotypical subspace $S_{d,p}$ of degree d polynomials on which G acts with character χ_p. In some examples, we have $I_{d,p} = S_{d,p}$, but more commonly, we have strict containment $I_{d,p} \subset S_{d,p}$. In this case we write elements of $I_{d,p}$ as generic linear combinations of the basis elements of $S_{d,p}$ and then seek coefficients that yield a smooth algebraic curve with the correct degree and genus.

The coefficients used to form these generic linear combinations form the base space $\mathbb{A}^n$ of the family $\mathcal{X}$ produced by the main algorithm.

1.4. Step 5: Flattening stratifications.

THEOREM 1.6 ([**17**, Lecture 8]). *Let $f : X \to S$ be a projective morphism with S a reduced Noetherian scheme. Then there exist locally closed subsets $S_1, \dots, S_n$ such that $S = \sqcup_{i=1}^n S_i$ and $f|_{f^{-1}(S_i)}$ is flat.*

The stratification $S = \sqcup_{i=1}^n S_i$ is called a flattening stratification for the map f. Since S is reduced, flatness implies that over each stratum, the Hilbert polynomial of the fibers is constant. We find the stratum with Hilbert polynomial $P(t) = (2g-2)t - g + 1$, then intersect this stratum with the locus where the fibers are smooth. This completes Algorithm 1.1.

Flattening stratifications have been an important tool in theoretical algebraic geometry for over 50 years. There exist Gröbner basis techniques for computing flattening stratifications; in the computational literature, these are typically called *comprehensive* or *parametric Gröbner bases*, or *Gröbner systems*. The foundational work on this problem was begun by Weispfenning, and many authors, including Manubens and Montes, Suzuki and Sato, Nabeshima, and Kapur, Sun, and Wang, have made important improvements on the original algorithm [**12, 18, 25**].

The size of a Gröbner basis can grow very quickly with the number of variables and generators of an ideal, and unfortunately, even the most recent software cannot compute flattening stratifications for the examples we consider. We discuss a strategy for circumventing this obstacle below.

The algorithms for comprehensive Gröbner bases described in Section 1.4 all begin with the same observation. Let $S = K[x_0, \dots, x_m]$ be a polynomial ring over a field. Let $\preceq$ be a multiplicative term order on S. Then a theorem of Macaulay states that the Hilbert function of I is the same as the Hilbert function of its initial ideal with respect to this term order (see [**8**, Theorem 15.26]).

Therefore, whenever two ideals in S have Gröbner bases with the same leading monomials with respect to some term order, they will have the same initial ideal for that term order, hence they must have the same Hilbert function and Hilbert polynomial, and therefore they will lie in the same stratum of a flattening stratification. To reach a different stratum of the flattening stratification, it is necessary to alter the leading terms of the Gröbner basis — for instance, by restricting to the locus where that coefficient vanishes.

Here is a brief example to illustrate this idea. Let $\mathbb{A}^2$ have coordinates c_1, c_2, and let $\mathbb{P}^3$ have coordinates x_0, x_1, x_2, x_3. The ideal

$$I = \langle c_1x_0x_2 - c_2x_1^2, c_1x_0x_3 - c_2x_1x_2, c_1x_1x_3 - c_2x_2^2 \rangle$$

defines a 2-parameter family of subschemes of $\mathbb{P}^3$. A Gröbner basis for I in $\mathbb{C}[c_1, c_2][x_0, x_1, x_2, x_3]$ with respect to the lexicographic term order is

$$\begin{gathered} c_1x_0x_2 - c_2x_1^2, c_1x_0x_3 - c_2x_1x_2, c_1x_1x_3 - c_2x_2^2, \\ (c_1c_2 - c_2^2)x_1x_2^2, c_2x_1^2x_3 - c_2x_1x_2^2, (c_1c_2 - c_2^2)x_1^2x_2, c_2x_0x_2^2 - c_2x_1^2x_2, \\ (c_1c_2^2 - c_2^3)x_1^4, (c_1c_2^2 - c_2^3)x_2^4, c_2^2x_1x_2^2x_3 - c_2^2x_2^4, c_2^2x_0x_1^2x_2 - c_2^2x_1^4. \end{gathered}$$

Over the locus where c_1, c_2, and $c_1 - c_2$ are invertible, the initial ideal is $\langle x_0x_2, x_0x_3, x_1x_3, x_1x_2^2, x_1^2x_2, x_1^4, x_2^4\rangle$ with Hilbert polynomial $P(t) = 8$. On the other hand, when $c_1 = 0$, or $c_2 = 0$, or $c_1 - c_2 = 0$, we get a different initial ideal and Hilbert polynomial. For example, the locus $c_1 = c_2 \neq 0$ yields the twisted cubic with $P(t) = 3t + 1$.

2. Example: A genus 7 Riemann surface with 54 automorphisms

Magaard, Shaska, Shpectorov, and Völklein's tables show that there exist two smooth, compact genus 7 Riemann surfaces with full automorphism group G given by the group labeled $(54, 6)$ in the `GAP` library of small finite groups [**15**]. For each of these two Riemann surfaces, $X/G \cong \mathbb{P}^1$. The quotient morphisms are branched over 3 points of $\mathbb{P}^1$, and the ramification indices over these points are 2, 6, and 9.

A polycyclic presentation of the group G is

$$\langle g_1, g_2, g_3, g_4 | g_1^2, g_2^3, g_3^3g_4^{-2}, g_4^3, g_1^{-1}g_3g_1(g_3^2g_4)^{-1}, g_2^{-1}g_3g_2(g_3g_4)^{-1}, g_1^{-1}g_4g_1(g_4)^{-2}\rangle.$$

Breuer's surface kernel generators for the first of these two surfaces are $a = g_1$, $b = g_1g_2g_3^2g_4$, and $c = g_2^2g_3g_4^2$. This Riemann surface is not cyclic or superelliptic.

STEP 1. We use `Magma` to compute the conjugacy classes and character table of G. There are 10 conjugacy classes, represented by the following elements: Id_G, g_1, g_4, g_2^2, g_2,g_1g_2, $g_1g_2^2$, $g_2g_3g_4$, g_3, $g_2^2g_3g_4^2$.

The irreducible characters are given below by their values on the ten conjugacy classes.

Class	1	2	3	4	5	6	7	8	9	10
χ_1	1	1	1	1	1	1	1	1	1	1
χ_2	1	-1	1	1	1	-1	-1	1	1	1
χ_3	1	-1	1	ζ_3	ζ_3^2	$-\zeta_3^2$	$-\zeta_3$	ζ_3^2	1	ζ_3
χ_4	1	1	1	ζ_3^2	ζ_3	ζ_3	ζ_3^2	ζ_3	1	ζ_3^2
χ_5	1	1	1	ζ_3	ζ_3^2	ζ_3^2	ζ_3	ζ_3^2	1	ζ_3
χ_6	1	-1	1	ζ_3^2	ζ_3	$-\zeta_3$	$-\zeta_3^2$	ζ_3	1	ζ_3^2
χ_7	2	0	2	2	2	0	0	-1	-1	-1
χ_8	2	0	2	$2\zeta_3$	$2\zeta_3^2$	0	0	$-\zeta_3^2$	-1	$-\zeta_3$
χ_9	2	0	2	$2\zeta_3^2$	$2\zeta_3$	0	0	$-\zeta_3$	-1	$-\zeta_3^2$
χ_{10}	6	0	-3	0	0	0	0	0	0	0

STEP 2. Let V_i be the irreducible G-module with character χ_i given by the table above. For any G-module V, let $V \cong \bigoplus_{i=1}^r V_i^{\oplus m_i}$ be its decomposition into irreducible G-modules.

We use the Eichler trace formula to compute these multiplicities m_i for several relevant G-modules. Let $S = \mathbb{C}[x_0, x_1, x_2, x_3, x_4, x_5, x_6]$, and let S_d denote polynomials of degree d. Let I_d be the kernel of $\to S_d \to \Gamma(\omega_X^{\otimes d})$. Then we have

$$S_1 \cong \Gamma(\omega_X) \cong V_3 \oplus V_{10}$$

$$I_2 \cong V_1 \oplus V_4 \oplus V_8 \oplus V_{10}$$

$$S_2 \cong V_1 \oplus V_4^{\oplus 2} \oplus V_5 \oplus V_7 \oplus V_8 \oplus V_9 \oplus V_{10}^{\oplus 3}$$
$$\Gamma(\omega_X^{\otimes 2}) \cong V_4 \oplus V_5 \oplus V_7 \oplus V_9 \oplus V_{10}^{\oplus 2}$$

STEP 3. We use GAP to obtain matrix representatives of a G action with character equal to the character of the G action on S_1. Such a representation is obtained by mapping the surface kernel generators $a = g_1$ and $b = g_1 g_2 g_3^2 g_4$ to the matrices below.

$$\begin{bmatrix} -1 & 0 & 0 & 0 & 0 & 0 & 0 \\ 0 & 0 & 0 & 0 & 0 & 0 & 1 \\ 0 & 0 & 0 & 0 & 1 & 0 & 0 \\ 0 & 0 & 0 & 0 & 0 & 1 & 0 \\ 0 & 0 & 1 & 0 & 0 & 0 & 0 \\ 0 & 0 & 0 & 1 & 0 & 0 & 0 \\ 0 & 1 & 0 & 0 & 0 & 0 & 0 \end{bmatrix}, \quad \begin{bmatrix} -\zeta_3^2 & 0 & 0 & 0 & 0 & 0 & 0 \\ 0 & 0 & 0 & 0 & 0 & \zeta_9^{-1} & 0 \\ 0 & 0 & 0 & 0 & 0 & 0 & \zeta_9^2 \\ 0 & 0 & 0 & 0 & \zeta_9^5 & 0 & 0 \\ 0 & \zeta_9^{-2} & 0 & 0 & 0 & 0 & 0 \\ 0 & 0 & \zeta_9^4 & 0 & 0 & 0 & 0 \\ 0 & 0 & 0 & \zeta_9 & 0 & 0 & 0 \end{bmatrix}$$

STEP 4. We use the projection formula to decompose the G-module of quadrics S_2 into its isotypical components. When an isotypical component has multiplicity greater than 1, we (noncanonically) choose ordered bases so that the G action is given by the same matrices on each ordered basis.

$$\begin{array}{lllll}
S_{2,1} & \cong & V_1 & = & \langle x_1x_6 + x_2x_4 + x_3x_5\rangle \\
S_{2,4} & \cong & V_4^{\oplus 2} & = & \langle x_0^2\rangle \oplus \langle x_1x_6 - \zeta_6x_2x_4 + \zeta_3x_3x_5\rangle \\
S_{2,8} & \cong & V_8 & = & \langle x_1x_4 + \zeta_3x_2x_5 - \zeta_6x_3x_6, x_1x_5 + \zeta_3x_2x_6 - \zeta_6x_3x_4\rangle \\
S_{2,10} & \cong & V_{10}^{\oplus 3} & = & \langle x_0x_1, x_0x_2, x_0x_3, x_0x_4, x_0x_5, x_0x_6\rangle \\
 & & & \oplus & \langle \zeta_6x_5^2, -x_6^2, -\zeta_3x_4^2, x_1^2, \zeta_3x_2^2, -\zeta_6x_3^2\rangle \\
 & & & \oplus & \langle -x_4x_6, -\zeta_3x_4x_5, \zeta_6x_5x_6, \zeta_3x_2x_3, -\zeta_6x_1x_3, x_1x_2\rangle
\end{array}$$

The second isotypical subspace yields a polynomial of the form

$$c_1(x_0^2) + c_2(x_1x_6 - \zeta_6x_2x_4 + \zeta_3x_3x_5).$$

Assuming that c_1 and c_2 are nonzero, we may scale x_0 to get $c_1 = c_2 = 1$.

The fourth isotypical subspace yields polynomials of the form

$$\begin{array}{l}
c_3x_0x_1 + c_4\zeta_6x_5^2 - c_5x_4x_6, \\
c_3x_0x_2 - c_4x_6^2 - c_5\zeta_3x_4x_5, \\
c_3x_0x_3 - c_4\zeta_3x_4^2 + c_5\zeta_6x_5x_6, \\
c_3x_0x_4 + c_4x_1^2 + c_5\zeta_3x_2x_3, \\
c_3x_0x_5 + c_4\zeta_3x_2^2 - c_5\zeta_6x_1x_3, \\
c_3x_0x_6 - c_4\zeta_6x_3^2 + c_5x_1x_2.
\end{array}$$

Assuming that the leading coefficient is nonzero, we may set it to 1.

Thus, this Riemann surface has a canonical ideal of the form

$$\begin{array}{l}
x_1x_6 + x_2x_4 + x_3x_5, \\
x_0^2 + x_1x_6 - \zeta_6x_2x_4 + \zeta_3x_3x_5, \\
x_1x_4 + \zeta_3x_2x_5 - \zeta_6x_3x_6, \\
x_1x_5 + \zeta_3x_2x_6 - \zeta_6x_3x_4, \\
x_0x_1 + c_4\zeta_6x_5^2 - c_5x_4x_6, \\
x_0x_2 - c_4x_6^2 - c_5\zeta_3x_4x_5, \\
x_0x_3 - c_4\zeta_3x_4^2 + c_5\zeta_6x_5x_6, \\
x_0x_4 + c_4x_1^2 + c_5\zeta_3x_2x_3, \\
x_0x_5 + c_4\zeta_3x_2^2 - c_5\zeta_6x_1x_3, \\
x_0x_6 - c_4\zeta_6x_3^2 + c_5x_1x_2
\end{array}$$

For generic values of c_4 and c_5, the intersection of these ten quadrics is empty.

STEP 5. To find values of the coefficients c_4, c_5 that yield a smooth curve, we partially compute a flattening stratification. Begin Buchberger's algorithm with respect to the lexicographic term order. We compute the S-pair reductions between the generators and find that

$$S(f_1, f_5) \to (\zeta_6 - 1)(c_4 + c_5)x_4^2x_5 + \cdots$$
$$S(f_2, f_5) \to (-c_4c_5 + 1)x_1^2x_6 + \cdots$$

Therefore, in Buchberger's algorithm, these polynomials are added to the Gröbner basis. This suggests that the locus given by the equations $c_4+c_5 = 0$ and $-c_4c_5+1 = 0$ may be an interesting stratum in the flattening stratification.

We may check in Magma that the values $c_4 = i$ and $c_5 = -i$ yield a smooth genus 7 curve in $\mathbb{P}^6$ with the desired automorphism group. (Notice that further scaling x_0 by i allows us to write these equations over a smaller field.)

From these equations, we can compute the minimal free resolution and Betti table of this ideal:

1					
	10	16			
			16	10	
					1

Schreyer has classified Betti tables of genus 7 canonical curves in [**21**]. This Betti table implies that the curve is not hyperelliptic, trigonal, or tetragonal, and it has no degree 6 morphism $C \to \mathbb{P}^2$.

3. Selected results

In the following table we give equations for the Riemann surfaces of genus 7 with large automorphism groups (that is, $|\operatorname{Aut}(X)| > 4(g_X - 1)$) that are isolated in moduli ($\delta = 0$, in the notation of [**15**]). Equations for the 1-dimensional families as well as many lower genus examples can be found at [**24**]. We order the examples as they appear in Table 4 in [**15**].

For hyperelliptic Riemann surfaces, we give an equation of the form $y^2 = f(x)$. Many of these are classically known, and all of them can be found in [**22**].

For nonhyperelliptic curves, we give equations of the canonical ideals and surface kernel generators as elements of $\mathrm{GL}(g, \mathbb{C})$. When a matrix $M \in \mathrm{GL}(g, \mathbb{C})$ is sufficiently sparse, we frequently write the product $[x_0, \ldots, x_{g-1}]M$ to save space. For the cyclic trigonal curves, we also give a cyclic trigonal equation.

Throughout the tables below, canonical ideals are shown in the polynomial ring $\mathbb{C}[x_0, \ldots, x_6]$. The symbol ζ_n denotes $e^{2\pi i/n}$, and we write i for ζ_4.

After computing the canonical equations of the nonhyperelliptic Riemann surfaces, we can compute the Betti tables of these ideals and use the results of [**21**] to classify the curve as having a g_4^1, g_6^2, g_3^1, or none of these.

Finally, note that some of the examples below are not cyclic or superelliptic.

Genus 7, Locus 1: Group $(504, 156)$, signature (2,3,7)

Ideal: Macbeath,[**14**]:

$$x_0^2 + x_1^2 + x_2^2 + x_3^2 + x_4^2 + x_5^2 + x_6^2,$$
$$x_0^2 + \zeta_7 x_1^2 + \zeta_7^2 x_2^2 + \zeta_7^3 x_3^2 + \zeta_7^4 x_4^2 + \zeta_7^5 x_5^2 + \zeta_7^6 x_6^2,$$
$$x_0^2 + \zeta_7^{-1} x_1^2 + \zeta_7^{-2} x_2^2 + \zeta_7^{-3} x_3^2 + \zeta_7^{-4} x_4^2 + \zeta_7^{-5} x_5^2 + \zeta_7^{-6} x_6^2,$$
$$(\zeta_7^{-3} - \zeta_7^3) x_0 x_6 - (\zeta_7^{-2} - \zeta_7^2) x_1 x_4 + (\zeta_7 - \zeta_7^{-1}) x_3 x_5,$$
$$(\zeta_7^{-3} - \zeta_7^3) x_1 x_0 - (\zeta_7^{-2} - \zeta_7^2) x_2 x_5 + (\zeta_7 - \zeta_7^{-1}) x_4 x_6,$$
$$(\zeta_7^{-3} - \zeta_7^3) x_2 x_1 - (\zeta_7^{-2} - \zeta_7^2) x_3 x_6 + (\zeta_7 - \zeta_7^{-1}) x_5 x_0,$$
$$(\zeta_7^{-3} - \zeta_7^3) x_3 x_2 - (\zeta_7^{-2} - \zeta_7^2) x_4 x_0 + (\zeta_7 - \zeta_7^{-1}) x_6 x_1,$$
$$(\zeta_7^{-3} - \zeta_7^3) x_4 x_3 - (\zeta_7^{-2} - \zeta_7^2) x_5 x_1 + (\zeta_7 - \zeta_7^{-1}) x_0 x_2,$$
$$(\zeta_7^{-3} - \zeta_7^3) x_5 x_4 - (\zeta_7^{-2} - \zeta_7^2) x_6 x_2 + (\zeta_7 - \zeta_7^{-1}) x_1 x_3,$$
$$(\zeta_7^{-3} - \zeta_7^3) x_6 x_5 - (\zeta_7^{-2} - \zeta_7^2) x_0 x_3 + (\zeta_7 - \zeta_7^{-1}) x_2 x_4$$

Maps: $(x_0, \ldots, x_6) \mapsto (x_0, -x_1, -x_2, -x_3, x_4, x_5, -x_6),$

$$\begin{bmatrix} 0 & \frac{1}{2} & \frac{1}{2} & -\frac{1}{2} & 0 & -\frac{1}{2} & 0 \\ -\frac{1}{2} & -\frac{1}{2} & \frac{1}{2} & 0 & -\frac{1}{2} & 0 & 0 \\ \frac{1}{2} & -\frac{1}{2} & 0 & -\frac{1}{2} & 0 & 0 & -\frac{1}{2} \\ -\frac{1}{2} & 0 & -\frac{1}{2} & 0 & 0 & -\frac{1}{2} & -\frac{1}{2} \\ 0 & \frac{1}{2} & 0 & 0 & -\frac{1}{2} & \frac{1}{2} & -\frac{1}{2} \\ \frac{1}{2} & 0 & 0 & \frac{1}{2} & -\frac{1}{2} & -\frac{1}{2} & 0 \\ 0 & 0 & \frac{1}{2} & \frac{1}{2} & \frac{1}{2} & 0 & -\frac{1}{2} \end{bmatrix}$$

Genus 7, Locus 2: Group $(144, 127)$, signature (2,3,12). Has g_6^2

Ideal:

$$x_0^2 + x_3 x_4 - \zeta_6 x_3 x_5 - \zeta_6 x_5 x_6,$$
$$2i x_1^2 + x_3 x_4 + \zeta_6 x_3 x_5 + 2 x_4 x_6 - \zeta_6 x_5 x_6,$$
$$2i x_1 x_2 + (-2\zeta_6 + 1) x_3 x_4 + \zeta_6 x_3 x_5 - 2\zeta_6 x_4 x_6 + \zeta_6 x_5 x_6,$$
$$2i x_2^2 - x_3 x_4 + (-\zeta_6 + 2) x_3 x_5 + (2\zeta_6 - 2) x_4 x_6 + (-\zeta_6 + 2) x_5 x_6,$$
$$x_1 x_3 - \zeta_6 x_2 x_6 + \zeta_{12} x_4^2 + (\zeta_{12}^3 - 2\zeta_{12}) x_4 x_5 + \zeta_{12} x_5^2,$$
$$x_1 x_4 - \zeta_3 x_2 x_5 - x_3 x_6 - x_6^2,$$
$$x_1 x_5 - x_2 x_5 + x_3^2 + (-\zeta_6 + 2) x_3 x_6,$$
$$x_1 x_6 + x_2 x_6 - \zeta_{12} x_4^2 + \zeta_{12} x_4 x_5 - \zeta_{12} x_5^2,$$
$$x_2 x_3 - \zeta_6 x_2 x_6 + \zeta_{12} x_4^2,$$
$$x_2 x_4 + (-\zeta_6 - 1) x_2 x_5 + \zeta_6 x_3^2 + 2\zeta_6 x_3 x_6 + \zeta_6 x_6^2$$

$$\text{Maps:}\quad \begin{bmatrix} -1 & 0 & 0 & 0 & 0 & 0 & 0 \\ 0 & \zeta_{12}^{-1} & -\zeta_{12} & 0 & 0 & 0 & 0 \\ 0 & -\zeta_{12} & -\zeta_{12}^{-1} & 0 & 0 & 0 & 0 \\ 0 & 0 & 0 & 0 & \zeta_6 & \zeta_6 & 0 \\ 0 & 0 & 0 & -\zeta_3 & 0 & 0 & \zeta_3 \\ 0 & 0 & 0 & 0 & 0 & 0 & -\zeta_3 \\ 0 & 0 & 0 & 0 & 0 & \zeta_6 & 0 \end{bmatrix},$$

$$\begin{bmatrix} \zeta_3 & 0 & 0 & 0 & 0 & 0 & 0 \\ 0 & -1 & \zeta_3 & 0 & 0 & 0 & 0 \\ 0 & \zeta_6 & 0 & 0 & 0 & 0 & 0 \\ 0 & 0 & 0 & \zeta_3 & 0 & 0 & 1 \\ 0 & 0 & 0 & 0 & \zeta_3 & 0 & 0 \\ 0 & 0 & 0 & 0 & -\zeta_3 & -\zeta_6 & 0 \\ 0 & 0 & 0 & 0 & 0 & 0 & 1 \end{bmatrix}$$

Genus 7, Locus 3: Group $(64,41)$, signature (2,3,16), tetragonal

Ideal: $x_0^2 + x_1x_2,$
$x_3x_4 - \zeta_8 x_5x_6,$
$x_1^2 + x_3x_6 + ix_4x_5,$
$x_2^2 + ix_3x_6 + x_4x_5,$
$x_0x_1 + \zeta_{16}^7 x_3^2 - \zeta_{16}^5 x_5^2,$
$x_0x_2 - \zeta_{16}^7 x_4^2 + \zeta_{16}^5 x_6^2,$
$x_0x_3 - \zeta_{16} x_2x_6,$
$x_0x_4 + \zeta_{16} x_1x_5,$
$x_0x_5 + \zeta_{16}^7 x_2x_4,$
$x_0x_6 - \zeta_{16}^7 x_1x_3$

Maps: $(x_0,\dots,x_6) \mapsto (-x_0, x_2, x_1, x_4, x_3, x_6, x_5),$
$(x_0,\dots,x_6) \mapsto (ix_0, -\zeta_8^3 x_2, -\zeta_8 x_1, x_6, -\zeta_8^3 x_5, -\zeta_8 x_4, x_3)$

Genus 7, Locus 4: Group $(64,38)$, signature (2,4,16), hyperelliptic
$y^2 = x^{16} - 1$

Genus 7, Locus 5: Group $(56,4)$, signature (2,4,28), hyperelliptic
$y^2 = x^{15} - x$

Genus 7, Locus 6: Group $(54,6)$, signature (2,6,9)

Ideal: $x_1x_6 + x_2x_4 + x_3x_5,$
$x_0^2 - x_1x_6 + \zeta_6 x_2x_4 - \zeta_3 x_3x_5,$
$x_1x_4 + \zeta_3 x_2x_5 - \zeta_6 x_3x_6,$
$x_1x_5 + \zeta_3 x_2x_6 - \zeta_6 x_3x_4,$
$x_0x_1 - \zeta_6 x_5^2 - x_4x_6,$
$x_0x_2 + x_6^2 - \zeta_3 x_4x_5,$
$x_0x_3 + \zeta_3 x_4^2 + \zeta_6 x_5x_6,$
$x_0x_4 - x_1^2 + \zeta_3 x_2x_3,$
$x_0x_5 - \zeta_3 x_2^2 - \zeta_6 x_1x_3,$
$x_0x_6 + \zeta_6 x_3^2 + x_1x_2$

Maps: $(x_0,\dots,x_6) \mapsto (-x_0, \zeta_9^5 x_6, \zeta_9^8 x_4, \zeta_9^2 x_5, \zeta_9 x_2, \zeta_9^7 x_3, \zeta_9^4 x_1),$
$(x_0,\dots,x_6) \mapsto (\zeta_6 x_0, \zeta_3^2 x_4, \zeta_3^2 x_5, \zeta_3^2 x_6, \zeta_3 x_3, \zeta_3 x_1, \zeta_3 x_2)$

Genus 7, Locus 7: Group $(54,6)$, signature (2,6,9)
Complex conjugate of the previous curve

Genus 7, Locus 8: Group $(54,3)$, signature (2,6,9) cyclic trigonal
Trigonal equation: $y^3 = x^9 - 1$

Ideal: 2×2 minors of $\begin{bmatrix} x_0 & x_2 & x_3 & x_4 & x_5 \\ x_1 & x_3 & x_4 & x_5 & x_6 \end{bmatrix}$, and

$x_0^3 - x_6^2x_3 + x_2^3,$
$x_0^2x_1 - x_6^2x_4 + x_2^2x_3,$
$x_0x_1^2 - x_6^2x_5 + x_2^2x_4,$
$x_1^3 - x_6^3 + x_2^2x_5$

Maps: $(x_0,\ldots,x_6) \mapsto (x_1, x_0, -x_6, -x_5, -x_4, -x_3, -x_2),$
$(x_0,\ldots,x_6) \mapsto (\zeta_9x_1, \zeta_9^2x_0, -\zeta_9x_6, -\zeta_9^2x_5, -\zeta_9^3x_4, -\zeta_9^4x_3, -\zeta_9^5x_2)$

Genus 7, Locus 9: Group $(48,32)$, signature (3,4,6). Has g_6^2
Ideal:
$x_0^2 + x_3x_5 + \zeta_6x_3x_6 - \zeta_3x_4x_6,$
$\sqrt{3}i(x_1x_3 - x_2x_4 + x_1x_5) - x_1x_6 + x_2x_5 - x_2x_6,$
$\sqrt{3}i(2x_1x_4 - x_2x_5 + x_2x_6) + x_1x_5 - x_1x_6,$
$\sqrt{3}i(2x_2x_3 + x_1x_6 - x_2x_5) + 3x_1x_5 + x_2x_6,$
$-3(x_1^2 + x_3x_5 - x_3x_6) + \sqrt{3}i(x_4x_5 - x_4x_6) + 2x_5^2 + 2(\zeta_6 - 1)x_5x_6 - 2\zeta_6x_6^2,$
$-3(x_1x_2 - x_4x_5) + \sqrt{3}i(x_3x_5 - x_3x_6 + x_4x_6 - x_5^2) + 2\zeta_6x_5x_6 - x_6^2,$
$-3(x_2^2 - x_3x_5 - x_4x_6) + \sqrt{3}i(x_3x_6 - 3x_4x_5) + 2(\zeta_6 + 1)x_5x_6 + 2(\zeta_6 - 1)x_6^2,$
$-3(2x_4^2 - x_3x_5 + x_3x_6) + \sqrt{3}i(-x_4x_5 + x_4x_6) + 2x_5^2 + 2(\zeta_6 - 1)x_5x_6 - 2\zeta_6x_6^2,$
$-3(-2x_3x_4 + x_4x_5) + \sqrt{3}i(-x_3x_5 + x_3x_6 - x_4x_6 - x_5^2) + 2\zeta_6x_5x_6 - x_6^2,$
$-3(2x_3^2 + x_3x_5 + x_4x_6) + \sqrt{3}i(-x_3x_6 + 3x_4x_5) + 2(\zeta_6 + 1)x_5x_6 + 2(\zeta_6 - 1)x_6^2$
Maps:
$(x_0,\ldots,x_6) \mapsto (\zeta_3x_0, -\zeta_6x_2, -\zeta_3x_1 - x_2, -x_3 + \zeta_3x_4, \zeta_6x_3, \zeta_3x_5, -\zeta_6x_5 + x_6),$
$(x_0,\ldots,x_6) \mapsto (-x_0, -x_2, x_1, x_4, -x_3, \zeta_3x_5 - \zeta_6x_6, -\zeta_6x_5 - \zeta_3x_6)$

Genus 7, Locus 10: Group $(42,4)$, signature (2,6,21) cyclic trigonal
Trigonal equation: $y^3 = x^8 - x$

Ideal: 2×2 minors of $\begin{bmatrix} x_0 & x_2 & x_3 & x_4 & x_5 \\ x_1 & x_3 & x_4 & x_5 & x_6 \end{bmatrix}$, and

$x_0^3 - x_6^2x_2 + x_2^2x_3$
$x_0^2x_1 - x_6^2x_3 + x_2^2x_4,$
$x_0x_1^2 - x_6^2x_4 + x_2^2x_5,$
$x_1^3 - x_6^2x_5 + x_2^2x_6$

Maps:
$(x_0,\ldots,x_6) \mapsto (\zeta_7x_1, \zeta_7^6x_0, -\zeta_7^4x_6, -\zeta_7^2x_5, -x_4, -\zeta_7^5x_3, -\zeta_7^3x_2),$
$(x_0,\ldots,x_6) \mapsto (\zeta_{21}^{-2}x_1, \zeta_{21}^{-5}x_0, -\zeta_{21}^{-1}x_6, -\zeta_{21}^{-4}x_5, (\zeta_3 + 1)x_4, -\zeta_{21}^{11}x_3, -\zeta_{21}^{8}x_2)$

Genus 7, Locus 11: Group $(32, 11)$, signature (4,4,8). Has g^2_6
Ideal: $x_3x_5 + x_4x_6,$
$x_0^2 + x_1x_5 + ix_2x_6,$
$x_1x_4 + ix_2x_3 + x_5x_6,$
$x_1x_2 + x_3x_4,$
$x_1x_6 + \zeta_8^3 x_4x_5,$
$x_2x_5 + \zeta_8 x_3x_6,$
$x_1^2 - ix_3^2 - \zeta_8^3 x_5^2,$
$x_2^2 + ix_4^2 + \zeta_8^3 x_6^2,$
$-ix_2x_4 + \zeta_8^3 x_3^2,$
$x_1x_3 - \zeta_8^3 x_4^2$
Maps: $(x_0, \ldots, x_6) \mapsto (-x_0, x_2, -x_1, -ix_4, -ix_3, ix_6, ix_5),$
$(x_0, \ldots, x_6) \mapsto (ix_0, x_1, ix_2, -x_3, -ix_4, -x_5, ix_6)$

Genus 7, Locus 12: Group $(32, 10)$, signature (4,4,8). Has g^2_6
Ideal: $x_1x_6 + \zeta_{16}^6 x_2x_5 + x_3x_4,$
$x_1x_2 + x_5x_6,$
$x_0^2 + x_1x_6 - \zeta_{16}^6 x_2x_5,$
$x_3x_6 - \zeta_{16}^4 x_4x_5,$
$x_1^2 - \zeta_{16}^7 x_4^2 - \zeta_{16}^6 x_5^2,$
$x_2^2 + \zeta_{16}^3 x_3^2 - \zeta_{16}^{10} x_6^2,$
$-\zeta_{16}^2 x_2x_6 - \zeta_{16}^7 x_5^2,$
$x_1x_5 + (-\zeta_{16}^{12} - \zeta_{16}^4)x_3^2 - \zeta_{16}^{11} x_6^2,$
$x_1x_3 + \zeta_{16}^7 x_4x_6,$
$x_2x_4 + \zeta_{16} x_3x_5$
Maps: $(x_0, \ldots, x_6) \mapsto (-x_0, x_2, -x_1, -\zeta_{16}^6 x_4, -\zeta_{16}^2 x_3, -\zeta_{16}^2 x_6, -\zeta_{16}^6 x_5),$
$(x_0, \ldots, x_6) \mapsto (ix_0, -\zeta_{16}^2 x_2, -\zeta_{16}^6 x_1, ix_4, -ix_3, -ix_6, -ix_5)$

Genus 7, Locus 13: Group $(30, 4)$, signature (2,15,30), hyperelliptic
$y^2 = x^{15} - 1$

Acknowledgements. This work was partially supported by the University of Georgia's NSF VIGRE grant DMS-03040000, a Simons Foundation Travel Grant, and a Fordham Faculty Research Grant. The author acknowledges contributions from the students of the Fall 2010 Introductory Fall 2010 VIGRE Research Group at the University of Georgia and Darcy Chanin, who performed many calculations for genus 4, 5, and 6 surfaces. The main algorithm was programmed in `Magma` [**16**] during a visit to the computational algebra group at the University of Sydney, hosted by John Cannon and Mark Watkins. Valery Alexeev and James McKernan suggested the algorithm for matrix generators of representations cited in Section 1.2. The author thanks Jennifer Paulhus for sharing early versions of the `Magma` and `GAP` code in [**19**]; Tony Shaska, for sharing data on automorphism groups of genus 4 curves; John Voight for supplying references; and the anonymous referees for many helpful comments on the first version of this manuscript.

References

[1] L. Beshaj, T. Shaska, and E. Zhupa, *The case for superelliptic curves*, Advances on superelliptic curves and their applications, NATO Sci. Peace Secur. Ser. D Inf. Commun. Secur., vol. 41, IOS, Amsterdam, 2015, pp. 1–14. MR3525570

[2] Thomas Breuer, *Characters and automorphism groups of compact Riemann surfaces*, London Mathematical Society Lecture Note Series, vol. 280, Cambridge University Press, Cambridge, 2000. MR1796706

[3] E. Bujalance, F. J. Cirre, and G. Gromadzki, *Groups of automorphisms of cyclic trigonal Riemann surfaces*, J. Algebra **322** (2009), no. 4, 1086–1103, DOI 10.1016/j.jalgebra.2009.05.017. MR2537674

[4] E. Bujalance, J. J. Etayo, and E. Martínez, *Automorphism groups of hyperelliptic Riemann surfaces*, Kodai Math. J. **10** (1987), no. 2, 174–181, DOI 10.2996/kmj/1138037412. MR897252

[5] M. Conder, *Lists of regular maps*. `https://www.math.auckland.ac.nz/ conder/`.

[6] Vahid Dabbaghian-Abdoly, *An algorithm for constructing representations of finite groups*, J. Symbolic Comput. **39** (2005), no. 6, 671–688, DOI 10.1016/j.jsc.2005.01.002. MR2168613

[7] Vahid Dabbaghian and John D. Dixon, *Computing matrix representations*, Math. Comp. **79** (2010), no. 271, 1801–1810, DOI 10.1090/S0025-5718-10-02330-6. MR2630014

[8] David Eisenbud, *Commutative algebra*, Graduate Texts in Mathematics, vol. 150, Springer-Verlag, New York, 1995. With a view toward algebraic geometry. MR1322960

[9] H. M. Farkas and I. Kra, *Riemann surfaces*, 2nd ed., Graduate Texts in Mathematics, vol. 71, Springer-Verlag, New York, 1992. MR1139765

[10] William Fulton and Joe Harris, *Representation theory*, Graduate Texts in Mathematics, vol. 129, Springer-Verlag, New York, 1991. A first course; Readings in Mathematics. MR1153249

[11] The GAP Group, `GAP`*: Groups, Algorithms, and Programming, a system for computational discrete algebra* (2015), available at `http://www.gap-system.org`. Version 4.7.8.

[12] Deepak Kapur, Yao Sun, and Dingkang Wang, *An efficient method for computing comprehensive Gröbner bases*, J. Symbolic Comput. **52** (2013), 124–142, DOI 10.1016/j.jsc.2012.05.015. MR3018131

[13] Dan Grayson and Mike Stillman, `Macaulay 2`*: a software system for research in algebraic geometry* (2015), available at `http://www.math.uiuc.edu/Macaulay2/`. Version 1.8.

[14] A. M. Macbeath, *On a curve of genus* 7, Proc. London Math. Soc. (3) **15** (1965), 527–542, DOI 10.1112/plms/s3-15.1.527. MR0177342

[15] K. Magaard, T. Shaska, S. Shpectorov, and H. Völklein, *The locus of curves with prescribed automorphism group*, Sūrikaisekikenkyūsho Kōkyūroku **1267** (2002), 112–141. Communications in arithmetic fundamental groups (Kyoto, 1999/2001). MR1954371

[16] School of Mathematics and Statistics Computational Algebra Research Group University of Sydney, `MAGMA` *computational algebra system* (2015), available at `http://magma.maths.usyd.edu.au/magma/`. Version 2.21-7.

[17] David Mumford, *Lectures on curves on an algebraic surface*, With a section by G. M. Bergman. Annals of Mathematics Studies, No. 59, Princeton University Press, Princeton, N.J., 1966. MR0209285

[18] Katsusuke Nabeshima, *On the computation of parametric Gröbner bases for modules and syzygies*, Jpn. J. Ind. Appl. Math. **27** (2010), no. 2, 217–238, DOI 10.1007/s13160-010-0003-z. MR2720554

[19] Jennifer Paulhus, *Branching data for curves up to genus 48* (2015). `http://arxiv.org/abs/1512.07657`.

[20] R. Sanjeewa and T. Shaska, *Determining equations of families of cyclic curves*, Albanian J. Math. **2** (2008), no. 3, 199–213. MR2492096

[21] Frank-Olaf Schreyer, *Syzygies of canonical curves and special linear series*, Math. Ann. **275** (1986), no. 1, 105–137, DOI 10.1007/BF01458587. MR849058

[22] Tanush Shaska, *Determining the automorphism group of a hyperelliptic curve*, Proceedings of the 2003 International Symposium on Symbolic and Algebraic Computation, ACM, New York, 2003, pp. 248–254, DOI 10.1145/860854.860904. MR2035219

[23] Manfred Streit, *Homology, Belyĭ functions and canonical curves*, Manuscripta Math. **90** (1996), no. 4, 489–509, DOI 10.1007/BF02568321. MR1403719

[24] D. Swinarski, *Supplementary files for "Equations of Riemann surfaces with large automorphism groups."* (2016). `http://faculty.fordham.edu/dswinarski/RiemannSurfaceAutomorphisms/`.

[25] Volker Weispfenning, *Comprehensive Gröbner bases*, J. Symbolic Comput. **14** (1992), no. 1, 1–29, DOI 10.1016/0747-7171(92)90023-W. MR1177987

Department of Mathematics, Fordham University, 113 W 60th St. Room 813, New York, New York 10023

Email address: dswinarski@fordham.edu

Contemporary Mathematics
Volume **703**, 2018
http://dx.doi.org/10.1090/conm/703/14130

On the field of moduli of superelliptic curves

Ruben Hidalgo and Tony Shaska

Abstract. A superelliptic curve $\mathcal{X}$ of genus $g \geq 2$ is not necessarily defined over its field of moduli but it can be defined over a quadratic extension of it. While a lot of work has been done by many authors to determine which hyperelliptic curves are defined over their field of moduli, less is known for superelliptic curves.

In this paper we observe that if the reduced group of a genus $g \geq 2$ superelliptic curve $\mathcal{X}$ is different from the trivial or cyclic group, then $\mathcal{X}$ can be defined over its field of moduli; in the cyclic situation we provide a sufficient condition for this to happen. We also determine those families of superelliptic curves of genus at most 10 which might not be definable over their field of moduli.

1. Introduction

Let k be an algebraically closed field of characteristic zero and $\mathcal{X}$ a genus $g \geq 2$, smooth projective, irreducible algebraic curve defined over k. The *field of moduli* of $\mathcal{X}$ is the field of definition of the corresponding moduli point $\mathfrak{p} = [\mathcal{X}]$ in the moduli space $\mathcal{M}_g$ (see Section 2.1.2 for a formal definition of the field of moduli). In general, to determine the field of moduli and to decide if it is a field of definition is a difficult task and it is an active research topic. Examples of algebraic curves (for $k = \mathbb{C}$) which cannot be defined over their field of moduli have been provided by Earle [**8**], Huggins [**13**] and Shimura [**24**] for the hyperelliptic situation and by the first author [**10**] and Kontogeorgis [**15**] in the non-hyperelliptic situation. In other words, $\mathcal{M}_g$ is not a *fine* moduli space.

By results due to Koizumi [**14**], there is always a field of definition that is a finite extension of the field of moduli and if, moreover, the field k has transcendental numbers over the field of moduli (for instance $k = \mathbb{C}$), then the field of moduli coincides with the intersection of all the fields of definitions of the curve. On the other hand, if $k = \overline{\mathbb{Q}}$, then it might be that the intersection of all fields of definitions (inside k) is different from the field of moduli.

Investigating the obstruction for the field of moduli to be a field of definition is part of descent theory for fields of definition and has many consequences in arithmetic geometry. Many works have been devoted to this problem, most notably

2010 *Mathematics Subject Classification.* Primary 30F10, 14H37, 14H45.
Key words and phrases. Field of moduli, superelliptic curves.
The first author was partially supported by Project Fondecyt 1150003 and Project Anillo ACT 1415 PIA-CONICYT.

by Weil [**25**], Shimura [**24**] and Grothendieck, among many others. Weil's criterion [**25**] assures that if a curve has no non-trivial automorphisms then its field of moduli is a field of definition. On the other extreme, if the curve $\mathcal{X}$ is quasiplatonic (that is, when the quotient orbifold $\mathcal{X}/\mathrm{Aut}(\mathcal{X})$ has genus zero and exactly three cone points), then Wolfart [**26**] proved that the field of moduli is also a field of definition. Hence, the real problem occurs when the curve has non-trivial automorphism group but the quotient orbifold $\mathcal{X}/\mathrm{Aut}(\mathcal{X})$ has non-trivial moduli.

Mestre [**18**] provided the first algorithm which determines if the field of moduli is a field of definition for genus two curves with automorphism group of order 2. The algorithm gives explicitly an equation of the curve over its field of moduli when such equation exists. In [**22**] it is shown that the field of moduli is a field of definition for genus two curves with automorphism group isomorphic to the dihedral groups D_8 or D_{12}. Cardona and Quer [**6**] shows that this is true also for genus two curves with automorphism group the Klein 4-group $D_4 = V_4$. Most recently Malmendier and Shaska [**17**] are able to construct a pair of universal genus two curves defined over a quadratic number field. In all these papers the moduli point $\mathfrak{p} \in \mathcal{M}_2$ was computed and an explicit equation of a curve was determined via the invariants of the binary sextics.

Shaska conjectured in [**23**] and [**9**] that all hyperelliptic curves $\mathcal{X}$ with extra automorphisms (i.e. $|\mathrm{Aut}\,(\mathcal{X})| > 2$) are defined over their field of moduli. This was proved to be incorrect by Huggins [**13**] who showed explicit examples of hyperelliptic curves with reduced automorphism group isomorphic to a cyclic group which cannot be defined over their field of moduli (curves defined over $k = \mathbb{C}$ which cannot be defined over $\mathbb{R}$ and whose field of moduli is contained inside $\mathbb{R}$). In the same paper Huggins proved that every hyperelliptic curve whose reduced automorphism group is different from a cyclic group (including the trivial situation) is definable over its field of moduli; even they are hyperelliptically defined over it. Kontogeorgis [**15**] generalized the above result to cyclic p-gonal curves (where p is a prime integer) and in [**12**] the first author and Quispe generalized the above for curves admitting a subgroup of automorphisms being unique up to conjugation (note that in case of cyclic p-gonal curves, in the case that it is definable over its field of moduli, a rational model over its field of moduli may not be in a cyclic p-gonal form).

A genus $g \geq 2$ smooth algebraic curve (or closed Riemann surface) $\mathcal{X}$ is called *superelliptic of level* n if there exists an element $\tau \in \mathrm{Aut}\,(\mathcal{X})$ of order n such that (i) τ is central, (ii) the quotient $\mathcal{X}/\langle\tau\rangle$ has genus zero, (iii) every cone point has order n and (iv) the rotation number of τ at each of its fixed points is the same (these are very special types of cyclic n-gonal curves); in which case τ is called a *superelliptic automorphism of level* n and $H = \langle\tau\rangle$ a *superelliptic group of level* n. In Proposition 1 we observe that the superelliptic group of level n is unique.

These superelliptic Riemann surfaces are natural generalizations of hyperelliptic Riemann surfaces (in which case $n = 2$ and τ is given by the hyperelliptic involution), but they are in general non-hyperelliptic ones. Superelliptic curves are the natural cases where one could try to extend the results from hyperelliptic curves; see [**4**]. These are, with some exceptions, the only classes of curves for which we can easily write down equations starting from the automorphism group of the curve. These curves have equations of the form

$$\mathcal{X}: \quad y^n = \prod_{i=1}^{r}(x - a_i)$$

so that either $r \equiv 0 \mod (n)$ or $r+1 \equiv 0 mod(n)$, and we may assume that the generator of the superelliptic group of level n is generated by $\tau : (x,y) \mapsto (x, \omega y)$, where $\omega^n = 1$ (recall that we are assuming τ to have the same rotation number about all of its fixed points, so we may choose a power of it, of the same order n, for which the rotation number is equal to one).

Beshaj and Thompson [5] have proved that a superelliptic curve can always be defined over a quadratic extension of its field of moduli. Moreover, they have provided an equation over an at most quadratic extension of its field of moduli using the Shaska invariants which has the above particular cyclic gonal form. At this point, we should remark that if a superelliptic curve S is definable over its field of moduli, then it might be that it is not definable in a cyclic gonal form over it; but it will be so definable over a quadratic extension of it.

In general, the field of moduli of a superelliptic curve $\mathcal{X}$ is not a field of definition, but we will see (cf. Thm. 3) that if $|\operatorname{Aut}(\mathcal{X})| > n$, where n is the order of τ, for the majority of cases, the field of moduli is a field of definition.

A superelliptic descent over a field k means that we may find a defining curve with equation $y^n = f(x)$ where $f \in k[x]$, which is a slightly more restricted class than the typical descent that ask for a curve over k not necessarily of the above form; see for example [16] for the typical descent for hyperelliptic situation for $g = 3$ (in there there is an example of a hyperelliptic curve of genus three, which is definable over $\mathbb{Q}$, but it cannot be hyperelliptically defined over $\mathbb{Q}$).

The main goal of this paper is to (partially) describe which superelliptic curves of genus $g \leq 10$ are definable over their field of moduli. In each genus (see Tables 1, 2 and 3), we describe those ones which might not be definable over their field of moduli; all cases which are not in blue are defined over their field of moduli and the blue ones might or not be definable over the field of moduli.

The results of this paper can be easily extended to positive characteristic in the case when the covering $\mathcal{X} \to \mathcal{X}/\langle\tau\rangle$ is tame. The list of full automorphisms groups in positive characteristic ($\operatorname{char} k = p > 0$) was determined in [20] and the rest of the proofs should follow easily when $(n,p) = 1$. We only focus in characteristic zero.

The paper is organized as follows. In Section 2 we give a brief review of superelliptic curves and their automorphism groups. Such groups were fully classified in [20] and the equations of each parametric curve for any given group are given in [21] for any given genus $g \geq 2$; in Section 5 we provide such a list for $5 \leq g \leq 10$ (see also [3] and [19]). In Section 3 we recall some known results which provide sufficient conditions for a curve to be definable over its field of moduli. Using such conditions, we give the list of all possible superelliptic curves of genus $2 \leq g \leq 10$ which might not be definable over the field of moduli (in particular, the complementary ones are definable over them). Let us note that the case of genus $g = 2, 3$ are already well known and $g = 4$ can be obtained by [4].

NOTATION. Throughout this paper, k denotes an algebraically closed field of characteristic zero and $\mathcal{X}$ a genus $g \geq 2$, smooth, projective, irreducible, algebraic curve defined over k. We will keep the notation used in [20] to denote certain groups (i.e, C_n is the cyclic group of order n, D_m the dihedral group of order m, $V_4 = D_4$ the 4-Klein group, etc).

2. Preliminaries

In this section, k will be a fixed algebraically closed field of characteristic zero and we denote by $\mathrm{Gal}(k)$ its group of field automorphisms.

2.1. The field of moduli and fields of definition. Let $\mathcal{X}$ be a genus g projective, irreducible, algebraic curve defined over k, say given as the common zeroes of the polynomials $P_1, \ldots, P_r$, and let us denote by $G = \mathrm{Aut}(\mathcal{X})$ the full automorphism group of $\mathcal{X}$.

If $\sigma \in \mathrm{Gal}(k)$, then X^σ will denote the curve defined as the common zeroes of the polynomials $P_1^\sigma, \ldots, P_r^\sigma$, where P_j^σ is obtained from P_j by applying σ to its coefficients. In particular, if τ is also a field automorphism of k, then $X^{\tau\sigma} = (X^\tau)^\sigma$.

2.1.1. *Field of definition.* A subfield k_0 of k is called a *field of definition* of $\mathcal{X}$ if there is a curve $\mathcal{Y}$, defined over k_0, which is isomorphic to $\mathcal{X}$. It is clear that every subfield of k containing k_0 is also a field of definition of it. In the other direction, a subfield of k_0 might not be a field of definition of $\mathcal{X}$.

Weil's descent theorem [**25**] provides sufficient conditions for a subfield k_0 of k to be a field of definition. Let us denote by $\mathrm{Gal}(k/k_0)$ the group of field automorphisms of k acting as the identity on k_0.

THEOREM 1 (Weil's descent theorem [**25**]). *Assume that for every element* $\sigma \in \mathrm{Gal}(k/k_0)$ *there is an isomorphism* $f_\sigma : \mathcal{X} \to \mathcal{X}^\sigma$ *so that*

$$f_{\tau\sigma} = f_\sigma^\tau \circ f_\tau, \quad \forall \sigma, \tau \in \mathrm{Gal}(k/k_0).$$

Then there is a curve $\mathcal{Y}$*, defined over* k_0*, and there is an isomorphism* $R : \mathcal{X} \to \mathcal{Y}$*, defined over a finite extension of* k_0*, so that* $R = R^\sigma \circ f_\sigma$*, for every* $\sigma \in \mathrm{Gal}(k/k_0)$.

Clearly, the sufficient conditions in Weil's descent theorem are trivially satisfied if $\mathcal{X}$ has no non-trivial automorphisms (a generic situation for $\mathcal{X}$ of genus at least three).

COROLLARY 1. *If* $\mathcal{X}$ *has trivial group of automorphisms and for every* $\sigma \in \mathrm{Gal}(k/k_0)$ *there is an isomorphism* $f_\sigma : \mathcal{X} \to \mathcal{X}^\sigma$*, then* $\mathcal{X}$ *can be defined over* k_0.

2.1.2. *Field of moduli.* The notion of field of moduli was originally introduced by Shimura for the case of abelian varieties and later extended to more general algebraic varieties by Koizumi. If $G_\mathcal{X}$ is the subgroup of $\mathrm{Gal}(k)$ consisting of those σ so that $\mathcal{X}^\sigma$ is isomorphic to $\mathcal{X}$, then the fixed field $M_\mathcal{X}$ of $G_\mathcal{X}$ is called *the field of moduli* of $\mathcal{X}$.

As we are assuming that k is algebraically closed and of characteristic zero, we have that $G_\mathcal{X}$ consists of all automorphisms of $\mathrm{Gal}(k)$ acting as the identity on $M_\mathcal{X}$.

It is known that every curve of genus $g \leq 1$ can be defined over its field of moduli. If $g \geq 2$, then (as already said in the Introduction) there are known examples of curves which cannot be defined over their field of moduli. A direct consequence of Corollary 1 is the following.

COROLLARY 2. *Every curve with trivial group of automorphisms can be defined over its field of moduli.*

As a consequence of Belyi's theorem [**2**], every quasiplatonic curve $\mathcal{X}$ can be defined over $\overline{\mathbb{Q}}$ (so over a finite extension of $\mathbb{Q}$).

THEOREM 2 (Wolfart [**26**]). *Every quasiplatonic curve can be defined over its field of moduli (which is a number field).*

2.2. Two practical sufficient conditions. When the curve $\mathcal{X}$ has a non-trivial group of automorphisms, then Weil's conditions (in Weil's descent theorem) are in general not easy to check. Next we consider certain cases for which it is possible to check for $\mathcal{X}$ to be definable over its field of moduli.

2.2.1. *Sufficient condition 1: unique subgroups.* Let H be a subgroup of the automorphism group $\mathrm{Aut}\,(\mathcal{X})$. In general it might be another different subgroup K which is isomorphic to H and with $\mathcal{X}/K$ and $\mathcal{X}/H$ having the same signature. For instance, the genus two curve $\mathcal{X}$ defined by $y^2 = x(x-1/2)(x-2)(x-1/3)(x-3)$ has two conformal involutions, τ_1 and τ_2, whose product is the hyperelliptic involution. The quotient $\mathcal{X}/\langle \tau_j \rangle$ has genus one and exactly two cone points (of order two).

We say that H is is *unique* in $\mathrm{Aut}\,(\mathcal{X})$ if it is the unique subgroup of $\mathrm{Aut}\,(\mathcal{X})$ isomorphic to H and with quotient orbifold of same signature as $\mathcal{X}/H$. Typical examples are (i) $H = \mathrm{Aut}\,(\mathcal{X})$ and (ii) H being the cyclic group generated by the hyperelliptic involution for the case of hyperelliptic curves.

If H is unique in $\mathrm{Aut}\,(\mathcal{X})$, then it is a normal subgroup; so we may consider the reduced group $\overline{\mathrm{Aut}}\,(\mathcal{X}) = \mathrm{Aut}\,(\mathcal{X})/H$, which is a group of automorphisms of the quotient orbifold $\mathcal{X}/H$. In [**12**] the following sufficient condition for a curve to definable over its field of moduli was obtained.

THEOREM 3 (Hidalgo and Quispe [**12**]). *Let $\mathcal{X}$ be a curve of genus $g \geq 2$ admitting a subgroup H which is unique in* $\mathrm{Aut}\,(\mathcal{X})$ *and so that $\mathcal{X}/H$ has genus zero. If the reduced group of automorphisms* $\overline{\mathrm{Aut}}\,(\mathcal{X}) = \mathrm{Aut}\,(\mathcal{X})/H$ *is different from trivial or cyclic, then $\mathcal{X}$ is definable over its field of moduli.*

If $\mathcal{X}$ is a hyperelliptic curve, then a consequence of the above is the following result (originally due to Huggins [**13**]).

COROLLARY 3. *Let $\mathcal{X}$ be a hyperelliptic curve with extra automorphisms and reduced automorphism group* $\overline{\mathrm{Aut}}\,(\mathcal{X})$ *not isomorphic to a cyclic group. Then, the field of moduli of $\mathcal{X}$ is a field of definition.*

2.2.2. *Sufficient condition 2: Odd signature.* Another sufficient condition of a curve $\mathcal{X}$ to be definable over its field of moduli, which in particular contains the case of quasiplatonic curves, was provided in [**1**]. We say that $\mathcal{X}$ has *odd signature* if $\mathcal{X}/\mathrm{Aut}(\mathcal{X})$ has genus zero and in its signature one of the cone orders appears an odd number of times.

THEOREM 4 (Artebani and Quispe [**1**]). *Let $\mathcal{X}$ be a curve of genus $g \geq 2$. If $\mathcal{X}$ has odd signature, then it can be defined over its field of moduli.*

2.3. The locus of curves with prescribed group action, moduli dimension of families. Fix an integer $g \geq 2$ and a finite group G. Let $C_1, \ldots, C_r$ be nontrivial conjugacy classes of G. Let $\mathbf{C} = (C_1, \ldots, C_r)$, viewed as an unordered tuple, where repetitions are allowed. We allow r to be zero, in which case $\mathbf{C}$ is empty. Consider pairs $(\mathcal{X}, \mu)$, where $\mathcal{X}$ is a curve and $\mu : G \to \mathrm{Aut}\,(\mathcal{X})$ is an injective homomorphism. We will suppress μ and just say $\mathcal{X}$ is a curve with G-action, or a G-curve. Two G-curves $\mathcal{X}$ and $\mathcal{X}'$ are called equivalent if there is a G-equivariant conformal isomorphism $\mathcal{X} \to \mathcal{X}'$.

We say a G-curve $\mathcal{X}$ is **of ramification type** $(g, G, \mathbf{C})$ (for short, of type $(g, G, \mathbf{C})$) if

i) g is the genus of $\mathcal{X}$,
ii) $G < \mathrm{Aut}(\mathcal{X})$,
iii) the points of the quotient $\mathcal{X}/G$ that are ramified in the cover $\mathcal{X} \to \mathcal{X}/G$ can be labeled as $p_1, \dots, p_r$ such that C_i is the conjugacy class in G of distinguished inertia group generators over p_i (for $i = 1, \dots, r$).

If $\mathcal{X}$ is a G-curve of type $(g, G, \mathbf{C})$, then the genus g_0 of $\mathcal{X}/G$ is given by the Riemann-Hurwitz formula

$$2(g-1) = 2|G|(g_0 - 1) + |G| \sum_{j=1}^{r} (1 - |C_j|^{-1}).$$

Define $\mathcal{H} = \mathcal{H}(g, G, \mathbf{C})$ to be the set of equivalence classes of G-curves of type $(g, G, \mathbf{C})$. By covering space theory, $\mathcal{H}$ is non-empty if and only if G can be generated by elements $\alpha_1, \beta_1, ..., \alpha_{g_0}, \beta_{g_0}, \gamma_1, ..., \gamma_r$ with $\gamma_i \in C_i$ and $\prod_j [\alpha_j, \beta_j] \prod_i \gamma_i = 1$, where $[\alpha, \beta] = \alpha^{-1}\beta^{-1}\alpha\beta$.

Let $\mathcal{M}_g$ be the moduli space of genus g curves, and $\mathcal{M}_{g_0,r}$ the moduli space of genus g_0 curves with r distinct marked points, where we view the marked points as unordered. Consider the map

$$\Phi : \mathcal{H} \to \mathcal{M}_g,$$

forgetting the G-action, and the map $\Psi : \mathcal{H} \to \mathcal{M}_{g_0,r}$ mapping (the class of) a G-curve $\mathcal{X}$ to the class of the quotient curve $\mathcal{X}/G$ together with the (unordered) set of branch points $p_1, \dots, p_r$.

If $\mathcal{H} \neq \emptyset$, then Ψ is surjective and has finite fibers, by covering space theory. Also Φ has finite fibers, since the automorphism group of a curve of genus ≥ 2 is finite. The set $\mathcal{H}$ carries a structure of quasi-projective variety (over $\mathbb{C}$) such that the maps Φ and Ψ are finite morphisms. If $\mathcal{H} \neq \emptyset$, then all (irreducible) components of $\mathcal{H}$ map surjectively to $\mathcal{M}_{g_0,r}$ (through a finite map), hence they all have the same dimension

$$\delta(g, G, \mathbf{C}) := \dim \mathcal{M}_{g_0,r} = 3g_0 - 3 + r$$

Let $\mathcal{M}(g, G, \mathbf{C})$ denote the image of Φ, i.e., the locus of genus g curves admitting a G-action of type $(g, G, \mathbf{C})$. Since Φ is a finite map, if this locus is non-empty, each of its (irreducible) components has dimension $\delta(g, G, \mathbf{C})$.

Theorem 2 can be written as follows.

THEOREM 5. *If $\delta(g, G, \mathbf{C}) = 0$, then every curve in $\mathcal{M}(g, G, \mathbf{C})$ is defined over its field of moduli.*

The above is due to the fact that $\delta = 0$ ensures that the quotient orbifold $\mathcal{X}/G$ must be of genus zero and with exactly three conical points, that is, $\mathcal{X}$ is a quasiplatonic curve.

3. Field of moduli of superelliptic curves

3.1. Automorphism groups of superelliptic curves. Let $\mathcal{X}$ be a superelliptic curve of level n with $G = \mathrm{Aut}(\mathcal{X})$. By the definition, there is some $\tau \in G$, of order n and central, so that the quotient $\mathcal{X}/\langle\tau\rangle$ has genus zero, that is, it can be identified with the projective line, and all its cone points have order n. As, in this case, the cyclic group $H = \langle\tau\rangle \cong C_n$ is normal subgroup of G, we may consider the quotient group $\bar{G} := G/H$, called the *reduced automorphism group of $\mathcal{X}$ with respect to H*; so G is a degree n central extension of $\bar{G}$.

In the particular case that $n = p$ is a prime integer, Castelnuovo-Severi's inequality [**7**] asserts that for $g > (p-1)^2$ the cyclic group H is unique in $\mathrm{Aut}(\mathcal{X})$. In [**11**] it is observed that if $n = p \geq 5(r-1)$ is prime, where $r \geq 3$ denotes the number of cone points of the quotient $\mathcal{X}/H$, then H is again unique. The following result shows that the superelliptic group of level n is unique.

PROPOSITION 1. *A superelliptic curve of level n and genus $g \geq 2$ has a unique superelliptic group of level n.*

PROOF. Let $\mathcal{X}$ be a superelliptic curve of level n and assume that $\langle \tau \rangle$ and $\langle \eta \rangle$ are two different superelliptic groups of level n. The condition that the cone points of both quotient orbifolds $\mathcal{X}/\langle \tau \rangle$ and $\mathcal{X}/\langle \eta \rangle$ are of order n asserts that a fixed point of a non-trivial power of τ (respectively, of η) must also be a fixed point of τ (respectively, η). In this way, our previous assumption asserts that no non-trivial power of η has a common fixed point with a non-trivial power of τ. In this case, the fact that τ and η are central asserts that $\eta\tau = \tau\eta$ and that $\langle \tau, \eta \rangle \cong C_n^2$ (see also [**20**]). Let $\pi : \mathcal{X} \to \mathbb{P}_k^1$ be a regular branched cover with $\langle \tau \rangle$ as deck group. Then the automorphism η induces a automorphism $\rho \in \mathrm{PGL}_2(k)$ (also of order n) so that $\pi\eta = \rho\pi$. As ρ is conjugated to a rotation $x \mapsto \omega_n x$, where $\omega_n^n = 1$, we observe that it has exactly two fixed points. This asserts that η must have either n or $2n$ fixed points (forming two orbits under the action of $\langle \tau \rangle$). As this is also true by interchanging the roles of τ and η, the same holds for the fixed points of τ. It follows that the cone points of π consists of (i) exactly two sets of cardinality n each one or (ii) exactly one set of cardinality n, and each one being invariant under the rotation ρ. Up to post-composition by a suitable transformation in $\mathrm{PGL}_2(k)$, we may assume these in case (i) the $2n$ cone points are given by the n roots of unity and the n roots of a point different from 1 and 0 and in case (ii) that the n cone points are the n roots of unity. In other words, $\mathcal{X}$ can be given either as

$$\mathcal{X}_1 : y^n = (x^n - 1)(x^n - a^n), \quad a \in k - \{0, 1\}$$

or as the classical Fermat curve

$$\mathcal{X}_2 : y^n = x^n - 1$$

and, in these models,

$$\tau(x, y) = (x, \omega_n y), \quad \eta(x, y) = (\omega_n x, y).$$

As the genus of $\mathcal{X}_1$ is at least two, we must have that $n \geq 3$. But such a curve also admits the order two automorphism

$$\gamma(x, y) = \left(\frac{a}{x}, \frac{ay}{x^2} \right)$$

which does not commute with η, a contradiction to the fact that η was assumed to be central.

In the Fermat case, the full group of automorphisms is $C_n^2 \rtimes S_3$ and it may be checked that it is not superelliptic. □

3.2. Most of superelliptic curves are definable over their field of moduli. The group $\bar{G}$ is a subgroup of the group of automorphisms of a genus zero field, so $\bar{G} < PGL_2(k)$ and $\bar{G}$ is finite. It is a classical result that every finite subgroup of $PGL_2(k)$ (since we are assuming k of characteristic zero) is either the trivial group or isomorphic to one of the following: C_m, D_m, A_4, S_4, A_5. All automorphisms groups of superelliptic curves and their equations were determined in [**20**] and [**21**]. Determining the automorphism groups G, the signature $\mathbf{C}$ of the covering $\mathcal{X} \to \mathcal{X}/G$, and the dimension of the locus $\mathcal{M}(g, G, \mathbf{C})$ for superelliptic curves is known (see, for instance, [**20**]).

We have seen in Theorem 1 that its superelliptic group of level n is unique. As a consequence of Theorem 3, we obtain the following fact concerning the field of moduli of superelliptic curves.

THEOREM 6. *Let $\mathcal{X}$ be a superelliptic curve of genus $g \geq 2$ with superelliptic group $H \cong C_n$. If the reduced group of automorphisms $\overline{\mathrm{Aut}}\,(\mathcal{X}) = \mathrm{Aut}\,(\mathcal{X})/H$ is different from trivial or cyclic, then $\mathcal{X}$ is definable over its field of moduli.*

As a consequence of the above, we only need to take care of the case when the reduced group $\bar{G} = G/H$ is either trivial or cyclic. As a consequence of Theorem 4 we have the following fact.

THEOREM 7. *Let $\mathcal{X}$ be a superelliptic curve of genus $g \geq 2$ with superelliptic group $H \cong C_n$ so that $\bar{G} = G/H$ is either trivial or cyclic. If $\mathcal{X}$ has odd signature, then it can be defined over its field of moduli.*

As a consequence, the only cases we need to take care are those superelliptic curves with reduced group $\bar{G} = G/H$ being either trivial or cyclic and with $\mathcal{X}/G$ having not an odd signature.

4. Superelliptic curves of genus at most 10

Using the list provided in Section 5 and the previous results, we proceed, in each genus $2 \leq g \leq 10$, to describe those superelliptic curves which are definable over their field of moduli. Observe that in the left cases (which might or might not be definable over their field of moduli) the last column provides an algebraic model $y^n = f(x)$, where $f(x)$ is defined over the algebraic closure and not necessarily over a minimal field of definition. The branched regular covering $\pi : \mathcal{X} \to \mathbb{P}^1_k$, defined by $\pi(x, y) = x$, has deck group $H = \langle \tau(x, y) = (x, \epsilon_n y) \rangle \cong C_n$.

4.0.1. **Genus 2.** The case of genus $g = 2$ is well known since in this case every curve $\mathcal{X}$ such that $|\,\mathrm{Aut}\,(\mathcal{X})| > 2$ the field of moduli is a field of definition. There are examples of genus two curves, whose reduced group is trivial, which are not definable over their field of moduli.

4.0.2. **Genus 3.** There are 21 signatures for genus $g = 3$ from which 12 of them are hyperelliptic and 3 are trigonal.

LEMMA 1. *Every superelliptic curve of genus 3, other than Nr. 1 and 2 in Table 1, is definable over its field of moduli.*

PROOF. If $\overline{\mathrm{Aut}}\,(\mathcal{X})$ is isomorphic to A_4 or S_4 then the corresponding locus consists of the curves $y^4 = x^4 + 2x^2 + \frac{1}{3}$ and $y^2 = x^8 + 14x^4 + 1$ which are both defined over their field of moduli.

If $\overline{\mathrm{Aut}}\,(\mathcal{X})$ is isomorphic to a dihedral group and $\mathcal{X}$ is not hyperelliptic, then $\mathrm{Aut}\,(\mathcal{X})$ is isomorphic to $V_4 \times C_4$, G_5, $D_6 \times C_3$, and G_8. These cases G_5, $D_6 \times C_3$, and G_8 correspond to $y^4 = x^4 - 1$, $y^3 = x(x^3 - 1)$, and $y^4 = x(x^2 - 1)$, which are all defined over the field of moduli.

If $\overline{\mathrm{Aut}}\,(\mathcal{X})$ is isomorphic to a cyclic group, then in the cases when it is isomorphic to C_{14}, C_{12} there are two cases which correspond to the curves $y^2 = x^7 + 1$ and $y^3 = x^4 + 1$. The left cases are given in Table 1. The curve No. 5 is definable

TABLE 1. Genus 3 curves No. 1 and 2 are the only one whose field of moduli is not necessarily a field of definition

Nr.	$\overline{G}$	G	n	m	sig.	δ	Equation $y^n = f(x)$
1	$\{I\}$	C_2	2	1	2^8	5	$x\left(x^6 + \sum_{i=1}^{5} a_i x^i + 1\right)$
2	C_2	V_4	2	2	2^6	3	$x^8 + a_1x^2 + a_2x^4 + a_3x^6 + 1$
3	C_2	C_4	2	2	$2^3, 4^2$	2	$x\left(x^6 + a_1x^2 + a_2x^4 + 1\right)$
4	C_2	C_6	3	2	$2, 3^2, 6$	1	$x^4 + a_1x^2 + 1$
5	V_4	$V_4 \times C_4$	4	2	$2^3, 4$	1	$x^4 + a_1x^2 + 1$

over its field of moduli by Theorem 6. All the other cases, with the exception of Nr. 1 and 2, the curves are of odd signature, so they are definable over their field of moduli by Theorem 7. □

4.0.3. Genus 4.

LEMMA 2. *Every superelliptic curve of genus 4, other than Nr. 1, 3 and 5 in Table 2, is definable over its field of moduli.*

PROOF. There is only one case when the reduced automorphism group $\overline{\mathrm{Aut}}\,(\mathcal{X})$ is not isomorphic to a cyclic or a dihedral group, namely $\overline{G} \cong S_4$. In this case, the curve is $y^3 = x(x^4 - 1)$ and is defined over the field of moduli. If $\overline{G}$ is isomorphic to a dihedral group, then there are only 6 signatures which give the groups $D_6 \times C_3$, $D_4 \times C_3$, $D_{12} \times C_3$, $D_4 \times C_3$, $D_8 \times C_3$, and $D_4 \times C_5$. The groups $D_{12} \times C_3$, $D_8 \times C_3$, and $D_4 \times C_5$ correspond to curves $y^3 = x^6 - 1$, $y^3 = x(x^4 - 1)$, and $y^5 = x(x^2 - 1)$ respectively. The remaining three cases are given by Nrs. 7, 8 and 9 in Table 2 which are definable over their field of moduli by Theorem 6.

TABLE 2. Genus 4 curves No. 1, 3 and 5 are the only ones whose field of moduli is not necessarily a field of definition

Nr.	$\overline{G}$	G	n	m	sig.	δ	Equation $y^n = f(x)$
1		C_2	2	1	2^{10}	7	$x\left(x^8 + \sum_{i=1}^{7} a_i x^i + 1\right)$
2		V_4	2	2	2^7	4	$x^{10} + \sum_{i=1}^{4} a_i x^{2i} + 1$
3	C_m	C_4	2	2	$2^4, 4^2$	3	$x(x^8 + a_3x^6 + a_2x^4 + a_1x^2 + 1)$
4		C_6	2	3	$2^3, 3, 6$	2	$x^9 + a_1x^3 + a_2x^6 + 1$
5		C_3	3	1	3^6	3	$x(x^4 + a_1x + a_2x^2 + a_3x^3 + 1)$
6		$C_2 \times C_3$	3	2	$2^2, 3^3$	2	$x^6 + a_2x^4 + a_1x^2 + 1$
7		$D_6 \times C_3$	3	3	$2^2, 3^2$	1	$x^6 + a_1x^3 + 1$
8	D_{2m}	$V_4 \times C_3$	3	2	$2^2, 3, 6$	1	$(x^2 - 1)(x^4 + a_1x^2 + 1)$
9		$V_4 \times C_3$	3	2	$2^2, 3, 6$	1	$x(x^4 + a_1x^2 + 1)$

If $\overline{\mathrm{Aut}}\,(\mathcal{X})$ is isomorphic to a cyclic group, then there are two signatures for each of the groups C_{18} and C_{15}. In each case, both signatures give the same curve, namely $y^2 = x^9 + 1$ and $y^3 = x^5 + 1$ respectively. The left cases are given by Nrs. 1 to 6 in Table 2. As all cases, with the exception of Nrs. 1, 3 and 5, the curves are of odd signature; so definable over their field of moduli by Theorem 7. □

4.0.4. **Genus $5 \leq g \leq 10$.** We proceed to indicate which cases in the table provided in Section 5 are definable over the field of moduli.

(1) Genus 5: We may see from the table in Section 5 that, for $g = 5$, there are 20 cases to consider. By Theorem 6 all cases, from Nr. 8 to Nr. 20, are definable over their field of moduli. The left cases, with the exception of Nrs. 1, 2 and 6, are of odd signature, so they are definable over their field of moduli.
(2) Genus 6: We see from the table in Section 5 that, for $g = 6$, there are 36 cases to consider. By Theorem 6 all cases, from Nr. 17 to Nr. 36, are definable over their field of moduli. The left cases, with the exception of Nrs. 9, 10, 13 and 15, are of odd signature, so they are definable over their field of moduli.
(3) Genus 7: We see from the table in Section 5 that, for $g = 7$, there are 27 cases to consider. By Theorem 6 all cases, from Nr. 14 to Nr. 27, are definable over their field of moduli. The left cases, with the exception of Nrs. 1, 2 and 11, are of odd signature, so they are definable over their field of moduli.
(4) Genus 8: We see from the table in Section 5 that, for $g = 8$, there are 22 cases to consider. By Theorem 6 all cases, from Nr. 9 to Nr. 22, are definable over their field of moduli. The left cases, with the exception of Nrs. 2, 6, 7 and 8, are of odd signature, so they are definable over their field of moduli.
(5) Genus 9: We see from the table in Section 5 that, for $g = 9$, there are 50 cases to consider. By Theorem 6 all cases, from Nr. 23 to Nr. 50, are definable over their field of moduli. The left cases, with the exception of Nrs. 1, 3, 4, 14, 16 and 20, are of odd signature, so they are definable over their field of moduli.
(6) Genus 10: We see from the last table in Section 5 that, for $g = 10$, there are 55 cases to consider. From them, there are 18 hyperelliptic, 18 trigonal, and 4 quintagonal. By Theorem 6 all cases, from Nr. 24 to Nr. 55, are definable over their field of moduli. The left cases, with the exception of Nrs. 2, 3, 16, 17, 19, 20 and 23, are of odd signature, so they are definable over their field of moduli.

5. Tables of superelliptic curves of genus $5 \leq g \leq 10$

The following tables are taken from [**19**, Table 1]. The first column of the tables is simply a counter for each genus $5 \leq g \leq 10$. The second column is the reduced automorphism group and the third column some information about the full automorphism group. Notice that such column is left mostly empty, but a presentation of the group via its generators can be found in [**20**] for all the cases. In the fourth column is the level n of the superelliptic curve. Hence, the equation of the curve is given by $y^n = f(x)$, where $f(x)$ is the polynomial displayed in the last column. Columns 5 and 6 respectively represent the order of an automorphism in the reduced automorphism group and the signature of the covering $\mathcal{X} \to \mathcal{X}/G$. The seventh column represents the dimension of the corresponding locus in the moduli space $\mathcal{M}_g$.

TABLE 3. Superelliptic curves for genus $5 \leq g \leq 10$

Nr.	$\overline{G}$	G	n	m	sig.	δ	Equation $y^n = f(x)$
				Genus 5			
1		V_4	2	2	2^8	5	$x^{12}+\sum_{i=1}^{5} a_i x^{2i}+1$
2	C_m	$C_3\times C_2$	2	3	$2^4, 3^2$	3	$x^{12}+\sum_{i=1}^{3} a_i x^{3i}+1$
3		$C_2\times C_4$	2	4	$2^3, 4^2$	2	$x^{12}+a_2\,x^8+a_1\,x^4+1$
4		C_{22}	2	11	2, 11, 22	0	$x^{11}+1$
5		C_{22}	11	2	2, 22, 22	0	x^2+1
6		C_2	2	1	2^{12}	9	$x(x^{10}+\sum_{i=1}^{9} a_i x^i+1)$
7		C_4	2	2	$2^5, 4^2$	4	$x(x^{10}+\sum_{i=1}^{4} a_i x^{2i}+1)$
8			2	2	2^6	3	$\prod_{i=1}^{3}(x^4+a_i x^2+1)$
9	D_{2m}		2	3	2^4, 3	2	$(x^6+a_1x^3+1)(x^6+a_2x^3+1)$
10			2	6	2^3, 6	1	$x^{12}+a_1x^6+1$
11			2	4	$2^2, 4^2$	1	$(x^4-1)(x^8+a_1x^4+1)$
12			2	12	2, 4, 12	0	$x^{12}-1$
13			2	5	2^3, 10	1	$x(x^{10}+a_1x^5+1)$
14			2	2	$2^3, 4^2$	2	$(x^4-1)(x^4+a_1x^2+1)(x^4+a_2x^2+1)$
15			2	3	2, 3, 4^2	1	$(x^6-1)(x^6+a_1x^3+1)$
16			2	2	$2^3, 4^2$	2	$x(x^2-1)(x^4+a_1x^2+1)(x^4+a_2x^2+1)$
17			2	10	2, 4, 20	0	$x(x^{10}-1)$
18	A_4		2		$2^2, 3^2$	1	$f_1(x)$
19	S_4		2	0	3, 4^2	0	$x^{12}-33x^8-33x^4+1$
20	A_5		2		2,3,10	0	$x(x^{10}+11x^5-1)$
				Genus 6			
1		V_4	2	2	2^9	6	$x^{14}+\sum_{i=1}^{6} a_i x^{2i}+1$
2	C_m	C_{26}	2	13	2, 13, 26	0	$x^{13}+1$
3		C_{21}	3	7	3, 7, 21	0	x^7+1
4		C_{20}	4	5	4, 5, 20	0	x^5+1
5		C_{10}	5	2	2, 5, 5, 10	1	$x^4+a_1x^2+1$
6		C_{20}	5	4	4, 5, 20	0	x^4+1
7		C_{21}	7	3	3, 7 , 21	0	x^3+1
8		C_{26}	13	2	2, 13, 26	0	x^2+1
9		C_2	2	1	2^{14}	11	$x(x^{12}+\sum_{i=1}^{11} a_i x^i+1)$
10		C_4	2	2	$2^6, 4^2$	5	$x(x^{12}+\sum_{i=1}^{5} a_i x^{2i}+1)$
11		C_6	2	3	$2^3, 3^2, 6^2$	3	$x(x^{12}+\sum_{i=1}^{3} a_i x^{3i}+1)$
12		C_8	2	4	$2^3, 8^2$	2	$x(x^{12}+\sum_{i=1}^{2} a_i x^{4i}+1)$
13		C_3	3	1	3^8	5	$x^6+\sum_{i=1}^{5} a_i x^i+1$
14		C_6	3	2	$3^3, 6^2$	2	$x^6+a_2x^4+a_1x^2+1$
15		C_4	4	1	4^6	3	$x^4+\sum_{i=1}^{3} a_i x^i+1$
16		C_5	5	1	5^5	2	$x^3+a_1x+a_2x^2+1$
17		$D_{14}\times C_2$	2	7	2^3, 7	1	$x^{14}+a_1x^7+1)$
18		G_5	2	2	2^5, 4	3	$(x^2-1)\prod_{i=1}^{3}(x^4+a_ix^2+1)$
19		G_5	2	14	2,4,14	0	$x^{14}-1$
20		$D_{10}\times C_2$	5	5	2,5,10	0	x^5-1
21		D_8	2	2	2^5,4	3	$x\cdot\prod_{i=1}^{3}(x^4+a_ix^2+1)$
22		$D_6\times C_2$	2	3	2^4, 6	2	$x\cdot\prod_{i=1}^{2}(x^4+a_ix^2+1)$
23		D_{24}	2	6	2^3, 12	1	$x(x^{12}+a_1x^6+1)$
24		$D_6\times C_3$	3	3	2^2,3,9	1	$x(x^6+a_1x^3+1)$
25	D_{2m}	D_{16}	4	2	2^2, 4, 8	1	$x(x^4+a_1x^2+1)$
26		G_8	2	4	2^2, 4, 8	1	$x(x^4-1)(x^8+a_1x^4+1)$
27		G_8	2	12	2, 4, 24	0	$x(x^{12}-1)$
28		$V_4\times C_3$	3	2	2, 3, 6^2	1	$x(x^2-1)(x^4+a_1x^2+1)$
29		$D_{12}\times C_3$	3	6	2, 6, 18	0	$x(x^6-1)$
30		G_8	4	4	2, 8, 16	0	$x(x^4-1)$
31		$D_6\times C_5$	5	3	2, 10, 15	0	$x(x^3-1)$
32		$V_4\times C_7$	7	2	2, 14^2	0	$x(x^2-1)$
33		G_9	2	2	$2^2, 4^3$	2	$x(x^4-1)\cdot\prod_{i=1}^{2}(x^4+a_ix^2+1)$
34		G_9	2	3	$2, 4^2, 6$	1	$x(x^6-1)(x^6+a_1x^3+1)$
35	S_4	G_{18}	4	0	2, 3, 16	0	$x(x^4-1)$
36		G_{19}	2	0	2, 6, 8	0	$x(x^4-1)(x^8+14x^4+1)$

TABLE 3. (Cont.)

Nr.	$\overline{G}$	G	n	m	sig.	δ	Equation $y^n = f(x)$
Genus 7							
1		V_4	2	2	2^{10}	7	$x^{16} + \sum_{i=1}^{7} a_i x^{2i} + 1$
2	C_m	$C_2 \times C_4$	2	4	$2^4, 4^2$	3	$x^{16} + \sum_{i=1}^{3} a_i x^{4i} + 1$
3		${C_3}^2$	3	3	3^5	2	$x^9 + a_2 x^6 + a_1 x^3 + 1$
4		C_6	2	3	2^5, 3, 6	4	$x^{15} + \sum_{i=1}^{4} a_1 x^{3i} + 1$
5		C_{10}	2	5	2^3, 5, 10	2	$x^{15} + a_1 x^5 + a_2 x^{10} + 1$
6		C_{30}	2	15	2, 15, 30	0	$x^{15} + 1$
7		C_6	3	2	2, 3^4, 6	3	$x^8 + a_3 x^6 + a_2 x^4 + a_1 x^2 + 1$
8		C_{12}	3	4	3^2, 4, 12	1	$x^8 + a_1 x^4 + 1$
9		C_{24}	3	8	3, 8, 24	0	$x^8 + 1$
10		C_{30}	15	2	2, 15, 30	0	$x^2 + 1$
11		C_2	2	1	2^{16}	13	$x(x^{14} + \sum_{i=1}^{13} a_i x^i + 1)$
12		C_4	2	2	$2^7, 4^2$	6	$x(x^{14} + \sum_{i=1}^{6} a_i x^{2i} + 1)$
13		C_3	3	1	3^9	6	$x^7 + \sum_{i=1}^{6} a_i x^i + 1$
14	D_{2m}	$V_4 \times C_2$	2	2	2^7	4	$\prod_{i=1}^{4}(x^4 + a_i x^2 + 1)$
15		$D_8 \times C_2$	2	4	2^4, 4	2	$(x^8 + a_1 x^4 + 1)(x^8 + a_2 x^4 + 1)$
16		$D_{16} \times C_2$	2	8	2^3, 8	1	$x^{16} + a_1 x^8 + 1$
17	D_{2m}	G_5	2	16	2, 4, 16	0	$x^{16} - 1$
18		$D_6 \times C_3$	3	3	2, 3^2, 6	1	$(x^3 - 1)(x^6 + a_1 x^3 + 1)$
19		$D_{18} \times C_3$	3	9	2, 6, 9	0	$x^9 - 1$
20		$D_{14} \times C_2$	2	7	2^3, 14	1	$x(x^{14} + a_1 x^7 + 1)$
21		G_7	2	2	2^4, 4^2	3	$(x^4 - 1) \prod_{i=1}^{3}(x^4 + a_i x^2 + 1)$
22		G_7	2	4	2, 4^3	1	$(x^8 - 1)(x^8 + a_1 x^4 + 1)$
23		G_8	2	2	2^4, 4^2	3	$x(x^2 - 1) \prod_{i=1}^{3}(x^4 + a_i x^2 + 1)$
24		G_8	2	14	2, 4, 28	0	$x(x^{14} - 1)$
25		$D_{14} \times C_3$	3	7	2, 6, 21	0	$x(x^7 - 1)$
26		G_8	8	2	2,16^2	0	$x(x^2 - 1)$
27	A_4	K	2	0	2^2, 3, 6	1	$(x^4 + 2\sqrt{-3}x^2 + 1)\, f_1(x)$
Genus 8							
1		V_4	2	2	2^{11}	8	$x^{18} + \sum_{i=1}^{8} a_i x^{2i} + 1$
2	C_m	$C_2 \times C_3$	2	3	$2^6, 3^2$	5	$x^{18} + \sum_{i=1}^{5} a_i x^{3i} + 1$
3		$C_2 \times C_6$	2	6	$2^3, 6^2$	2	$x^{18} + a_1 x^6 + a_2 x^{12} + 1$
4		C_{34}	2	17	2, 17, 34	0	$x^{17} + 1$
5		C_{34}	17	2	2, 17 , 34	0	$x^2 + 1$
6		C_2	2	1	2^{18}	15	$x(x^{16} + \sum_{i=1}^{1} 5a_i x^i + 1)$
7		C_4	2	2	$2^8, 4^2$	7	$x(x^{16} + \sum_{i=1}^{7} a_i x^{2i} + 1)$
8		C_8	2	4	$2^4, 8^2$	3	$x(x^{16} + a_1 x^4 + a_2 x^8 + a_3 x^{12} + 1)$
9		$D_6 \times C_2$	2	3	2^5, 3	3	$\prod_{i=1}^{3}(x^6 + a_i x^3 + 1)$
10	D_{2m}	$D_{18} \times C_2$	2	9	2^3, 9	1	$x^{18} + a_1 x^9 + 1$
11		G_5	2	2	2^6, 4	4	$(x^2 - 1) \prod_{i=1}^{4}(x^4 + a_i x^2 + 1)$
12		G_5	2	6	2^2, 4, 6	1	$(x^6 - 1)(x^{12} + a_1 x^6 + 1)$
13		G_5	2	18	2, 4, 18	0	$x^{18} - 1$
14		D_8	2	2	2^6, 4	4	$x \prod_{i=1}^{4}(x^4 + a_i x^2 + 1)$
15		D_{16}	2	4	2^4, 8	2	$x(x^8 + a_1 x^4 + 1)(x^8 + a_2 x^4 + 1)$
16		D_{32}	2	8	2^3, 16	1	$x(x^{16} + a_1 x^8 + 1)$
17		G_9	2	3	2^2, 3, 4^2	2	$(x^6 - 1)(x^6 + a_1 x^3 + 1)(x^6 + a_2 x^3 + 1)$
18		G_8	2	16	2, 4, 32	0	$x(x^{16} - 1)$
19		G_9	2	2	$2^3, 4^3$	3	$x \prod_{i+1}^{3}(x^6 + a_i x^3 + 1)$
20		G_9	2	4	2, 4^2, 8	1	$x(x^8 - 1)(x^8 + a_1 x^4 + 1)$
21	A_4	K	2	0	2, 3^2, 4	1	$x(x^4 - 1)\, f_1(x)$
22	S_4	G_{22}	2	0	3, 4, 8	0	$x(x^4 - 1)(x^{12} - 33x^8 - 33x^4 + 1)$

TABLE 3. (Cont.)

Nr.	$\overline{G}$	G	n	m	sig.	δ	Equation $y^n = f(x)$
					Genus 9		
1		V_4	2	2	2^{12}	9	$x^{20} + \sum_{i=1}^{9} a_i x^{2i} + 1$
2	C_m	$C_2 \times C_4$	2	4	$2^5, 4^2$	4	$x^{20} + \sum_{i=1}^{4} a_i x^{4i} + 1$
3		$C_2 \times C_5$	2	5	$2^4, 5^2$	3	$x^{20} + a_1 x^5 + a_2 x^{10} + a_3 x^{15} + 1$
4		$C_2 \times C_4$	4	2	$2^2, 4^4$	3	$x^8 + a_1 x^2 + a_2 x^4 + a_3 x^6 + 1$
5		C_{38}	2	19	2, 19, 38	0	$x^{19} + 1$
6		C_6	3	2	2, 3^5, 6	4	$x^{10} + a_1 x^2 + a_2 x^4 + a_3 x^6 + a_4 x^8 + 1$
7		C_{15}	3	5	3^2, 5, 15	1	$x^{10} + a_1 x^5 + 1$
8		C_{30}	3	10	3, 10^2	0	$x^{10} + 1$
9		C_{28}	4	7	4, 7^2	0	$x^7 + 1$
10		C_{14}	7	2	$2, 7^2, 14$	1	$x^4 + a_1 x^2 + 1$
11		C_{28}	7	4	4^2, 7	0	$x^4 + 1$
12		C_{30}	10	3	3^2, 10	0	$x^3 + 1$
13		C_{38}	19	2	2^2, 19	0	$x^2 + 1$
14		C_2	2	1	2^{20}	17	$x(x^{18} + \sum_{i=1}^{17} a_i x^i + 1)$
15		C_4	2	2	$2^9, 4^2$	8	$x(x^{18} + \sum_{i=1}^{8} a_i x^{2i} + 1)$
16		C_6	2	3	$2^6, 6^2$	5	$x(x^{18} + \sum_{i=1}^{5} a_i x^{3i} + 1)$
17		C_{12}	2	6	$2^3, 12^2$	2	$x(x^{18} + a_1 x^6 + a_2 x^{12} + 1)$
18		C_3	3	1	3^{11}	8	$x^9 + \sum_{i=1}^{8} a_i x^i + 1$
19		C_9	3	3	$3^3, 9^2$	2	$x^9 + a_2 x^6 + a_1 x^3 + 1$
20		C_4	4	1	4^8	5	$x^6 + \sum_{i=1}^{5} a_i x^i + 1$
21		C_8	4	2	$4^3, 8^2$	2	$x^6 + a_2 x^4 + a_1 x^2 + 1$
22		C_7	7	1	7^5	2	$x^3 + a_1 x + a_2 x^2 + 1$
23		$V_4 \times C_2$	2	2	2^8	5	$\prod_{i=1}^{5} (x^4 + a_i x^2 + 1))$
24	D_{2m}	$D_{10} \times C_2$	2	5	2^4, 5	2	$(x^{10} + a_1 x^5 + 1)(x^{10} + a_2 x^5 + 1)$
25		$D_{20} \times C_2$	2	10	2^3, 10	1	$x^{20} + a_1 x^{10} + 1$
26		$V_4 \times C_4$	4	2	$2^3, 4^2$	2	$(x^4 + a_1 x^2 + 1)(x^4 + a_2 x^2 + 1)$
27		$D_8 \times C_4$	4	4	$2^2, 4^2$	1	$x^8 + a_1 x^4 + 1$
28		G_5	2	4	2^3, 4^2	2	$(x^4 - 1)(x^8 + a_1 x^4 + 1)(x^8 + a_2 x^4 + 1)$
29		G_5	2	20	2, 4, 20	0	$x^{20} - 1$
30		G_5	4	8	2, 8^2	0	$x^8 - 1$
31		$D_6 \times C_2$	2	3	2^5, 6	3	$x \prod_{i=1}^{3} (x^6 + a_i x^3 + 1)$
32		$D_{18} \times C_2$	2	9	2^3, 18	1	$x(x^{18} + a_1 x^9 + 1)$
33		$D_6 \times C_4$	4	3	2^2, 4, 12	1	$x(x^6 + a_1 x^3 + 1)$
34		G_7	2	2	2^5, 4^2	4	$(x^4 - 1) \prod_{i=1}^{4} (x^4 + a_i x^2 + 1)$
35		G_9	2	5	2, 4^2, 5	1	$(x^{10} - 1)(x^{10} + a_1 x^5 + 1)$
36		G_7	4	2	2, 4, 8^2	1	$(x^4 - 1)(x^4 + a_1 x^2 + 1)$
37		G_8	2	2	2^5, 4^2	4	$x(x^2 - 1) \prod_{i=1}^{4} (x^4 + a_i x^2 + 1)$
38		G_8	2	6	2^2, 4, 12	1	$x(x^6 - 1)(x^{12} + a_1 x^6 + 1)$
39		G_8	2	18	2, 4, 36	0	$x(x^{18} - 1)$
40		$D_6 \times C_3$	3	3	2, 3, 6, 9	1	$x(x^3 - 1)(x^6 + a_1 x^3 + 1)$
41		$D_{18} \times C_3$	3	9	2, 6, 27	0	$x(x^9 - 1)$
42		G_8	4	2	2, 4, 8^2	1	$x(x^2 - 1)(x^4 + a_1 x^2 + 1)$
43		G_8	4	6	2, 8, 24	0	$x(x^6 - 1)$
44		$D_6 \times C_7$	7	3	2, 14, 21	0	$x(x^3 - 1)$
45		G_8	10	2	2, 20^2	0	$x(x^2 - 1)$
46		G_9	2	3	$2^2, 4^2, 6$	2	$x(x^6 - 1)(x^6 + a_1 x^3 + 1)(x^6 + a_2 x^3 + 1)$
47	A_4	K	2	0	2^2, 6^2	1	$(x^8 + 14x^4 + 1)\, f_1(x)$
48	S_4	G_{17}	4	0	2, 4, 12	0	$x^8 + 14x^4 + 1$
49		G_{21}	2	0	4^2, 6	0	$(x^8 + 14x^4 + 1)(x^{12} - 33x^8 - 33x^4 + 1)$
50	A_5		2		2, 5, 6	0	$x^{20} - 228x^{15} + 494x^{10} + 228x^5 + 1$

TABLE 3. (Cont.)

Nr.	$\overline{G}$	G	n	m	sig.	δ	Equation $y^n = f(x)$
				Genus 10			
1		V_4	2	2	2^{13}	10	$x^{22} + \sum_{i=1}^{10} a_i x^{2i} + 1$
2	C_m	$C_2 \times C_3$	3	2	$2^2, 3^6$	5	$x^{12} + \sum_{i=1}^{5} a_i x^{2i} + 1$
3		C_3^2	3	3	3^6	3	$x^{12} + a_1x^3 + a_2x^6 + a_3x^9 + 1$
4		$C_3 \times C_4$	3	4	$3^3, 4^2$	2	$x^{12} + a_1x^4 + a_2x^8 + 1$
5		$C_2 \times C_6$	6	2	$2^2, 6^3$	2	$x^6 + a_1x^2 + a_2x^4 + 1$
6		C_6	2	3	2^7, 3, 6	6	$x^{21} + \sum_{i=1}^{6} a_i x^{3i} + 1$
7		C_{14}	2	7	2^3, 7, 14	2	$x^{21} + a_1x^7 + a_2x^{14} + 1$
8		C_{42}	2	21	2, 4, 21	0	$x^{21} + 1$
9		C_{33}	3	11	3, 11^2	0	$x^{11} + 1$
10		C_{10}	5	2	2, 5^3, 10	2	$x^6 + a_2x^4 + a_1x^2 + 1$
11		C_{15}	5	3	3, 5^2, 15	1	$x^6 + a_1x^3 + 1$
12		C_{30}	5	6	5, 6^2	0	$x^6 + 1$
13		C_{30}	6	5	5^2, 6	0	$x^5 + 1$
14		C_{33}	11	3	3^2, 11	0	$x^3 + 1$
15		C_{42}	21	2	2, 21, 42	0	$x^2 + 1$
16		C_2	2	1	2^{22}	19	$x(x^{20} + \sum_{i=1}^{19} a_i x^i + 1)$
17		C_4	2	2	$2^{10}, 4^2$	9	$x(x^{20} + \sum_{i=1}^{9} a_i x^{2i} + 1)$
18		C_8	2	4	$2^5, 8^2$	4	$x(x^{20} + a_1x^4 + a_2x^8 + a_3x^{12} + a_4x^{16} + 1)$
19		C_{10}	2	5	$2^4, 10^2$	3	$x(x^{20} + a_1x^5 + a_2x^{10} + a_3x^{15} + 1)$
20		C_3	3	1	3^{12}	9	$x^{10} + \sum_{i=1}^{9} a_i x^i + 1$
21		C_6	3	2	$3^5, 6^2$	4	$x^{10} + a_1x^2 + a_2x^4 + a_3x^6 + a_4x^8 + 1$
22		C_5	5	1	5^7	4	$x^5 + \sum_{i=1}^{4} a_i x^i + 1$
23		C_6	6	1	6^6	3	$x^4 + a_1x + a_2x^2a_3x^3 + 1$
24		$D_{22} \times C_2$	2	11	2^3, 11	1	$x^{22} + a_1x^{11} + 1$
25	D_{2m}	$V_4 \times C_3$	3	2	$2^3, 3^3$	3	$\prod_{i=1}^{3}(x^4 + a_ix^2 + 1)$
26		$D_6 \times C_3$	3	3	$2^2, 3^3,$	2	$(x^6 + a_1x^3 + 1)(x^6 + a_2x^3 + 1)$
27		$D_{12} \times C_3$	3	6	$2^2, 3, 6$	1	$(x^{12} + a_1x^6 + 1$
28		$D_6 \times C_6$	6	3	$2^2, 3, 6$	1	$x^6 + a_1x^3 + 1$
29		G_5	2	2	2^7, 4	5	$(x^2 - 1)\prod_{i=1}^{5}(x^4 + a_ix^2 + 1)$
30		G_5	2	22	2, 4, 22	0	$x^{22} - 1$
31		$D_8 \times C_3$	3	4	2, 3, 4, 6	1	$(x^4 - 1)(x^8 + a_1x^4 + 1)$
32		$D_{24} \times C_3$	3	12	2, 6, 12	0	$x^{12} - 1$
33		G_5	6	2	2^2, 6, 12	1	$(x^2 - 1)(x^4 + a_1x^2 + 1)$
34		G_5	6	6	2, 6, 12	0	$x^6 - 1$
35		D_8	2	2	2^7, 4	5	$x\prod_{i=1}^{5}(x^4 + a_ix^2 + 1)$
36		$D_{10} \times C_2$	2	5	2^4, 10	2	$x(x^{10} + a_1x^5 + 1)(x^{10} + a_2x^5 + 1)$
37		D_{40}	2	10	2^3, 20	1	$x(x^{20} + a_1x^{10} + 1)$
38		$D_{10} \times C_3$	3	5	2^2, 3, 15	1	$x(x^{10} + a_1x^5 + 1)$
39		D_{24}	6	2	2^2, 6, 12	1	$x(x^4 + a_1x^2 + 1)$
40		$V_4 \times C_3$	3	2	2, 3^2, 6^2	2	$(x^2 - 1)(x^4 + a_1x^2 + 1)(x^4 + a_2x^2 + 1)$
41		$D_6 \times C_3$	3	3	$3^2, 6^2$	1	$(x^6 - 1)(x^6 + a_1x^3 + 1)$
42		G_8	2	4	2^3, 4, 8	2	$x(x^4 - 1)(x^8 + a_1x^4 + 1)(x^8 + a_2x^4 + 1)$
43		G_8	2	20	2, 4, 40	0	$x(x^{20} - 1)$
44		$V_4 \times C_3$	3	2	$2, 3^2, 6^2$	2	$x(x^2 - 1)(x^4 + a_1x^2 + 1)(x^4 + a_2x^2 + 1)$
45		$D_{20} \times C_3$	3	10	2, 6, 30	0	$x(x^{10} - 1)$
46		$D_{10} \times C_5$	5	5	2, 10, 25	0	$x(x^5 - 1)$
47		G_8	6	4	2, 12, 24	0	$x(x^4 - 1)$
48		$V_4 \times C_{11}$	11	2	2, 22^2	0	$x(x^2 - 1)$
49		G_9	2	2	$2^4, 4^3$	4	$x(x^4 - 1)\prod_{i=1}^{4}(x^4 + a_ix^2 + 1)$
50		G_9	2	5	$2, 4^2, 10$	1	$x(x^{10} - 1)(x^{10} + a_1x^5 + 1)$
51	A_4		3	0	2, 3^3	1	$f_1(x)$
52			2	0	2, 3, 4, 6	1	$x(x^4 - 1)(x^4 + 2\sqrt{-3}\,x^2 + 1)\,f_1(x)$
53	S_4	G_{18}	6	0	2, 3, 24	0	$x(x^4 - 1)$
54		$S_4 \times C_3$	3	0	3, 4, 6	0	$x^{12} - 33x^8 - 33x^4 + 1$
55	A_5	$A_5 \times C_3$	3	0	2, 3, 15	0	$x(x^{10} + 11x^5 - 1)$

In [**19**], the authors create a database of superelliptic curves. Moreover, they display all curves of genus $g \leq 10$ in [**19**, Table 1]. We present such tables below. The first column of the table represents the case from Table 1 of [**20**], the second column is the reduced automorphism group. In the third column is the full automorphism group. Such groups are well known and only the 'obvious' cases are displayed, for full details one can check [**20**] and [**21**].

In the fourth column is the level n of the superelliptic curve; see [**3**]. Hence, the equation of the curve is given by $y^n = f(x)$, where $f(x)$ is the polynomial displayed in the last column. Columns 5 and 6, respectively, represent the order of the superelliptic automorphism in the reduced automorphism group and the signature of the covering $\mathcal{X} \to \mathcal{X}/G$. The sixth column represents the dimension of the corresponding locus in the moduli space $\mathcal{M}_g$. Throughout these tables $f_1(x)$ is as follows

$$f_1(x) = x^{12} - a_1x^{10} - 33x^8 + 2a_1x^6 - 33x^4 - a_1x^2 + 1$$

for $a_1 \in \mathbb{C}$. In [**19**, Table 1] the signatures of the coverings are not fully given. Indeed, for a full signature $(\sigma_1, \dots, \sigma_r)$, we know that $\sigma_r = \sigma_{r-1}^{-1} \cdots \sigma_1^{-1}$. Hence, the authors present only $(\sigma_1, \dots, \sigma_{r-1})$. In our Table 3 that follows, we present the full signature $(\sigma_1, \dots, \sigma_r)$.

For example, in the case $g = 5$, Nr.1, we have a signature of eight branch points, each corresponding to a double transposition. Such signature is presented as 2^8. Another clue for the reader must be that the moduli dimension is always $\delta = r - 3$, where r is the number of branch points of the covering.

REMARK 1. It is worth mentioning, to avoid any confusion, that the equations of the curves in column eight are not over the field of moduli of the corresponding curve. They are used only to help identify the corresponding family.

Acknowledgements. The authors want to thanks Jeroen Sijsling for his comments and suggestions to a previous version which help us to correctly state some points of this paper.

References

[1] Michela Artebani and Saúl Quispe, *Fields of moduli and fields of definition of odd signature curves*, Arch. Math. (Basel) **99** (2012), no. 4, 333–344, DOI 10.1007/s00013-012-0427-6. MR2990152

[2] G. V. Belyĭ, *Galois extensions of a maximal cyclotomic field* (Russian), Izv. Akad. Nauk SSSR Ser. Mat. **43** (1979), no. 2, 267–276, 479. MR534593

[3] Lubjana Beshaj, Valmira Hoxha, and Tony Shaska, *On superelliptic curves of level n and their quotients, I*, Albanian J. Math. **5** (2011), no. 3, 115–137. MR2846162

[4] L. Beshaj, T. Shaska, and E. Zhupa, *The case for superelliptic curves*, Advances on superelliptic curves and their applications, NATO Sci. Peace Secur. Ser. D Inf. Commun. Secur., vol. 41, IOS, Amsterdam, 2015, pp. 1–14. MR3525570

[5] Lubjana Beshaj and Fred Thompson, *Equations for superelliptic curves over their minimal field of definition*, Albanian J. Math. **8** (2014), no. 1, 3–8. MR3253208

[6] Gabriel Cardona and Jordi Quer, *Field of moduli and field of definition for curves of genus 2*, Computational aspects of algebraic curves, Lecture Notes Ser. Comput., vol. 13, World Sci. Publ., Hackensack, NJ, 2005, pp. 71–83, DOI 10.1142/9789812701640_0006. MR2181874

[7] G. Castelnuovo, *Sulle serie algebriche di gruppi di punti appartenenti ad una curva algebrica.* (Italian), Rom. Acc. L. Rend. (5) **15** (1906), no. 1, 337–344.

[8] Clifford J. Earle, *On the moduli of closed Riemann surfaces with symmetries*, Advances in the theory of riemann surfaces (Proc. Conf., Stony Brook, N.Y., 1969), Princeton Univ. Press, Princeton, N.J., 1971, pp. 119–130. Ann. of Math. Studies, No. 66. MR0296282

[9] J. Gutierrez and T. Shaska, *Hyperelliptic curves with extra involutions*, LMS J. Comput. Math. **8** (2005), 102–115, DOI 10.1112/S1461157000000917. MR2135032

[10] Rubén A. Hidalgo, *Non-hyperelliptic Riemann surfaces with real field of moduli but not definable over the reals*, Arch. Math. (Basel) **93** (2009), no. 3, 219–224, DOI 10.1007/s00013-009-0025-4. MR2540787

[11] Rubén A. Hidalgo, *Genus zero p-groups of automorphisms of riemann surfaces*, 2016. In preparation.

[12] Rubén A. Hidalgo and Saúl Quispe, *Fields of moduli of some special curves*, J. Pure Appl. Algebra **220** (2016), no. 1, 55–60, DOI 10.1016/j.jpaa.2015.05.042. MR3393450

[13] Bonnie Huggins, *Fields of moduli of hyperelliptic curves*, Math. Res. Lett. **14** (2007), no. 2, 249–262, DOI 10.4310/MRL.2007.v14.n2.a8. MR2318623

[14] Shoji Koizumi, *The fields of moduli for polarized abelian varieties and for curves*, Nagoya Math. J. **48** (1972), 37–55. MR0352095

[15] Aristides Kontogeorgis, *Field of moduli versus field of definition for cyclic covers of the projective line* (English, with English and French summaries), J. Théor. Nombres Bordeaux **21** (2009), no. 3, 679–692. MR2605539

[16] Reynald Lercier, Christophe Ritzenthaler, and Jeroen Sijsling, *Explicit Galois obstruction and descent for hyperelliptic curves with tamely cyclic reduced automorphism group*, Math. Comp. **85** (2016), no. 300, 2011–2045, DOI 10.1090/mcom3032. MR3471117

[17] Andreas Malmendier and Tony Shaska, *A universal pair of genus-two curves* (2016), 1–17 pp., available at 1607.08294.

[18] Jean-François Mestre, *Construction d'une courbe elliptique de rang* ≥ 12 (French, with English summary), C. R. Acad. Sci. Paris Sér. I Math. **295** (1982), no. 12, 643–644. MR688896

[19] Rezart Muço, Nejme Pjero, Ervin Ruci, and Eustrat Zhupa, *Classifying families of superelliptic curves*, Albanian J. Math. **8** (2014), no. 1, 23–35. MR3270074

[20] R. Sanjeewa, *Automorphism groups of cyclic curves defined over finite fields of any characteristics*, Albanian J. Math. **3** (2009), no. 4, 131–160. MR2578064

[21] R. Sanjeewa and T. Shaska, *Determining equations of families of cyclic curves*, Albanian J. Math. **2** (2008), no. 3, 199–213. MR2492096

[22] T. Shaska, *Genus 2 curves with* $(3,3)$*-split Jacobian and large automorphism group*, Algorithmic number theory (Sydney, 2002), 2002, pp. 205–218. MR2041085

[23] T. Shaska, *Computational aspects of hyperelliptic curves*, Computer mathematics, Lecture Notes Ser. Comput., vol. 10, World Sci. Publ., River Edge, NJ, 2003, pp. 248–257. MR2061839

[24] Goro Shimura, *On the field of rationality for an abelian variety*, Nagoya Math. J. **45** (1972), 167–178. MR0306215

[25] André Weil, *The field of definition of a variety*, Amer. J. Math. **78** (1956), 509–524, DOI 10.2307/2372670. MR0082726

[26] J. Wolfart, *ABC for polynomials, dessins d'enfants and uniformization—a survey*, Elementare und analytische Zahlentheorie, Schr. Wiss. Ges. Johann Wolfgang Goethe Univ. Frankfurt am Main, vol. 20, Franz Steiner Verlag Stuttgart, Stuttgart, 2006, pp. 313–345. MR2310190

Departamento de Matemática y Estadística, Universidad de La Frontera, Temuco, Chile

Email address: ruben.hidalgo@ufrontera.cl

Department of Mathematics and Statistics, Oakland University, Rochester, Michigan 48386

Email address: shaska@oakland.edu

Contemporary Mathematics
Volume **703**, 2018
http://dx.doi.org/10.1090/conm/703/14131

Minimal integral Weierstrass equations for genus 2 curves

Lubjana Beshaj

Abstract. We study the Weierstrass equations for genus 2 curves defined over a ring of integers $\mathcal{O}_{\mathbb{F}}$ which correspond to reduced binary sextics. This is done via reduction theory and Julia quadratic of binary sextics. We show that when a binary sextic has extra automorphisms then it is usually easier to compute its Julia quadratic. Moreover, we show that when the curve is given in the standard form $y^2z^4 = f(x^2, z^2)$, such that $f(0,1) = 1$, $f(1,0) = 1$, and defined over $\mathcal{O}_{\mathbb{F}}$, then the binary form f is reduced. Such curves have minimal height among integral models defined by sextics in x^2, z^2, even up to twist.

1. Introduction

Let $\mathcal{M}_2$ be the moduli space classifying algebraic curves of genus 2. Using the classical theory of invariants of binary forms, J. Igusa (1960) constructed an arithmetic model of the moduli space $\mathcal{M}_2$. Given a moduli point $\mathfrak{p} \in \mathcal{M}_2(\mathbb{Q})$ there are basically two main cases to get an equation of the curve defined over $\mathbb{Q}$ when such an equation exists.

If $\mathrm{Aut}(\mathfrak{p})$ has order 2, then one can use an algorithm of Mestre [**9**] to determine if there is a curve $\mathcal{X}$ defined over $\mathbb{Q}$ and construct its equation. If $\mathrm{Aut}(\mathfrak{p})$ has order > 2, then there always exists a curve $\mathcal{X}$ defined over $\mathbb{Q}$ and its equation can be found via the dihedral-invariants (u, v) defined in [**13**]. Such invariants determine uniquely the isomorphism class of a genus 2 curve with automorphism group isomorphic to the Klein 4-group V_4. The cases when the automorphism group of the curve is isomorphic to the dihedral group of order 8 or of order 12 correspond to the singular points of the V_4-locus in $\mathcal{M}_2$ and are treated differently; see [**1**] for such loci. From [**13**], and [**14**] we have a method which determines the equation of the curve when the curve has extra automorphisms. A more recent approach to recover an equation of a curve starting with any point $\mathfrak{p} \in \mathcal{M}_2$ regardless of the automorphism group can be found in the work of Malmendier and Shaska in [**8**]. In any case, for any number field K, when a point $\mathfrak{p} \in \mathcal{M}_2(K)$ is given we can determine a genus 2 curve at worst defined over a quadratic extension of K. The reader interested in such computational routines can check [**4**].

Let us now assume that $\mathbb{F}$ is a number field and $\mathcal{X}$ be a genus 2 curve with extra automorphisms defined over $\mathbb{F}$. Let $\mathcal{O}_{\mathbb{F}}$ be the ring of integers of $\mathbb{F}$. Then, without any loss of generality we can assume that $\mathcal{X}$ is defined over $\mathcal{O}_{\mathbb{F}}$. The height $\mathfrak{h}(\mathcal{X})$ of $\mathcal{X}$ over $\mathcal{O}_{\mathbb{F}}$ is defined in [**12**]. We further assume that F is a minimal field

2010 *Mathematics Subject Classification.* Primary 14Hxx, 14Qxx.

of definition of $\mathcal{X}$. The focus of this paper is to find a twist $\mathcal{X}'$ of $\mathcal{X}$ defined over $\mathcal{O}_\mathbb{F}$ such that the height $\mathfrak{h}(\mathcal{X}')$ is minimal; see [**2**] and [**12**] for further details.

Reduction theory was introduced by Julia in [**6**] and has been revisited recently by [**5**], [**15**], and [**2**]. For every binary form there is a corresponding positive definite binary quadratic called the *Julia quadratic*. Since the Julia quadratic is positive definite, then there is a unique root in the upper half-plane $\mathcal{H}_2$. Hence, there is a map from the set of semistable binary forms to the upper half-plane, which is called the *zero map* and denoted by ζ_0. The binary form f is called *reduced* when $\zeta_0(f)$ is in the fundamental domain of the modular group $\Gamma := SL_2(\mathcal{O}_\mathbb{F})/\{\pm I\}$. For binary quadratics it is shown in [**2**, Thm.13] that f is reduced implies that f has minimal height. It is expected that this occurs under some mild condition in higher degrees as well.

Hence, for any genus two curve $\mathcal{X}$ we write this curve in its Weierstrass form $y^2z^4 = f(x,z)$over some algebraically closed field k, where $f(x,z)$ is a binary sextic. Finding a twist of $\mathcal{X}$ with smaller coefficients is equivalent to finding the reduction red f as described in [**2**]. The main issue with this method is that determining red f involves solving a system via Gröbener bases. There is also a numerical approach suggested in [**15**], which of course it is open to numerical analysis.

In this paper we investigate whether determining a Weierstrass equation with minimal height is easier in the case that the curve has extra involutions. Any curve with extra involutions can be written in as $y^2\, z^4 = g(x^2, z^2)$, where $g(x,z)$ is a binary cubic form. We discover that if $g(x,z) \in \mathcal{O}_\mathbb{F}[x,z]$ and $g(x^2,z^2)$ has minimal discriminant over $\mathcal{O}_\mathbb{F}$, then $y^2\, z^4 = g(x^2, z^2)$ is reduced.

The paper is organized as follows. In section two we describe briefly reduction theory of binary quintics and sextics, see [**3**] for more details. The most delicate and difficult part of reduction theory is computing the Julia quadratic. We show that computing the Julia quadratic for binary quintics and sextics in a direct way using the system given in [**3**, Eq. 4.13] is too difficult. Hence, we investigate alternative methods, considering separately totally real and totally complex binary forms. At the end of this section we give an example where we show how numerical computations can be used successfully in implementing a reduction algorithm.

In Section three we give a quick review of how for any binary form $f(x,z) \in \mathcal{O}_\mathbb{F}$ we can minimize the discriminant over $\mathcal{O}_\mathbb{F}$. There is a detailed treatment of this in [**10**]. For minimizing the discriminant of genus 2 curves (over global fields) there is the more classical result of Liu [**7**]. In section four we tailor the reduction for binary quintics and sextics and study how this can be performed when applied to forms with extra automorphisms. We show that the curves with extra automorphisms in the standard form as in $y^2\, z^4 = f(x,z)$ where $f(x,z) = x^6 + ax^4z^2 + bx^2z^4 + z^6$ are reduced over $\mathcal{O}_\mathbb{F}$ when defined over $\mathcal{O}_\mathbb{F}$ and have minimal discriminant.

In the last section we build a database of all such curves with height $\mathfrak{h} \leq 100$ defined over the integers. There are 20 292 such curves (up to isomorphism over $\mathbb{C}$). For each height $1 \leq \mathfrak{h} \leq 100$ we also display the number of curves for that height. The number of such curves with automorphism group D_4 and D_6 are also displayed. From these 20 292 curves we check if they are all of minimal height. Of course not all of them are expected to have minimal discriminant. We check how many of them have minimal height $\mathfrak{h} \leq 3$, while $\max\{a,b\} > 3$. We found 57 such cases, and as expected all of them do not have minimal discriminant over $\mathbb{Z}$.

2. Reduction of binary quintics and sextics

In this section we will define reduction theory for binary forms $f(x,z) \in \mathbb{R}[x,z]$. A generalization of reduction theory to binary forms defined over $\mathbb{C}$ is explained in details in [**3**]. Let $f(x,z) \in \mathbb{R}[x,z]$ be a degree n binary form given as follows:

$$f(x,z) = a_0x^n + a_1x^{n-1}z + \cdots + a_nz^n$$

and suppose that $a_0 \neq 0$. Let the real roots of $f(x,z)$ be α_i, for $1 \leq i \leq r$ and the pair of complex roots β_j, $\bar{\beta}_j$ for $1 \leq j \leq s$, where $r + 2s = n$. The form can be factored as

$$f(x,1) = \prod_{i=1}^{r}(x-\alpha_i) \cdot \prod_{i=1}^{s}(x-\beta_i)(x-\bar{\beta}_i). \tag{1}$$

The ordered pair (r,s) of numbers r and s is called the **signature** of the form f. We associate to f the two quadratics $T_r(x,1)$ and $S_s(x,1)$ given by the formulas

$$T_r(x,1) = \sum_{i=1}^{r} t_i^2(x-\alpha_i)^2, \quad \textit{and} \quad S_s(x,1) = \sum_{j=1}^{s} 2u_j^2(x-\beta_j)(x-\bar{\beta}_j), \tag{2}$$

where t_i, u_j are to be determined. The quadratics T and S are positive definite binary quadratics with discriminants as follows

$$\begin{aligned} \Delta(T_r) &= -4\sum_{i<j}^{r} t_i^2t_j^2\,(\alpha_i-\alpha_j)^2, \\ \Delta(S_s) &= -16\left(\sum_{i<j} u_i^2u_j^2\left[(a_i-a_j)^2+(b_i^2+b_j^2)\right] + \sum_{j=1}^{s} u_j^4b_j^2\right). \end{aligned} \tag{3}$$

where $\beta_i = a_i + I \cdot b_i$, for $i = 1, \ldots, s$. To each binary form f with signature (r,s) we associate the quadratic Q_f which is defined as

$$Q_f(x,1) = T_r(x,1) + S_s(x,1). \tag{4}$$

The quadratic Q_f is a positive definite quadratic with discriminant $\mathfrak{D}_f$ given by

$$\mathfrak{D}_f = \Delta(T_r) + \Delta(S_s) - 8\sum_{i,j} t_i^2u_j^2\left((a_i-a_j)^2+b_j^2\right). \tag{5}$$

Define the θ_0 of a binary form as follows

$$\theta_0(f) = \frac{a_0^2 \cdot |\mathfrak{D}_f|^{n/2}}{\prod_{i=1}^{r} t_i^2 \prod_{j=1}^{s} u_j^4}.$$

Consider $\theta_0(t_1,\ldots,t_r,u_1,\ldots,u_s)$ as a multivariable function in the variables t_i, u_j for $1 \leq i \leq r, 1 \leq j \leq s$. We would like to pick these variables such that Q_f is a reduced quadratic. It is shown in [**3**] that this is equivalent to $\theta_0(t_1,\ldots,t_r,u_1,\ldots,u_s)$ obtaining a minimal value. The function $\theta_0 : \mathbb{R}^{r+s} \to \mathbb{R}$ obtains a minimum at a unique point $(\bar{t}_1,\ldots,\bar{t}_r,\bar{u}_1,\ldots,\bar{u}_s)$. Julia in his thesis [**6**] proves existence and Stoll, and Cremona prove uniqueness in [**15**].

Choosing $(\bar{t}_1,\ldots,\bar{t}_r,\bar{u}_1,\ldots,\bar{u}_s)$ that make θ_0 minimal gives a unique positive definite quadratic $Q_f(x,z)$. We call this unique quadratic $Q_f(x,z)$ for such a choice of $(\bar{t}_1,\ldots,\bar{t}_r,\bar{u}_1,\ldots,\bar{u}_s)$ the **Julia quadratic** of $f(x,z)$, denote it by $\mathcal{J}_f(x,z)$, and the quantity $\theta_f := \theta_0(\bar{t}_1,\ldots,\bar{t}_r,\bar{u}_1,\ldots,\bar{u}_s)$ the **Julia invariant**. The proof of the following lemma can be found in [**3**].

LEMMA 1. *Consider $GL_2(\mathbb{C})$ acting on $V_{n,\mathbb{R}}$. Then θ is an invariant of binary forms and $\mathcal{J}$ is a covariant of order 2.*

It is an open problem to express θ and $\mathcal{J}$ in terms of invariants and covariants of binary sextics.

To any form $f \in V_{n,\mathbb{R}}$ we associate a positive definite quadratic $\mathcal{J}_f \in V_{2,\mathbb{R}}^+$ as showed above. In [**2**] we show that positive definite binary quadratic forms are in one-to-one correspondence with points in the upper half plane $\mathcal{H}_2$. Hence, we have the following map $\zeta : V_{n,\mathbb{R}} \mapsto V_{2,\mathbb{R}} \to \mathcal{H}_2$, where

$$f \mapsto \mathcal{J}_f \mapsto \xi(\mathcal{J}_f).$$

where $\xi(\mathcal{J}_f)$ is the zero of the quadratic $\mathcal{J}_f$ in $\mathcal{H}_2$. We call this map the **zero map** and denote it by $\zeta(f) := \xi(\mathcal{J}_f)$. The map $\zeta : V_{n,\mathbb{R}} \to \mathcal{H}_2$ is Γ-equivariant, i.e. for any matrix M in the modular group we have $\zeta(f^M) = \zeta(f)^{M^{-1}}$, see [**3**] for the proof. A binary form $f \in V_{n,\mathbb{R}}$ is **reduced** if $\zeta(f) \in \mathcal{F}_2$, where $\mathcal{F}_2$ is the classical fundamental domain.

If the binary form is not reduced, i.e. $\zeta(f) \notin \mathcal{F}_2$ let $M \in \Gamma$ be such that $\zeta(f)^M \in \mathcal{F}_2$. Since ζ is a Γ-equivariant map then $f^{M^{-1}}$ is the reduced binary form which is Γ-equivalent to f.

2.1. Binary quintics. Consider the case when f is a binary quintic with sig $(f) = (5,0)$. Computations for binary forms with sig $(f) = (3,1)$ and sig $(f) = (1,2)$ are similar and are treated in [**3**]. Let f be given as follows:

$$f(x,z) = a_0x^5 + a_1x^4z + a_2x^3z^2 + a_3x^2z^3 + a_4xz^4 + a_5z^5 \tag{6}$$

and $\alpha_1, \dots, \alpha_5$ the roots of $f(x,1)$. We associate to f the form

$$Q(x,z) = \sum_{i=1}^{5} t_i^2(x - \alpha_i z)^2 \tag{7}$$

which is a positive definite quadratic with discriminant $\mathfrak{D}_f = \sum_{1=i<j=5} u_i\, u_j\, \alpha_{i,j}$, where $u_i = t_i^2$ and for $i < j$ we set $\alpha_{i,j} = (\alpha_i - \alpha_j)^2$. The system as in [**3**, Eq. 4.13] is

$$\begin{cases} \frac{n-2}{n} \sum_{i \neq j} u_i u_j \alpha_{i,j} = \frac{2}{n} \sum_{\substack{l,k=1 \\ l \neq k}}^{n} u_j u_k \alpha_{j,k} \\ u_1 \cdot u_2 \cdots u_6 = 1 \end{cases} \tag{8}$$

Hence, we get

$$\begin{cases} 3\,u_1u_2\alpha_{1,2} + 3\,u_1u_3\alpha_{1,3} + 3\,u_1u_4\alpha_{1,4} + 3\,u_1u_5\alpha_{1,5} - 2\,u_2u_3\alpha_{2,3} - \\ \quad 2\,u_2u_4\alpha_{2,4} - 2\,u_2u_5\alpha_{2,5} - 2\,u_3u_4\alpha_{3,4} - 2\,u_3u_5\alpha_{3,5} - 2\,u_4u_5\alpha_{4,5} = 0 \\ 3\,u_1u_2\alpha_{1,2} - 2\,u_1u_3\alpha_{1,3} - 2\,u_1u_4\alpha_{1,4} - 2\,u_1u_5\alpha_{1,5} + 3\,u_2u_3\alpha_{2,3} + \\ \quad 3\,u_2u_4\alpha_{2,4} + 3\,u_2u_5\alpha_{2,5} - 2\,u_3u_4\alpha_{3,4} - 2\,u_3u_5\alpha_{3,5} - 2\,u_4u_5\alpha_{4,5} = 0 \\ -2\,u_1u_2\alpha_{1,2} + 3\,u_1u_3\alpha_{1,3} - 2\,u_1u_4\alpha_{1,4} - 2\,u_1u_5\alpha_{1,5} + 3\,u_2u_3\alpha_{2,3} - \\ \quad 2\,u_2u_4\alpha_{2,4} - 2\,u_2u_5\alpha_{2,5} + 3\,u_3u_4\alpha_{3,4} + 3\,u_3u_5\alpha_{3,5} - 2\,u_4u_5\alpha_{4,5} = 0 \\ -2\,u_1u_2\alpha_{1,2} - 2\,u_1u_3\alpha_{1,3} + 3\,u_1u_4\alpha_{1,4} - 2\,u_1u_5\alpha_{1,5} - 2\,u_2u_3\alpha_{2,3} + \\ \quad 3\,u_2u_4\alpha_{2,4} - 2\,u_2u_5\alpha_{2,5} + 3\,u_3u_4\alpha_{3,4} - 2\,u_3u_5\alpha_{3,5} + 3\,u_4u_5\alpha_{4,5} = 0 \\ -2\,u_1u_2\alpha_{1,2} - 2\,u_1u_3\alpha_{1,3} - 2\,u_1u_4\alpha_{1,4} + 3\,u_1u_5\alpha_{1,5} - 2\,u_2u_3\alpha_{2,3} - \\ \quad 2\,u_2u_4\alpha_{2,4} + 3\,u_2u_5\alpha_{2,5} - 2\,u_3u_4\alpha_{3,4} + 3\,u_3u_5\alpha_{3,5} + 3\,u_4u_5\alpha_{4,5} = 0 \\ u_1u_2u_3u_4u_5 - 1 = 0 \end{cases}$$

We want to solve the above system for $u_1, \dots, u_5$.

LEMMA 2. *The degree of the field extension* $[k(u_1, \dots, u_5) : k(\alpha_{i,j})] = 5$.

The proof is computational using Maple; see [**3**] for further details.

REMARK 1. *For any given 5-tuple* $(\alpha_1, \dots, \alpha_5)$ *we have a unique positive real solution* $(u_1, u_2, u_3, u_4, u_5)$. *Hence, as expected the coefficients of the Julia quadratic are uniquely defined.*

One can express invariants of quintics in terms of the root differences $\alpha_{i,j}$ and then eliminate $u_1, \dots, u_5$ in order to get syzygies among $u_1, \dots, u_5$ and invariants of quintics. Solving the system of such syzygies for $u_1, \dots, u_5$ one expects to get a unique real solution $(u_1, \dots, u_5)$. Such a tuple would determine the Julia quadratic for the form f in terms of the invariants of the quintics. This seems quite a difficult problem computationally.

2.2. Binary sextics. Consider the case when f is a binary sextic with signature sig $(f) = (6,0)$. Computations for binary forms with sig $(f) = (4,1)$, sig $(f) = (2,2)$ and sig $(f) = (0,3)$ are similar and are treated in [**3**]. Let f be given as follows:

$$f(x,z) = a_0x^6 + a_1x^5z + a_2x^4z^2 + a_3x^3z^3 + a_4x^2z^4 + a_5xz^5 + a_6z^6$$

and α_i, $i = 1, \dots, 6$, its roots when $z = 1$. Associate to it the quadratic

$$Q_f = \sum_{i=1}^{6} t_i^2 (x - \alpha_i)^2,$$

where the t_i's are non zero real numbers. For convenience, as above we fix the following notation $u_i := t_i^2$ and $\alpha_{i,j} := (\alpha_i - \alpha_j)^2$. We can determine $u_1, \cdots, u_6$ by solving the system given as follows

$$\begin{cases} \frac{n-2}{n} \sum_{i \neq j} u_i u_j \alpha_{i,j} = \frac{2}{n} \sum_{\substack{l,k=1 \\ l \neq k}}^{n} u_j u_k \alpha_{j,k} \\ u_1 \cdot u_2 \cdots u_6 = 1 \end{cases}$$

more explicitly

$$\begin{cases} \frac{2}{3}\left(u_1u_2\alpha_{1,2} + u_1u_3\alpha_{1,3} + u_1u_4\alpha_{1,4} + u_1u_5\alpha_{1,5} + u_1u_6\alpha_{1,6}\right) - \frac{1}{3}\left(u_2u_3\alpha_{2,3}\right. \\ \quad + u_2u_4\alpha_{2,4} + u_2u_5\alpha_{2,5} + u_2u_6\alpha_{2,6} + u_3u_4\alpha_{3,4} + u_3u_5\alpha_{3,5} + u_3u_6\alpha_{3,6} \\ \quad \left. + u_4u_5\alpha_{4,5} + u_4u_6\alpha_{4,6} + u_5u_6\alpha_{4,6}\right) = 0 \\ \quad \vdots \\ \frac{2}{3}\left(u_1u_6\alpha_{1,6} + u_2u_6\alpha_{2,6} + u_3u_6\alpha_{3,6} + u_4u_6\alpha_{4,6} + u_5u_6\alpha_{5,6}\right) - \frac{1}{3}\left(u_1u_2\alpha_{1,2}\right. \\ \quad + u_1u_3\alpha_{1,3} + u_1u_4\alpha_{1,4} + u_1u_5\alpha_{1,5} + u_2u_3\alpha_{2,3} + u_2u_4\alpha_{2,4} + u_2u_5\alpha_{2,5} \\ \quad \left. + u_3u_4\alpha_{3,4} + u_3u_5\alpha_{3,5} + u_4u_5\alpha_{4,5}\right) = 0 \\ u_1 \cdot u_2 \cdots u_6 = 1 \end{cases}$$

We want to solve the above system for $u_1, \dots, u_6$. The proof of the following lemma is computational using Maple.

LEMMA 3. *The degree of the field extension* $[k(u_1, \dots, u_6) : k(\alpha_{i,j})] = 36$.

On the other hand, the invariants of a binary sextic are given in terms of the difference of the roots. We work with invariants J_2, J_4, J_6, and J_{10} as in [**4**]. Explicit formulas of J_2, J_4, J_6, and J_{10} in terms of $\alpha_{i,j}$, i.e. $J_k = g_k(\alpha_{i,j})$, $i, j = 1, \dots, 6$, $i < j$, and $k = 2, 4, 6, 10$, are well known.

We would like to express $u_1, \dots, u_6$, hence the coefficient of Julia's quadratic, in terms of the invariants J_2, J_4, J_6, and J_{10}. Computationally this is too difficult since for $i, j = 1, \dots, 6$ and $i < j$ we have 15 variables $\alpha_{i,j}$. Hence, we go back to the substitution, $\alpha_{i,j} = (\alpha_j - \alpha_i)^2$, where α_j, α_i are the roots of the binary sextic. Without loss of generality fix a coordinate, i.e. let $\alpha_1 = 0$, $\alpha_2 = 1$, and $\alpha_3 = -1$. This way we are left with the problem of eliminating α_4, α_5, and α_6.

Eliminating α_4, α_5, and α_6 we are coming down from the field $k(\alpha_1, \dots, \alpha_6)$ to $k(u_1, \dots, u_6)$. Hence, we get equations only in terms of $u_1, \dots, u_6$, and J_2, J_4, J_6, J_{10}. The next thing is to solve for $u_1, \dots, u_6$. Computations are very large and involve Gröbner bases.

Note that computing the Julia quadratic using the above system is too difficult, hence we develop new methods. We consider separately totally real forms, i.e. forms with all real roots, and totally complex forms i.e. forms with complex roots.

2.3. Totally real and totally complex forms. Consider first totally real forms. Let $f \in V_{n,\mathbb{R}}$ such that f has signature $(n, 0)$ given by

$$f(x,z) = a_n x^n + a_{n-1} x^{n-1} z + \cdots + a_1 x z^{n-1} + a_0 z^n$$

where $a_0, \dots, a_n$ are transcendentals. Identify $a_0, \dots, a_n$ respectively with $1, \dots, n+1$. Then the symmetric group S_{n+1} acts on $\mathbb{R}[a_0, \dots a_n][x, z]$ by permuting a_i's. For any permutation $\tau \in S_{n+1}$ and $f \in \mathbb{R}[a_0, \dots a_n][x, z]$ we denote by $\tau(f) = f^\tau$. Then,

$$f^\tau(x,z) = \tau(a_n)\, x^n + \tau(a_{n-1})\, x^{n-1} z + \cdots + \tau(a_1)\, x z^{n-1} + \tau(a_0)\, z^n.$$

Define $G_f(x, z)$ as follows

$$G_f(x,z) = \frac{x \cdot f_x(-f_z(x,z), f_x(x,z)) + z \cdot f_z(-f_z(x,z), f_x(x,z))}{n\, f(x,z)}. \tag{9}$$

In [**15**] Stoll and Cremona was proved that $G_f(x, y)$ is a degree $d = (n-1)(n-2)$ homogenous polynomial in $\mathbb{R}[a_0, \dots a_n][x, z]$ and $\mathcal{J}_f(x, z) \,|\, G_f(x, z)$; see [**15**] for details. Therefore, this polynomial can be used to reduce totally real binary forms.

Note that, for $\sigma \in S_{n+1}$ we have an involution

$$\sigma = \begin{cases} (1, n+1)(2, n) \cdots \left(\frac{n}{2}, \frac{n}{2} + 2\right), & \text{if } n \text{ is even} \\ (1, n+1)(2, n) \cdots \left(\frac{n+1}{2}, \frac{n+3}{2}\right), & \text{if } n \text{ is odd.} \end{cases}$$

In [**3**] is proved that the polynomial $G_f(x, z)$ satisfies the following.

THEOREM 1. *Let $f \in V_{n,\mathbb{R}}$ with distinct roots, sig $(f) = (n, 0)$, and G_f as above. Then*

i) $G_f(x, z)$ is a covariant of f of degree $(n-1)$ and order $(n-1)(n-2)$.

ii) $G_f(x, z)$ has a unique quadratic factor over $\mathbb{R}$, which is $\mathcal{J}_f$.

iii) $G_f^\sigma(x, z) = (-1)^{n-1} G_f(x, z)$. Moreover, if $G_f = \sum_{i=1}^d g_i\, x^i z^{d-i}$, then

$$g_i^\sigma = (-1)^{n-1} g_{d-i},$$

for all $i = 0, \dots, d$.

Hence, to each totally real form f we associate the G_f-polynomial which can be computed easily since is defined in terms of the derivatives of the form. Then, from Thm. 1 this polynomial has a unique quadratic factor which is the Julia quadratic associated to the given binary form f.

Next, we illustrate the theory for totally real binary quintics.

2.4. Totally real quintics. Let $f \in V_{5,\mathbb{R}}$ be a totally real binary quintic $f = \sum_{i=0}^{5} a_i x^i z^{5-i}$ with roots α_i for $i = 1, \dots 5$. In the notation of the previous section, we have $Q_f = T_5$. The discriminant of T_5 in terms of the roots α_i of the form is given by the formula

$$\begin{aligned}\Delta(T_5) = -4t_1^2 \cdots t_5^2 \Bigg(&\frac{(\alpha_1-\alpha_2)^2}{t_3^2t_4^2t_5^2} + \frac{(\alpha_1-\alpha_3)^2}{t_2^2t_4^2t_5^2} + \frac{(\alpha_1-\alpha_4)^2}{t_3^2t_2^2t_5^2} + \frac{(\alpha_1-\alpha_5)^2}{t_2^2t_3^2t_4^2} \\ &+ \frac{(\alpha_2-\alpha_3)^2}{t_1^2t_4^2t_5^2} + \frac{(\alpha_2-\alpha_4)^2}{t_1^2t_3^2t_5^2} + \frac{(\alpha_2-\alpha_5)^2}{t_1^2t_3^2t_4^2} + \frac{(\alpha_3-\alpha_4)^2}{t_1^2t_2^2t_5^2} \\ &+ \frac{(\alpha_3-\alpha_5)^2}{t_1^2t_2^2t_4^2} + \frac{(\alpha_4-\alpha_5)^2}{t_1^2t_2^2t_3^2}\Bigg).\end{aligned}$$

In this case $\sigma = (1,6)(2,5)(3,4)$ which correspond to $\sigma = (a_0, a_5)(a_1, a_4)(a_2, a_3)$. Then computing $G_f(x, z)$ as in Eq. (9) we have

$$G_f(x,z) = c_{12}x^{12} + c_{11}x^{11}z + \cdots c_1 x z^{11} + c_0 z^{12} \tag{10}$$

where the coefficients are given as follows:

$$\begin{aligned}
c_{12} &= 125\, a_1 a_5{}^3 - 50\, a_2 a_4 a_5{}^2 + 15\, a_3 a_4^2 a_5 - 3\, a_4^4 \\
c_{11} &= 625\, a_0 a_5{}^3 + 175\, a_1 a_4 a_5{}^2 - 100\, a_2 a_3 a_5{}^2 - 55\, a_2 a_4^2 a_5 + 60\, a_3^2 a_4 a_5 - 9\, a_3 a_4^3 \\
c_{10} &= 1375\, a_0 a_4 a_5{}^2 - 25\, a_1 a_3 a_5{}^2 + 65\, a_1 a_4^2 a_5 - 150\, a_2{}^2 a_5{}^2 - 30\, a_2 a_3 a_4 a_5 \\
&\quad - 41\, a_2 a_4^3 + 60\, a_3^3 a_5 + 3\, a_3^2 a_4^2 \\
c_9 &= 875\, a_0 a_3 a_5{}^2 + 1025\, a_0 a_4^2 a_5 - 425\, a_1 a_2 a_5{}^2 + 30\, a_1 a_3 a_4 a_5 - 33\, a_1 a_4^3 \\
&\quad - 130\, a_2{}^2 a_4 a_5 + 160\, a_2 a_3^2 a_5 - 86\, a_2 a_3 a_4^2 + 24\, a_3^3 a_4 \\
c_8 &= 125\, a_0 a_2 a_5{}^2 + 1500\, a_0 a_3 a_4 a_5 + 215\, a_0 a_4^3 - 425\, a_1^2 a_5{}^2 - 450\, a_1 a_2 a_4 a_5 \\
&\quad + 175\, a_1 a_3^2 a_5 - 128\, a_1 a_3 a_4^2 + 175\, a_2{}^2 a_3 a_5 - 97\, a_2{}^2 a_4^2 + 8\, a_2 a_3^2 a_4 + 12\, a_3^4 \\
c_7 &= -750\, a_0 a_1 a_5{}^2 + 500\, a_0 a_2 a_4 a_5 + 775\, a_0 a_3^2 a_5 + 430\, a_0 a_3 a_4^2 - 530\, a_1^2 a_4 a_5 \\
&\quad + 310\, a_1 a_2 a_3 a_5 - 322\, a_1 a_2 a_4^2 - 61\, a_1 a_3^2 a_4 + 105\, a_2{}^3 a_5 - 33\, a_2{}^2 a_3 a_4 + 32\, a_2 a_3^3 \\
c_6 &= -625\, a_0^2 a_5{}^2 - 800\, a_0 a_1 a_4 a_5 + 1200\, a_0 a_2 a_3 a_5 + 30\, a_0 a_2 a_4^2 + 365\, a_0 a_3^2 a_4 \\
&\quad + 30\, a_1^2 a_3 a_5 - 303\, a_1^2 a_4^2 + 365\, a_1 a_2{}^2 a_5 - 268\, a_1 a_2 a_3 a_4 + a_1 a_3^3 \\
&\quad + a_2{}^3 a_4 + 37\, a_2{}^2 a_3^2 \\
c_5 &= -750\, a_0^2 a_4 a_5 + 500\, a_0 a_1 a_3 a_5 - 530\, a_0 a_1 a_4^2 + 775\, a_0 a_2{}^2 a_5 + 310\, a_0 a_2 a_3 a_4 \\
&\quad + 105\, a_0 a_3{}^3 + 430\, a_1^2 a_2 a_5 - 322\, a_1^2 a_3 a_4 - 61\, a_1 a_2{}^2 a_4 - 33\, a_1 a_2 a_3^2 + 32\, a_2{}^3 a_3 \\
c_4 &= 125\, a_0^2 a_3 a_5 - 425\, a_0^2 a_4^2 + 1500\, a_0 a_1 a_2 a_5 - 450\, a_0 a_1 a_3 a_4 + 175\, a_0 a_2{}^2 a_4 \\
&\quad + 175\, a_0 a_2 a_3^2 + 215\, a_1^3 a_5 - 128\, a_1^2 a_2 a_4 - 97\, a_1^2 a_3^2 + 8\, a_1 a_2{}^2 a_3 + 12\, a_2{}^4
\end{aligned}$$

$$c_3 = 875\, a_0{}^2 a_2 a_5 - 425\, a_0^2 a_3 a_4 + 1025\, a_0 a_1^2 a_5 + 30\, a_0 a_1 a_2 a_4 - 130\, a_0 a_1 a_3^2$$
$$+ 160\, a_0 a_2{}^2 a_3 - 33\, a_1^3 a_4 - 86\, a_1^2 a_2 a_3 + 24\, a_1 a_2{}^3$$
$$c_2 = 1375\, a_0^2 a_1 a_5 - 25\, a_0^2 a_2 a_4 - 150\, a_0^2 a_3^2 + 65\, a_0 a_1^2 a_4 - 30\, a_0 a_1 a_2 a_3 + 60\, a_0 a_2{}^3$$
$$- 41\, a_1^3 a_3 + 3\, a_1^2 a_2{}^2$$
$$c_1 = 625\, a_0^3 a_5 + 175\, a_0^2 a_1 a_4 - 100\, a_0^2 a_2 a_3 - 55\, a_0 a_1^2 a_3 + 60\, a_0 a_1 a_2{}^2 - 9\, a_1^3 a_2$$
$$c_0 = 125\, a_0^3 a_4 - 50\, a_0^2 a_1 a_3 + 15\, a_0 a_1^2 a_2 - 3\, a_1^4$$

The following is an immediate consequence of Thm. 1.

COROLLARY 1. *Let $f \in V_{5,\mathbb{R}}$ with signature $(5,0)$. Then $G_f^\sigma = G_f$ the above coefficients give a computational confirmation that $G_f^\sigma = G_f$ and $c_i^\sigma = c_{5-i}$ for all $i = 1, \dots, 5$.*

When evaluated at a specific binary form the above polynomial G_f has a unique quadratic factor which is the Julia quadratic and can be used for reduction.

Similar computations are done for totally real sextics. The polynomial $G_f(x, y)$ is displayed in [**3**, Appendix A]. Next we consider totally complex forms.

2.5. Totally complex. For reduction of totally complex forms the Julia quadratic can be determined using the system given in [**3**, Eq. 3.14]. For such forms the degree of the system will drop to $\frac{n}{2}$ and the computations are possible.

Let f be a totally complex binary sextic factored as follows

$$f(x, z) = (x^2 + a_1 xz + b_1 z^2)(x^2 + a_2 xz + b_2 z^2)(x^2 + a_3 xz + b_3 z^2).$$

Associate to f the quadratic

$$Q(x, 1) = 2u_1^2(x^2 + a_1 x + b_1) + 2u_2^2(x^2 + a_2 x + b_2) + 2u_3^2(x^2 + a_3 x + b_3)$$

where the u_i's are real numbers that make θ_f minimal. To find u_1, u_2 and u_3 that satisfy this condition we need to set up the system in [**3**, Eq. 3.14] and solve for the u_i's. Compute first the discriminant of the quadratic which is as follows

$$\begin{aligned}\mathfrak{D}_f =& 4\, a_1{}^2 u_1{}^4 + 8\, a_1 a_2 u_1{}^2 u_2{}^2 + 8\, a_1 a_3 u_1{}^2 u_3{}^2 + 4\, a_2{}^2 u_2{}^4 + 8\, a_2 a_3 u_2{}^2 u_3{}^2 \\ &+ 4\, a_3{}^2 u_3{}^4 - 16\, b_1 u_1{}^4 - 16\, b_1 u_1{}^2 u_2{}^2 - 16\, b_1 u_1{}^2 u_3{}^2 - 16\, b_2 u_1{}^2 u_2{}^2 \\ &- 16\, b_2 u_2{}^4 - 16\, b_2 u_2{}^2 u_3{}^2 - 16\, b_3 u_1{}^2 u_3{}^2 - 16\, b_3 u_2{}^2 u_3{}^2 - 16\, b_3 u_3{}^4\end{aligned}$$

Next, compute the partial derivatives of $\mathfrak{D}_f$ with respect to u_1, u_2, and u_3 and then set up the system. This is done in Maple but we do not display the system here because is too big. The system is given in terms of u_i's, a_i's, b_i's and λ, the Lagrange multiplier. Solving for λ we get

$$\lambda = \frac{4(a_3 u_1{}^2 a_1 + a_3 u_2{}^2 a_2 + u_3{}^2 a_3{}^2 - 2\, u_1{}^2 b_1 - 2\, u_2{}^2 b_2 - 2\, b_3 u_1{}^2 - 2\, b_3 u_2{}^2 - 4\, u_3{}^2 b_3)}{u_3{}^2 u_1{}^4 u_2{}^4}$$

Substitute λ as computed in the [**3**, Eq. 3.14] and add to this system the equation $Q(x, 1) = 0$. Using this approach we can compute the point $\xi(f)$ in the upper half plane corresponding to the Julia quadratic. Eliminating all three of them we get a degree 8 polynomial

$$G_f = \sum_{i=0}^{8} c_i x^i z^{8-i}, \tag{11}$$

with coefficients given in [**3**, Appendix A]. This degree 8 polynomial has a unique quadratic factor which is the Julia quadratic.

As a special case, consider the case when we let $b_1 = b_2 = b_3 = 1$. This case is interesting since curves with such equations correspond to genus two curves with extra automorphism. Assume the binary form f is given as follows

$$f(x,z) = \prod_{i=1}^{3} (x^2 + a_i xz + z^2).$$

The polynomial $G_f(x,z)$ associated to this binary form has coefficients as follows

$$\begin{aligned}
c_8 &= -c_0 = 3\,(a_2 - a_3)\,(a_1 - a_3)\,(a_1 - a_2)\,(a_1a_2 + a_1a_3 + a_2a_3)\\
c_7 &= -c_1 = 3\,(a_2 - a_3)\,(a_1 - a_3)\,(a_1 - a_2)\,(3\,a_1a_2a_3 + 8\,a_1 + 8\,a_2 + 8\,a_3)\\
c_6 &= -c_2 = 6\,(a_2 - a_3)\,(a_1 - a_3)\,(a_1 - a_2)\,(5\,a_1a_2 + 5\,a_1a_3 + 5\,a_2a_3 + 24)\\
c_5 &= -c_3 = 9\,(a_2 - a_3)\,(a_1 - a_3)\,(a_1 - a_2)\,(a_1a_2a_3 + 8\,a_1 + 8\,a_2 + 8\,a_3)\\
c_4 &= 0.
\end{aligned}$$

Note that $G_f(x,z)$ is a palindromic polynomial.

REMARK 2. *If $f = \prod_{i=1}^{n}(x^2 + a_i xz + z^2)$, then f is a palindromic form. In this case, the Julia quadratic is a factor of G_f, where G_f is a self-inverse form.*

We give an explicit example that illustrates the theory and emphasizes how numerical computations can be used in implementing a reduction algorithm.

EXAMPLE 1. *Let $f \in V_{\mathbb{R},6}$ be a binary form given as follows:*

$$f(x,z) = x^6 - 12x^5z + 96x^4z^2 - 458x^3z^3 + 1489x^2z^4 - 3014xz^5 + 3770z^6$$

Let $z = 1$ then the binary form f can be factored as

$$f(x,1) = (x - \alpha_1)(x - \bar{\alpha}_1)(x - \alpha_2)(x - \bar{\alpha}_2)(x - \alpha_3)(x - \bar{\alpha}_3)$$

where $\alpha_1 = 1 + 3I$, $\alpha_2 = 2 + 5I$, and $\alpha_3 = 3 + 2I$. Associate to it the quadratic

$$\begin{aligned}
S(x,1) &= 2u_1^2(x - \alpha_1)(x - \bar{\alpha}_1) + 2u_2^2(x - \alpha_2)(x - \bar{\alpha}_2) + 2u_3^2(x - \alpha_3)(x - \bar{\alpha}_3)\\
&= 2u_1^2(x^2 - 2x + 10) + 2u_2^2(x^2 - 4x + 29) + 2u_3^2(x^2 - 6x + 13).
\end{aligned}$$

To find u_1, u_2 and u_3 we need to solve the following system

$$\begin{cases}
-4\lambda u_1^3u_2^4u_3^4 - 576u_1^3 - 1120u_1u_2^2 - 544u_1u_3^2 = 0\\
-4\lambda u_1^4u_2^3u_3^4 - 1120u_1^2u_2 - 1600u_2^3 - 960u_2u_3^2 = 0\\
-4\lambda u_1^4u_2^4u_3^3 - 544u_1^2u_3 - 960u_2^2u_3 - 256u_3^3 = 0\\
u_1^4 \cdot u_2^4 \cdot u_3^4 = 1.
\end{cases}$$

First solve the above for λ. We have

$$\lambda = -\frac{8(17u_1^2 + 30u_2^2 + 8u_3^2)}{u_3^2u_1^4u_2^4}.$$

and

$$\begin{cases}
144u_1^4 + 280u_1^2u_2^2 - 240u_2^2u_3^2 - 64u_3^4 = 0\\
280u_1^2u_2^2 - 136u_1^2u_3^2 + 400u_2^4 - 64u_3^4 = 0\\
u_1^4 \cdot u_2^4 \cdot u_3^4 = 1
\end{cases}$$

We want to find $\zeta(f)$ which satisfies the polynomial $S(x,1)$. Make the substitution $u_i^2 = w_i$, for $i = 1,2,3$ and solve the following system

$$\begin{cases} 144w_1^2 + 280w_1w_2 - 240w_2w_3 - 64w_3^2 = 0 \\ 280w_1w_2 - 136w_1w_3 + 400w_2^2 - 64w_3^2 = 0 \\ w_1^2 \cdot w_2^2 \cdot w_3^2 = 1 \\ 2w_1(x^2 - 2x + 10) + 2w_2(x^2 - 4x + 29) + 2w_3(x^2 - 6x + 13) = 0. \end{cases}$$

The w_1, w_2 and w_3 can be given as rational functions of x, where x satisfies

$$G_f(x) = 2169x^8 - 13194x^7 - 221877x^6 + 2519358x^5 - 9693795x^4 + \\ + 10471902x^3 + 54904153x^2 - 176055146x + 134002174$$

By Strum's theorem the above polynomial has exactly 6 real roots. Hence, it has only one quadratic factor which is the Julia quadratic. The zero of the Julia is the point

$$w = 2.12067656802142 + 3.26692991594356\,i.$$

The form is reduced if this root is in the fundamental domain, therefore send w to $w_0 = w - 2$. The reduced binary form is as follows

$$f(x+2,1) = x^6 + 36x^4 - 10x^3 + 325x^2 - 250x + 1250.$$

This is the reduced form in its Γ-orbit. To find the $GL_2(\mathbb{C})$-reduced form notice that $1250 = 2 \cdot 5^4$ and if we send x to $5x$ and then factor out the greatest common factor we get

$$f(x,1) = 2x^6 + 2x^5 + 13x^4 + 2x^3 + 36x^2 + 25$$

Thus, the minimal integral form is

$$f(x,1) = 2x^6 + 2x^5 + 13x^4 + 2x^3 + 36x^2 + 25$$

and $\mathfrak{h}(f) = 36$.

Note that the reduction algorithm can be made rather efficient using numerical methods, i.e. given a binary form f we can compute $\zeta(f)$ numerically and then reduce f as explained above. But we are interested in giving a precise formula of the Julia's invariant or Julia's quadratic in terms of the invariants of the binary form. This is quite difficult and can be done for binary forms under some assumptions as we will show next.

3. Julia quadratic of genus two curves with extra automorphisms

In this section we give the Julia quadratic of curves with extra automorphism in terms of the invariants of binary forms.

3.1. Genus 2 curve with $\mathrm{Aut}(\mathcal{X}) \cong V_4$. In this subsection we focus on genus 2 curves $\mathcal{X}$ with $\mathrm{Aut}(\mathcal{X}) \simeq V_4$ that can be written as $y^2 = f(x)$ for f totally real. If the form f is not defined over $\mathcal{O}_F$ then we find an equation of the curve over $\mathcal{O}_F$ as explained in [**4**].

For a totally real binary form we can perform reduction using the polynomial $G(x,z)$ defined in Eq. (9). Computations show that $G_f(x,z)$ is factored in three factors. One has degree 2, one degree 6, and one degree 12 as follows

$$G_f(x,1) = 1024\,u^6\,\left(4\,u^3 - v^2\right)^3 \cdot g_0(x,1) \cdot g_1(x,1) \cdot g_2(x,1)$$

where

$$
\begin{aligned}
g_0(x) =& x^2 - (v^2 - 4\,u^3)\\
g_1(x) =& \left(-u^2 - 3\,v\right) x^6 + \left(36\,u^3 - 2\,u^2 v - 18\,v^2\right) x^5\\
&+ \left(-4\,u^5 + 180\,u^3 v + u^2 v^2 - 45\,v^3\right) x^4\\
&+ \left(-480\,u^6 - 16\,u^5 v + 360\,u^3 v^2 + 4\,u^2 v^3 - 60\,v^4\right) x^3 +\\
&+ \left(16\,u^8 - 720\,u^6 v - 8\,u^5 v^2 + 360\,u^3 v^3 + u^2 v^4 - 45\,v^5\right) x^2 +\\
&+ \left(576\,u^9 - 32\,u^8 v - 576\,u^6 v^2 + 16\,u^5 v^3 + 180\,u^3 v^4 - 2\,u^2 v^5 - 18\,v^6\right) x +\\
&+ 64\,u^{11} + 192\,u^9 v - 48\,u^8 v^2 - 144\,u^6 v^3 + 12\,u^5 v^4 + 36\,u^3 v^5 - u^2 v^6 - 3\,v^7
\end{aligned}
\tag{12}
$$

while we don't display $g_2(x)$. Computing the discriminant of all of these factors we get

$$
\begin{aligned}
\Delta_{g_0} &= 4(v^2 - 4u^3)\\
\Delta_{g_1} &= 2^{36}\,3^2\,\left(u^2 + 162\,u + 12\,v - 2187\right)^2 \left(4\,u^3 - v^2\right)^{10} u^{30}\\
\Delta_{g_2} &= 2^{156}\,3^{18}\,(u-9)^6 \left(u^2 + 162\,u + 12\,v - 2187\right)^4 \cdot\\
&\cdot \left(u^2 + 18\,u - 4\,v - 27\right)^{14} \left(4\,u^3 - v^2\right)^{66} u^{132}
\end{aligned}
$$

We proved that $G_f(x,1)$ has a unique quadratic factor, i.e. if $v^2 - 4u^3 < 0$ then the Julia quadratic associated to f is

$$\mathcal{J}_f = x^2 - (v^2 - 4u^3).$$

If $v^2 - 4u^3 = 0$ the Julia quadratic is not defined, but that corresponds to the case when the $\mathrm{Aut}(\mathcal{X}) \geq V_4$. If $v^2 - 4u^3 > 0$ then for a generic binary form we can not determine the Julia quadratic precisely in terms of the invariants u and v.

3.2. Genus 2 curve with $\mathrm{Aut}(\mathcal{X}) \cong D_4$. Let $\mathcal{X}$ be a genus two curve with automorphism $\mathrm{Aut}(\mathcal{X}) \cong D_4$. The set of Weierstrass points in this case is $W = \left\{\pm 1, \pm\alpha, \pm\frac{1}{\alpha}\right\}$ i.e. $\alpha \in \mathbb{R}$. Therefore the binary form corresponding to $\mathcal{X}$ is a totally real form. Computing the polynomial $G_f(x,z)$ defined in Eq. (9) for this curves we have

$$
\begin{aligned}
G_f(x,1) = &-\left(3\,s - 5\,x^4 - x^2\right)\cdot\\
&\left(z^8 s^3 - 70\,s^2 x^4 + 14\,s^2 x^2 + 25\,s x^8 - 10\,s x^6 + 37\,s x^4 + 12\,x^6\right)
\end{aligned}
$$

and its discriminant is as follows

$$\Delta_G = -2^{44}\,3^5\,5^{21}\,s^{29}\,(60\,s+1)^2\,(20\,s-21)^6\,(-1+4\,s)^{14}\,(5\,s+1)^4.$$

For a given binary form the Julia quadratic will be the unique quadratic factor of G_f.

3.3. Genus 2 curve with $\mathrm{Aut}(\mathcal{X}) \cong D_6$. In an analogue way with the previous subsection, curves with automorphism group D_6 fall under the category of totally real form. The polynomial $G_f(x,z)$ for this form is

$$
\begin{aligned}
G_f(x,1) = &-972\,x\,\left(-x^6 + w\right)\\
&\left(8\,w x^{12} + 12\,w x^9 + x^9 + 12\,w x^6 + 12\,w^2 x^3 + w x^3 + 8\,w^3\right)
\end{aligned}
\tag{13}
$$

and its discriminant is

$$\Delta_G = 2^{114}\,3^{198}\cdot w^{48}\,(4\,w-1)^{39}\,(32\,w+1)^6.$$

Hence, for genus two curves with extra involution we can conclude the following.

THEOREM 2. *Let $\mathcal{X}$ be a genus two curve with* $\mathrm{Aut}(\mathcal{X}) > 2$, *affine equation* $y^2 = f(x)$, *and F its field of moduli. Then, the following are true*

i) If $\mathrm{Aut}(C) \simeq V_4$, *then the Julia quadratic is the unique quadratic factor of* $G_f(x, z)$ *as defined in Eq.* (12).

ii) If $\mathrm{Aut}(C) \simeq D_4$, *then the Julia quadratic is the unique quadratic factor of*

$$G_f(x,1) = -\left(3\,s - 5\,x^4 - x^2\right) \cdot \left(z^8 s^3 - 70\,s^2 x^4 + 14\,s^2 x^2 + 25\,s x^8 - 10\,s x^6 + 37\,s x^4 + 12\,x^6\right)$$

iii) If $\mathrm{Aut}(C) \simeq D_6$, *then the Julia quadratic is the unique quadratic factor of*

$$G_f(x,1) = -972\,x\,\left(-x^6 + w\right) \left(8\,w x^{12} + 12\,w x^9 + x^9 + 12\,w x^6 + 12\,w^2 x^3 + w x^3 + 8\,w^3\right)$$

Next we would like to determine minimal models of genus two curves with extra automorphism. First we will explain how one can get genus two curves with extra automorphism with minimal discriminant, given the equation of the curve.

3.4. Minimal discriminants. Minimal discriminants of genus two curves are discussed in detail in [**4**]. Here we discus the case of genus two curves with extra automorphism.

Recall that given a binary form $f(x, z)$ and a matrix $M = \begin{pmatrix} a & b \\ c & d \end{pmatrix}$, such that $M \in GL_2(K)$, we have that $f^M = f(ax + bz, cx + dz)$ has discriminant $\Delta(f^M) = (\det M)^{d(d-1)} \Delta(f)$. Hence, given $\mathcal{X}$ a genus two curve with equation $y^2 = f(x)$ and $M \in GL_2(K)$ such that $\det M = \lambda$ we have that $\Delta(f^M) = \lambda^{30}\,\Delta(f)$.

To reduce the discriminant we factor Δ as a product of primes, say $\Delta = p_1^{\alpha_1} \cdots p_r^{\alpha_r}$, and take u to be the product of those powers of primes with exponents $\alpha_i \geq 30$. Then, the transformation $M = \begin{pmatrix} 1/u & 0 \\ 0 & 1 \end{pmatrix}$ will reduce the discriminant, see [**4**] for more details. The following lemma is proved in [**4**]

LEMMA 4. *A genus 2 curve $\mathcal{X}_g$ with integral equation*

$$y^2 = a_6 x^6 + \cdots + a_1 x + a_0$$

has minimal discriminant if $\mathfrak{v}(\Delta) < 30$.

Now let us consider the case of curves with extra automorphism. Let $\mathcal{X}$ be hyperelliptic curve with an extra automorphism of order $n \geq 2$. Then, from [**11**] we know that the equation of $\mathcal{X}$ can be written as

$$y^2 = f(x^n)$$

If $\mathcal{X}$ is given with such equation over $\mathbb{Q}$ and discriminant Δ, then for any transformation $\tau : x \mapsto \frac{x}{u^{\frac{1}{n}}}$ would have $\mathcal{X}^\tau$ defined over $\mathbb{Q}$ and isomorphic to $\mathcal{X}$ over $\mathbb{C}$. Hence, $\mathcal{X}^\tau$ is a twist of $\mathcal{X}$ with discriminant

$$\Delta' = \frac{1}{u^{\frac{30}{n}}} \cdot \Delta$$

Hence, the following result is clear.

LEMMA 5. *Let $\mathcal{X}$ be a genus 2 curve with* $\text{Aut}(\mathcal{X}) \simeq V_4$. *Assume that $\mathcal{X}$ has equation over $\mathbb{Q}$ as $y^2 = f(x^2)$. Then, $\mathcal{X}$ has minimal discriminant Δ if $\mathfrak{v}(\Delta) < 15$.*

If $\text{Aut}(\mathcal{X}) \simeq D_6$ *and $\mathcal{X}$ has equation over $\mathbb{Q}$ as $y^2 = f(x^3)$. Then, $\mathcal{X}$ has minimal discriminant Δ if $\mathfrak{v}(\Delta) < 10$.*

PROOF. Assume $\Delta(f) = p^\alpha \cdot N$, where $\alpha > 30$, for some prime p and some integer N such that $(p, N) = 1$. Let $M = \begin{pmatrix} 1/p & 0 \\ 0 & 1 \end{pmatrix}$. Then, the discriminant of the form f^M is

$$\Delta' = \frac{1}{p^{30}} \cdot p^\alpha \cdot N = p^{\alpha - 30} \cdot N.$$

In the case of hyperelliptic curves with equation $y^2 = f(x^2)$ let $M = \begin{pmatrix} \frac{1}{\sqrt{p}} & 0 \\ 0 & 1 \end{pmatrix}$ then

$$\Delta' = \frac{1}{\sqrt{p}^{30}} \cdot p^\alpha \cdot N = p^{\alpha - 15} \cdot N.$$

The same way we can prove it for curves with equation $y^2 = f(x^3)$. □

Next we will determine minimal models of genus two curves with extra automorphism. We study only the loci in $\mathcal{M}_2$ of dimension ≥ 1, other cases are obvious.

4. Minimal models of curves with extra involutions

In this section we will focus on curves $\mathcal{X}$ with extra automrphisms. The following lemma gives a choice for the set of Weierstrass points for curves with extra automorphism; see [**13**] for the proof.

LEMMA 6. *Let $\mathcal{X}$ be a genus 2 curve defined over a field k such that char $k \neq 2$ and W be the set of Weierstrass points. Then the following hold:*

i) If $\text{Aut}(\mathcal{X}) \simeq V_4$, *then* $W = \{\pm\alpha, \pm\beta, \pm\frac{1}{\alpha\beta}\}$

ii) If $\text{Aut}(\mathcal{X}) \simeq D_4$, *then* $W = \{\pm 1, \pm\alpha, \pm\frac{1}{\alpha}\}$.

iii) If $\text{Aut}(\mathcal{X}) \simeq D_6$, *then* $W = \{1, \zeta_3, \zeta_3^2, \lambda, \lambda\zeta_3, \lambda\zeta_3^2\}$, *where λ is a parameter and ζ_3 is a primitive third root of unity.*

For each of the three cases above we already know how to find an equation of the curve over its field of moduli as shown in [**4**] amongst other places. In the next theorem we discus integral equations and their reducibility for curves $\mathcal{X}$ with $\text{Aut}(\mathcal{X}) \simeq V_4$.

THEOREM 3. *Let $\mathfrak{p} \in \mathcal{M}_2(\mathbb{Q})$ be such that* $\text{Aut}(\mathfrak{p}) \simeq V_4$. *There is a genus 2 curve $\mathcal{X}$ corresponding to $\mathfrak{p}$ with equation $y^2 z^4 = f(x^2, z^2)$, where*

$$f(x, z) = x^6 - s_1 x^4 z^2 + s_2 x^2 z^4 - z^6. \tag{14}$$

If $f \in \mathbb{Z}[x, z]$, then $f(x, z)$ or $f(-z, x)$ is a reduced binary form.

PROOF. From Lem. 6 we have that the set of Weierstrass points for such curves is $W = \{\pm\alpha, \pm\beta, \pm\frac{1}{\alpha\beta}\}$. The affine equation of the corresponding curve is

$$y^2 = (x^2 - \alpha^2)(x^2 - \beta^2)\left(x^2 - \frac{1}{(\alpha\beta)^2}\right). \tag{15}$$

Hence,

$$y^2 z^4 = x^6 - s_1 x^4 z^2 + s_2 x^2 z^4 - z^6.$$

where s_1 and s_2 are the symmetric polynomials. This proves the first part.

We assume now that $s_1, s_2 \in \mathbb{Z}$. Then, if α is a non-real root so is its conjugate $\bar{\alpha}$. Suppose that α and β are both purely complex. Then, $\frac{1}{\alpha\beta}$ is real. Geometrically this case is illustrated in Fig. (1).

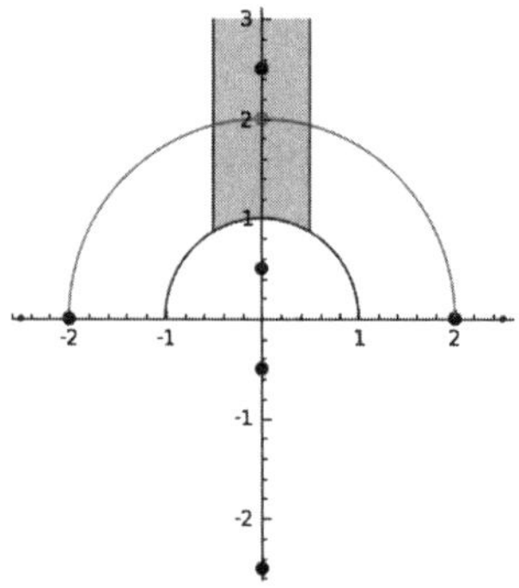

FIGURE 1. The image of the zero map $\zeta(f)$ is the red dot

We are denoting with black dots the roots of the form. In this case the zero map $\zeta(f)$ will be in the "middle" of the half semicircle connecting $\pm\frac{1}{\alpha\beta}$. Since, they are symmetric with respect to the y-axis this obviously will be in the y-axis, i.e. $\zeta(f)$ is purely complex. Next, assume α is purely complex and β is real. Then $\frac{1}{\alpha\beta}$ is purely complex and the proof is as above.

Now, let us assume that α and β are complex roots with real and imaginary part nonzero. Then, $\beta = \bar{\alpha}$ and the set of Weierstrass points for the curve is $\{\pm\alpha, \pm\bar{\alpha}, \pm\frac{1}{\alpha\bar{\alpha}}\}$. Then, the centroid of the rectangle with vertices$\{\pm\alpha, \pm\bar{\alpha}\}$ is the origin O. Finding the zero map $\zeta(f)$ is equivalent to finding the "middle" of the half semicircle connecting the real roots $\pm\frac{1}{\alpha\beta}$ which will be a point in the y-axis, illustrated in Fig. (2).

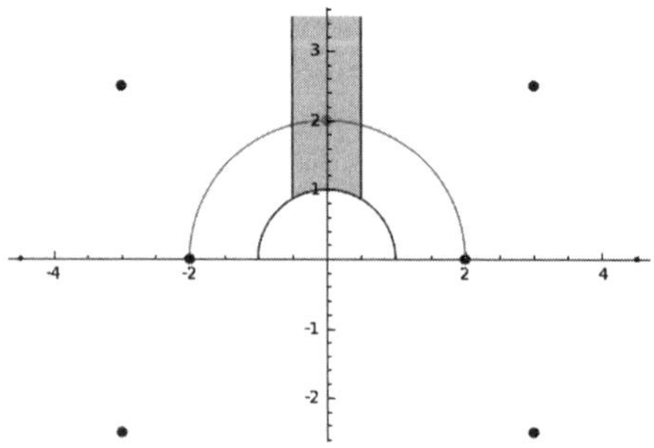

FIGURE 2. The image of the zero map $\zeta(f)$ is the red dot

Above we proved that $\zeta(F)$ is purely complex, i.e. $\zeta(F) = ci$ for some $c \in \mathbb{R}^+$. Then f is reduced if and only if $c \geq 1$. Assume f is not reduced. Then, there exists a binary form $g(x, z)$ such that $g(x, z) = f(-z, x) = f(z, x)$. The form $g(x, z)$ will be reduced since $\zeta(g) = -\frac{1}{\zeta(f)} = \frac{i}{c}$. Hence, either f or g (or possibly both, when $c = 1$) will be reduced.

Lastly, if α and β are both real the form f is totally real. In [**3**]Prop. 6.1 we proved that a superelliptic curve with such Weierstrass points is reduced in its orbit. This completes the proof. □

REMARK 3. *After this proof was completed M. Stoll pointed out that $\zeta(f)$ must be fixed by the extra involution $\sigma : (x, y) \to (-x, y)$. Notice that $\zeta(f)$ is uniquely determined by the coefficients s_1, s_2. Such pair of coefficients are fixed by the extra involution σ. Hence, $\zeta(f)$ is also fixed by such involution. Thus, $\zeta(f)$ is purely complex.*

Notice that in general a binary form being reduced doesn't necessarily mean that it has minimal height; see [**2**] for details. And an integral model of the form given as in Eq. (14) is not necessarily of minimal height among all integral models. But it is of minimal height among integral models defined by polynomials in x^2. This is proved in the following lemma.

LEMMA 7. *Let $\mathcal{X}$ be a genus 2 curve with* $\mathrm{Aut}(\mathfrak{p}) \simeq V_4$ *and equation $y^2 = f(x^2)$, where*

$$f(x) = x^6 - s_1 x^4 + s_2 x^2 - 1. \tag{16}$$

Then, integral models of this form have minimal height among integral models defined by polynomials in x^2, even up to twist.

PROOF. Let $f(x)$ be integral given as in Eq. (16) and consider another $\mathbb{Q}$-isomorphic integral model that is a polynomial in x^2. Any such model has coefficient vector as follows

$$(\lambda^6\mu^2, \lambda^4\mu^2 s_1, \lambda^2\mu^2 s_2, \mu^2)$$

for some rational λ and μ. Now, let us proceed prime by prime. Let v be the valuation of λ and let w be that of μ. For the equation to have smaller height, we would need that one of the valuation jumps

$$(6v + 2w, 4v + 2w, 2v + 2w, 2w)$$

to be negative. But because the new model is also integral, we have $2w \geq 0$ and $6v+2w \geq 0$. However, if those two inequalities hold, then all the jumps are positive, so $4v + 2w$ and $2v + 2w$ as well. Therefore, our integral model of the form

$$f(x) = x^6 - s_1 x^4 + s_2 x^2 - 1. \tag{17}$$

has minimal height among integral models defined by polynomials in x^2. □

Note that lots of curves with geometric automorphism group V_4 and field of moduli $\mathbb{Q}$ do not admit a model over $\mathbb{Q}$ defined by a polynomial in x^2. All of them descend, and in fact they even all descend to a hyperelliptic model instead of a cover of a conic. But they do not all admit that special form as given in Eq. (15).

The natural question is what are other additional conditions could force $f(x, z)$, where $f(x, z)$ represents the equation of a curve with extra automorphism, to be of minimal absolute height? We will elaborate more on this question in the next section.

5. Some heuristics for curves with extra involutions defined over $\mathbb{Q}$

In [**4**] we display a database of genus 2 curves defined over $\mathbb{Q}$. The curves in the database are ordered based on their minimal absolute height, therefore they provide a perfect case for us to check how many of our curves are in that database. In addition for each isomorphism class is given a minimal equation over the field of moduli, the automorphism group, and all the twists. All the computations involved in the database are done based on the absolute invariants i_1, i_2, i_3; see [**4**] for details. The database is explained in more details in [**4**].

We have added to that database even the curves discussed here defined over $\mathbb{Q}$. We have found all such curves of height $\mathfrak{h} \leq 101$ defined over $\mathbb{Q}$. The number of such curves for each height $\mathfrak{h}$ is displayed in the following Table 1.

In the first column is the height of the curve, the second column contains the number of tuples $(1, 0, a, 0, b, 0, 1)$ which gives a genus 2 curves (i.e. $J_{10} \neq 0$). Not all such tuples give a new moduli point. In the third column it is the number of such moduli points. The fourth and fifth column contain the number of curves with automorphism group D_4 and D_6, and the last column contains the number of all points in the moduli space of height $\leq h$.

TABLE 1. Number of curves with height $\mathfrak{h} \leq 100$

$\mathfrak{h}$	$J_{10} \neq 0$	in $\mathcal{M}_2$	D_4	D_6	# pts	$\mathfrak{h}$	$J_{10} \neq 0$	in $\mathcal{M}_2$	D_4	D_6	# pts
1	8	5	1	0	5	51	10607	205	2	0	5347
2	24	9	2	0	14	52	11023	209	2	0	5556
3	47	12	1	0	26	53	11447	213	2	0	5769
4	79	17	2	0	43	54	11879	217	2	0	5986
5	119	20	1	0	63	55	12319	221	2	0	6207
6	167	25	2	0	88	56	12767	225	2	0	6432
7	223	28	1	0	116	57	13223	229	2	0	6661
8	287	33	2	0	149	58	13687	233	2	0	6894
9	359	36	1	0	185	59	14159	237	2	0	7131
10	439	41	2	0	226	60	14639	241	2	0	7372
11	527	45	2	0	271	61	15127	245	2	0	7617
12	623	49	2	0	320	62	15623	249	2	0	7866
13	727	53	2	0	373	63	16127	253	2	0	8119
14	839	57	2	0	430	64	16639	257	2	0	8376
15	959	58	1	0	488	65	17159	261	2	0	8637
16	1087	65	2	0	553	66	17687	265	2	0	8902
17	1223	68	1	0	621	67	18223	269	2	0	9171
18	1367	73	2	0	694	68	18767	273	2	0	9444
19	1519	77	2	0	771	69	19319	277	2	0	9721
20	1679	81	2	0	852	70	19879	281	2	0	10002
21	1847	85	2	0	937	71	20447	285	2	0	10287
22	2023	89	2	0	1026	72	21023	289	2	0	10576
23	2207	93	2	0	1119	73	21607	293	2	0	10869
24	2399	97	2	0	1216	74	22199	297	2	0	11166
25	2599	101	2	0	1317	75	22799	301	2	0	11467
26	2807	105	2	0	1422	76	23407	305	2	0	11772
27	3023	109	2	0	1531	77	24023	309	2	0	12081
28	3247	113	2	0	1644	78	24647	313	2	0	12394

continued on next page

TABLE 1. (Continued)

$\mathfrak{h}$	$J_{10} \neq 0$	in $\mathcal{M}_2$	D_4	D_6	# pts	$\mathfrak{h}$	$J_{10} \neq 0$	in $\mathcal{M}_2$	D_4	D_6	# pts
29	3479	117	2	0	1761	79	25279	317	2	1	12711
30	3719	121	2	0	1882	80	25919	321	2	0	13032
31	3967	125	2	0	2007	81	26567	325	2	0	13357
32	4223	129	2	0	2136	82	27223	329	2	0	13686
33	4487	133	2	0	2269	83	27887	333	2	1	14019
34	4759	137	2	0	2406	84	28559	337	2	0	14356
35	5039	141	2	0	2547	85	29239	341	2	0	14697
36	5327	145	2	0	2692	86	29927	345	2	0	15042
37	5623	149	2	0	2841	87	30623	349	2	0	15391
38	5927	153	2	0	2994	88	31327	353	2	0	15744
39	6239	157	2	0	3151	89	32039	357	2	0	16101
40	6559	161	2	0	3312	90	32759	361	2	0	16462
41	6887	165	2	0	3477	91	33487	365	2	0	16827
42	7223	169	2	0	3646	92	34223	369	2	0	17196
43	7567	173	2	0	3819	93	34967	373	2	0	17569
44	7919	177	2	0	3996	94	35719	377	2	0	17946
45	8279	181	2	0	4177	95	36479	381	2	0	18327
46	8647	185	2	0	4362	96	37247	385	2	0	18712
47	9023	189	2	0	4551	97	38023	389	2	0	19101
48	9407	193	2	0	4744	98	38807	393	2	0	19494
49	9799	197	2	0	4941	99	39599	397	2	0	19891
50	10199	201	2	0	5142	100	40399	401	2	0	20292

Some interesting questions that can be addressed analyzing Table 1 are as follows. How many genus two curves with extra involutions are there with a fixed height $\mathfrak{h}$? How many isomorphism classes of genus two curves with extra involutions are there for a fixed height $\mathfrak{h}$? In other words, how many twists are for such curves with fixed height? We intend to further explore some of these questions in further work.

The main question that comes from the previous section was how many of these curves are of minimal absolute height. From 14523 = 20292 - 5769 binary forms of the form given in Eq. (14) we check how many of them have minimal absolute height ≤ 3 even though $r := \min\{|s_1|, |s_2|\} > 3$. Out of 14523 forms only for 57 of them

$$r = \max\{|s_1|, |s_2|\} \neq \mathfrak{h}\,.$$

We display all such forms in the Table 2. In the third column is the equation of the curve given the 7-tuple $(a_0, \dots, a_6)$ corresponding to the equation

$$y^2 = \sum_{i=0}^{6} a_i x^i = x^6 - s_1 x^4 + s_2 x^2 + 1.$$

In the fifth column is the twist with height $\mathfrak{h} \leq 4$ which is isomorphic over $\overline{\mathbb{Q}}$ with the corresponding curve in the first column. In the last column is given the automorphism group of the curve over $\overline{\mathbb{Q}}$.

There are a few questions which arise from Table 2. First, can the curves of column five be obtained from reducing curves of column three? Secondly, are they in the same Γ-orbit as the curves from column 2?

TABLE 2. Curves which have twists with height ≤ 4

#	r	$(1, 0, s_1, 0, s_2, 0, 1)$	$\mathfrak{h}$	$(a_0, \dots, a_6)$	Aut($\mathfrak{p}$)
1	7	[1, 0, 1, 0, -7, 0, 1]	3	[1, -3, -1, -2, -1, -3, 1]	[4, 2]
2	5	[1, 0, 5, 0, 1, 0, 1]	3	[1, -1, 3, 2, 3, -1, 1]	[4, 2]
3	17	[1, 0, 15, 0, -17, 0, 1]	2	[1, -1/2, -1, -1, -1, -1/2, 1]	[4, 2]
4	29	[1, 0, -29, 0, -29, 0, 1]	3	[0, 1, 0, -3/2, 0, 1]	[8, 3]
5	9	[1, 0, 9, 0, 5, 0, 1]	2	[1, -1/2, 1, 1, 1, -1/2, 1]	[4, 2]
6	41	[1, 0, -25, 0, -41, 0, 1]	3	[1, -1/2, -3/2, 1, -3/2, -1/2, 1]	[4, 2]
7	13	[1, 0, 3, 0, -13, 0, 1]	2	[1, -2, -1, 0, -1, -2, 1]	[4, 2]
8	51	[1, 0, 51, 0, -45, 0, 1]	3	[1, 0, -1, -2/3, -1, 0, 1]	[4, 2]
9	9	[1, 0, 7, 0, -9, 0, 1]	2	[1, -1, -1, -2, -1, -1, 1]	[4, 2]
10	19	[1, 0, 19, 0, -13, 0, 1]	2	[1, 0, -1, -2, -1, 0, 1]	[4, 2]
11	61	[1, 0, 35, 0, -61, 0, 1]	3	[1, -2/3, -1, 2/3, -1, -2/3, 1]	[4, 2]
12	7	[1, 0, -7, 0, -7, 0, 1]	4	[1, 4, -3, 0, -3, -4, 1]	[8, 3]
13	61	[1, 0, 3, 0, -61, 0, 1]	3	[1, -2, -1, 3, -1, -2, 1]	[4, 2]
14	17	[1, 0, -1, 0, -17, 0, 1]	3	[1, -3, -1, 2, -1, -3, 1]	[4, 2]
15	6	[1, 0, 6, 0, 6, 0, 1]	2	[-1, 1, 1/2, 0, -1/2, -1, 1]	[8, 3]
16	13	[1, 0, -5, 0, -13, 0, 1]	3	[1, -1, -3, 2, -3, -1, 1]	[4, 2]
17	29	[1, 0, 19, 0, -29, 0, 1]	3	[1, -2/3, -1, 0, -1, -2/3, 1]	[4, 2]
18	19	[1, 0, 19, 0, 19, 0, 1]	3	[0, 1, 0, -3, 0, 1]	[8, 3]
19	5	[1, 0, -5, 0, -5, 0, 1]	1	[0, -1, 0, 0, 0, 1]	[48, 5]
20	47	[1, 0, 47, 0, 47, 0, 1]	3	[1, 0, -2/3, 0, -2/3, 0, 1]	[8, 3]
21	39	[1, 0, 39, 0, 23, 0, 1]	2	[1, -1/2, -1/2, 1, -1/2, -1/2, 1]	[4, 2]
22	5	[1, 0, 3, 0, -5, 0, 1]	2	[0, 1, -2, -2, -2, 1]	[2, 1]
23	8	[1, 0, 8, 0, 8, 0, 1]	3	[1, 1, 3/2, 0, 3/2, -1, 1]	[8, 3]
24	19	[1, 0, 19, 0, 11, 0, 1]	2	[1, -1/2, 0, 1, 0, -1/2, 1]	[4, 2]
25	21	[1, 0, -5, 0, -21, 0, 1]	4	[1, -4, -1, 4, -1, -4, 1]	[4, 2]
26	53	[1, 0, 43, 0, -53, 0, 1]	3	[1, -1/3, -1, 0, -1, -1/3, 1]	[4, 2]
27	37	[1, 0, 27, 0, -37, 0, 1]	2	[1, -1/2, -1, 0, -1, -1/2, 1]	[4, 2]
28	35	[1, 0, 35, 0, -29, 0, 1]	1	[1, 0, -1, -1, -1, 0, 1]	[4, 2]
29	93	[1, 0, 35, 0, -93, 0, 1]	3	[1, -1, -1, 3/2, -1, -1, 1]	[4, 2]
30	21	[1, 0, 11, 0, -21, 0, 1]	1	[1, -1, -1, 0, -1, -1, 1]	[4, 2]
31	55	[1, 0, 55, 0, 39, 0, 1]	3	[1, -1/3, -2/3, 2/3, -2/3, -1/3, 1]	[4, 2]
32	77	[1, 0, 51, 0, -77, 0, 1]	2	[1, -1/2, -1, 1/2, -1, -1/2, 1]	[4, 2]
33	9	[1, 0, -1, 0, -9, 0, 1]	4	[1, -4, -1, 0, -1, -4, 1]	[4, 2]
34	37	[1, 0, -37, 0, -37, 0, 1]	3	[1, 3/2, -1, 0, -1, -3/2, 1]	[8, 3]
35	11	[1, 0, 11, 0, 11, 0, 1]	3	[1, 2, 3, 0, 3, -2, 1]	[8, 3]
36	29	[1, 0, 3, 0, -29, 0, 1]	2	[1, -2, -1, 2, -1, -2, 1]	[4, 2]
37	23	[1, 0, 23, 0, 23, 0, 1]	3	[1, 0, -1/3, 0, -1/3, 0, 1]	[8, 3]
38	11	[1, 0, 5, 0, -11, 0, 1]	3	[1, -3/2, -1, -1, -1, -3/2, 1]	[4, 2]
39	11	[1, 0, 11, 0, 3, 0, 1]	2	[1, -1, 1, 2, 1, -1, 1]	[4, 2]
40	25	[1, 0, -9, 0, -25, 0, 1]	2	[1, -1, -2, 2, -2, -1, 1]	[4, 2]
41	99	[1, 0, 99, 0, -93, 0, 1]	3	[1, 0, -1, -1/3, -1, 0, 1]	[4, 2]
42	15	[1, 0, 9, 0, -15, 0, 1]	3	[1, -1, -1, -2/3, -1, -1, 1]	[4, 2]
43	23	[1, 0, 23, 0, 7, 0, 1]	2	[1, -1, 0, 2, 0, -1, 1]	[4, 2]
44	45	[1, 0, 19, 0, -45, 0, 1]	1	[1, -1, -1, 1, -1, -1, 1]	[4, 2]
45	31	[1, 0, 31, 0, 31, 0, 1]	2	[1, 0, -1/2, 0, -1/2, 0, 1]	[8, 3]
46	9	[1, 0, 9, 0, 9, 0, 1]	3	[-1, -3, -3, -2, 3, -3, 1]	[8, 3]
47	25	[1, 0, 7, 0, -25, 0, 1]	3	[1, -3/2, -1, 1, -1, -3/2, 1]	[4, 2]
48	83	[1, 0, 83, 0, -45, 0, 1]	3	[1, -1/2, -1, 3/2, -1, -1/2, 1]	[4, 2]
49	13	[1, 0, 13, 0, 9, 0, 1]	3	[1, -1/3, 1/3, 2/3, 1/3, -1/3, 1]	[4, 2]

continued on next page

TABLE 2. (Continued)

50	25	[1, 0, 23, 0, -25, 0, 1]	3	[1, -1/3, -1, -2/3, -1, -1/3, 1]	[4, 2]
51	33	[1, 0, 15, 0, -33, 0, 1]	3	[1, -1, -1, 2/3, -1, -1, 1]	[4, 2]
52	27	[1, 0, 27, 0, 19, 0, 1]	3	[1, -1/3, -1/3, 2/3, -1/3, -1/3, 1]	[4, 2]
53	51	[1, 0, 51, 0, -13, 0, 1]	3	[1, -1, -1, 3, -1, -1, 1]	[4, 2]
54	13	[1, 0, -13, 0, -13, 0, 1]	1	[0, 1, 0, -1, 0, 1]	[8, 3]
55	27	[1, 0, 27, 0, 27, 0, 1]	3	[1, 3, 3, 0, 3, -3, 1]	[8, 3]
56	67	[1, 0, 67, 0, -61, 0, 1]	2	[1, 0, -1, -1/2, -1, 0, 1]	[4, 2]
57	33	[1, 0, -33, 0, -33, 0, 1]	3	[1, 0, -3/2, 0, -3/2, 0, 1]	[8, 3]

In response to this question, we found that twenty of the curves displayed in Table 2 can be reduced further using the reduction algorithm. They are displayed in Table 3. In the second column is displayed the curve from Table 2, in the third column the curve obtained by the reduction algorithm and the last column the automorphism group of the curve. Some of the reduced curves are isomorphic to the original curves over $\mathbb{Q}$.

TABLE 3. Curves which have twists with height ≤ 4 and reduced curves.

case #	(s_1, s_2)-curve	reduced curve	Group
57	(1, 0, -33, 0, -33, 0, 1)	(-2, 0, 3, 0, 3, 0, -2)	[8, 3]
50	(1 , 0, 23, 0, -25, 0, 1)	(0 , 3, 1, -6, 1, 3, 0)	[4, 2]
21	(1 , 0, 39, 0, 23, 0, 1)	(2, 1, -1, -2, -1, 1, 2)	[4, 2]
16	(1 , 0, -5, 0, -13, 0, 1)	(-1, -1, 3, 2, 3, -1, -1)	[4, 2]
20	(1 , 0, 47, 0, 47, 0, 1)	(3, 0, -2, 0, -2, 0, 3)	[8, 3]
39	(1 , 0, 11, 0, 3, 0, 1)	(1 , -1, 1, 2, 1, -1, 1)	[4, 2]
3	(1 , 0, 15, 0, -17, 0, 1)	(0, 2, 1, -4, 1, 2, 0)	[4, 2]
47	(1 , 0, 7, 0, -25, 0, 1)	(1, -4, 3, 8, 3, -4, -1)	[4, 2]
40	(1 , 0, -9, 0, -25, 0, 1)	(-1, -1, 2, 2, 2, -1, -1)	[4, 2]
24	(1 , 0, 19, 0, 11, 0, 1)	(2, 1, 0, -2, 0, 1, 2)	[4, 2]
9	(1 , 0, 7, 0, -9, 0, 1)	(0, -1, 1, 2, 1, -1, 0)	[4, 2]
37	(1 , 0, 23, 0, 23, 0, 1)	(3, 0, -1, 0, -1, 0, 3)	[8, 3]
43	(1 , 0, 23, 0, 7, 0, 1)	(1 , -1, 0, 2, 0, -1, 1)	[4, 2]
31	(1 , 0, 55, 0, 39, 0, 1)	(3, 1, -2, -2, -2, 1, 3)	[4, 2]
6	(1 , 0, -25, 0, -41, 0, 1)	(-2, 1, 3, -2, 3, 1, -2)	[4, 2]
45	(1 , 0, 31, 0, 31, 0, 1)	(2, 0, -1, 0, -1, 0, 2)	[8, 3]
52	(1 , 0, 27, 0, 19, 0, 1)	(3, 1, -1, -2, -2, 1, 3)	[4, 2]
14	(1 , 0, -1, 0, -17, 0, 1)	(-1, 2, 3, -4, 3, 2, -1)	[4, 2]
51	(1 , 0, 15, 0, -33, 0, 1)	(-1, -6, 3, 12, 3, -6, 1)	[4, 2]

It is worth noting that in each case the reduction algorithm does find a curve with minimal absolute height. It is also interesting to see that all 57 curves from Table 2 have one thing in common, their discriminant can be further reduced as explained in Section (3.4).

We believe that a generalization of Thm. 3 to higher degree binary forms $f(x^2, y^2)$ and in more general for forms $f(x^n, y^n)$ is possible. Hopefully, this will be the focus of investigation of another paper.

Acknowledgments

The author would like to thank M. Stoll, T. Shaska, and J. Sijsling for valuable comments and help. Furthermore, the author would like to thank the anonymous referees for all the comments they provided during the review.

References

[1] Lubjana Beshaj, *Singular locus on the space of genus 2 curves with decomposable Jacobians*, Albanian J. Math. **4** (2010), no. 4, 147–160. MR2755393

[2] Lubjana Beshaj, *Reduction theory of binary forms*, Advances on superelliptic curves and their applications, NATO Sci. Peace Secur. Ser. D Inf. Commun. Secur., vol. 41, IOS, Amsterdam, 2015, pp. 84–116. MR3525574

[3] Lubjana Beshaj, *Integral binary forms with minimal height*, ProQuest LLC, Ann Arbor, MI, 2016. Thesis (Ph.D.)–Oakland University. MR3579531

[4] L. Beshaj, R. Hidalgo, A. Malmendier, S. Kruk, S. Quispe, and T. Shaska, *Rational points on the moduli space of genus two*, Contemporary Mathematics (AMS) (2016).

[5] J. E. Cremona, *Reduction of binary cubic and quartic forms*, LMS J. Comput. Math. **2** (1999), 64–94. MR1693411

[6] Gaston Julia, *Étude sur les formes binaires non quadratiques à indéterminées réelles, ou complexes, ou à indéterminées conjuguées* (French), NUMDAM, [place of publication not identified], 1917. MR3532882

[7] Qing Liu, *Conducteur et discriminant minimal de courbes de genre* 2 (French), Compositio Math. **94** (1994), no. 1, 51–79. MR1302311

[8] Andreas Malmendier and Tony Shaska, *A universal pair of genus-two curves* (201607), available at 1607.08294.

[9] Jean-François Mestre, *Construction de courbes de genre* 2 *à partir de leurs modules* (French), Effective methods in algebraic geometry (Castiglioncello, 1990), Progr. Math., vol. 94, Birkhäuser Boston, Boston, MA, 1991, pp. 313–334. MR1106431

[10] Rachel Shaska, *Equations of curves with minimal discriminant* (201407), available at 1407.7064.

[11] T. Shaska, *Determining the automorphism group of a hyperelliptic curve*, Proceedings of the 2003 International Symposium on Symbolic and Algebraic Computation, 2003, pp. 248–254 (electronic). MR2035219 (2005c:14037)

[12] T. Shaska and L. Beshaj, *Heights on algebraic curves*, Advances on superelliptic curves and their applications, NATO Sci. Peace Secur. Ser. D Inf. Commun. Secur., vol. 41, IOS, Amsterdam, 2015, pp. 137–175. MR3525576

[13] Tanush Shaska and Helmut Völklein, *Elliptic subfields and automorphisms of genus 2 function fields*, Algebra, arithmetic and geometry with applications (West Lafayette, IN, 2000), Springer, Berlin, 2004, pp. 703–723. MR2037120

[14] Tony Shaska, *Genus 2 curves with* $(3,3)$*-split Jacobian and large automorphism group*, Algorithmic number theory (Sydney, 2002), 2002, pp. 205–218. MR2041085 (2005e:14048)

[15] Michael Stoll and John E. Cremona, *On the reduction theory of binary forms*, J. Reine Angew. Math. **565** (2003), 79–99. MR2024647 (2005e:11091)

Department of Mathematics, The University of Texas at Austin, 1 University Station C1200, Austin, Texas 78712

Email address: beshaj@math.utexas.edu

Contemporary Mathematics
Volume **703**, 2018
http://dx.doi.org/10.1090/conm/703/14132

Rational points in the moduli space of genus two

L. Beshaj, R. Hidalgo, S. Kruk, A. Malmendier, S. Quispe, and T. Shaska

ABSTRACT. We build a database of genus 2 curves defined over $\mathbb{Q}$ which contains all curves with minimal absolute height $\mathrm{h} \leq 5$, all curves with moduli height $\mathfrak{h} \leq 20$, and all curves with extra automorphisms in standard form $y^2 = f(x^2)$ defined over $\mathbb{Q}$ with height $\mathrm{h} \leq 101$. For each isomorphism class in the database, an equation over its minimal field of definition is provided, the automorphism group of the curve, Clebsch and Igusa invariants. The distribution of rational points in the moduli space $\mathcal{M}_2$ for which the field of moduli is a field of definition is discussed and some open problems are presented.

1. Introduction

In **[22]** were introduced concepts of *minimum absolute height* of a binary form, the *moduli height*, and discussed relations between the two. Moreover, some computations were performed for binary sextics of minimum absolute height one. A natural problem in that paper was to check whether a large database of binary forms (equivalently genus 2 curves) of relatively small minimum absolute height could be constructed. Comparing the minimum absolute height with the moduli height would be the main point of this database. Moreover, such a database would shed some light to other problems related to the rational points of the moduli space $\mathcal{M}_2$. For example, not every rational point $\mathfrak{p} \in \mathcal{M}_2$ has a representative genus 2 curve $\mathcal{X}$ defined over $\mathbb{Q}$. What is the percentage of points of bounded height for which the field of moduli is not a field of definition? How does this ratio is affected when the height increases?

From another point of view, we can list genus two curves based on their moduli height. For example, traditionally there have been plenty of effort to count the curves with bounded discriminant. That is not an easy problem, but would it make more sense to have an estimate on the number of curves with bounded moduli height? After all, the moduli height is the most natural way of sorting points in $\mathcal{M}_2$.

And then, there are also the curves with automorphisms. In [**2**] it was shown that such curves can always be written in an equation such that the corresponding binary sextic is reduced; see [**1**] for details. Reduced usually means minimal absolute height for the binary form. Is this really supported computationally?

Our goal was to construct a database of genus 2 curves which addresses some of these questions.

2010 *Mathematics Subject Classification.* Primary 14Hxx, 14Qxx.

In this paper we construct three main databases:

i) integral binary sextics of minimum absolute height ≤ 4,
ii) integral binary sextics with moduli height $\mathfrak{h} \leq 20$,
iii) integral binary sextics $f(x^2, z^2)$ of height $\mathrm{h} \leq 101$.

All combined we have over 1 million isomorphism classes of genus two curves (equivalently rational points in $\mathcal{M}_2$) without counting twists. We compute the minimal absolute height, the moduli height, the discriminant, automorphism group, field of definition, and an equation over the field of definition for all such points. We discuss all the technical details of organizing the data and some open questions on the distribution of the rational points with non-trivial obstruction in the moduli space $\mathcal{M}_2$.

Let k be an algebraically closed field of characteristic zero and $\mathcal{M}_g$ the moduli space of smooth, projective, genus $g \geq 2$ algebraic curves defined over k. The moduli space $\mathcal{M}_2$ of genus 2 curves is the most understood moduli space among all moduli spaces. This is mostly due to two main facts; first all genus two curves are hyperelliptic and therefore studying them it is easier than general curves, secondly even among hyperelliptic curves the curves of genus two have a special place since they correspond to binary sextics which, from the computational point of view, are relatively well understood compared to higher degree binary forms.

Some of the main questions related to $\mathcal{M}_2$ have been to recover a nice equation for any point $\mathfrak{p} \in \mathcal{M}_2$. Since $\mathcal{M}_2$ is a coarse moduli space, such equation is not always defined over the field of moduli of $\mathfrak{p}$. Can we find a universal equation for genus two curves over their minimal field of definition? Can such equation provide a minimal model for the curve? Does the height of this minimal model has any relation to the projective height of the corresponding moduli point $\mathfrak{p} \in \mathcal{M}_2$? What is the distribution in $\mathcal{M}_2$ of points $\mathfrak{p}$ for which the field of moduli is not a field of definition? The answers to these questions are still unknown.

In [**3**] we provide a computational package for computing with genus 2 curves and a database of genus 2 curves which contains all curves with height $\mathrm{h} \leq 5$, curves with moduli height $\mathfrak{h} \leq 20$, and curves with automorphism and height ≤ 101. They are organized in three Python directories $\mathcal{L}_i$, $i = 1, 2, 3$ as explained in Section 2.

The database is build with the idea of better understanding $\mathcal{M}_2$, the distribution of points in $\mathcal{M}_2$ based on the moduli height, the distribution of points for which the field of moduli is not a field of definition.

The goal of this paper is twofold: to provide the mathematical background for most of the algorithms in [**3**] and to discuss some of the open questions and problems raised there. Most of the material of the first part has already appeared in the vast literature on genus two curves some of which is previous work of these authors. For sake of completeness and straightening out some notational confusion we define all the basic invariants of genus two curves in this paper.

The current database and all the functions are implemented in Sage. It improves and expands a previous Maple genus 2 computational algebra package as in [**21**]. There is another database of genus 2 curves in [**5**] which collects all genus 2 curves with discriminants ≤ 1000. Some remarks on how the two databases overlap can be found in the last section.

There is a lot of confusion in the literature about the invariants of genus two curves. We go to great lengths to make sure that all the invariants are defined explicitly and there is no room for misunderstanding. We like to warn the reader that our invariants are different from the ones used by Magma and all the papers which use Magma in their computations.

2. A database of integral binary sextics

Our main goal is to create an extensive database of integral binary sextics with minimal absolute height and all the twists with minimal height. We will use the definitions of minimal absolute height, moduli height, and the basic properties of heights of polynomials from [**22**] and will provide details in the next coming sections. The database will be organized in a Sage/Python dictionary, where the key will be the moduli point

$$\mathfrak{p} = \begin{cases} (-1, i_1, i_2, i_3), & if\ J_2 \neq 0 \\ (0, t_1, t_2, t_3), & if\ J_2 = 0, \end{cases}$$

see Eq. (15) for definitions of such moduli point. The data is organized in three main dictionaries:

i) integral binary sextics with minimum absolute height $\mathrm{h} \leq 10$,
ii) decomposable integral binary sextics $f(x^2, z^2)$ with minimum absolute $\mathrm{h} \leq 101$ and,
iii) integral binary sextics with moduli height $\mathfrak{h} \leq 20$

Each point in the database has the following invariants

$$\begin{aligned} \mathfrak{p} = (r, i_1, i_2, i_3): \quad & \mathrm{h} && = \textit{minimal absolute height} \\ & \mathfrak{h} && = \textit{moduli height} \\ & \Delta && = \textit{minimal discriminant} \\ & \mathrm{Aut}(\mathfrak{p}) && = \textit{automorphism group} \\ & C && = \textit{Conductor} \\ & M_{\mathfrak{p}} && = \textit{field of definition of the universal curve} \\ & \text{Twist} && = \textit{List of twists} \end{aligned}$$

An entry in each dictionary looks as the following:

$$(r, i_1, i_2, i_3) : \Big[\mathrm{h},\ \mathfrak{h},\ \Delta,\ \mathrm{Aut}(\mathfrak{p}),\ C,\ M_{\mathfrak{p}},\ \Big[[a_0, \ldots, a_6], \ldots, [b_0, \ldots, b_6]\Big]\ \Big]$$

We illustrate with an example.

EXAMPLE 1. Let $\mathcal{X}$ be the curve with equation

$$y^2 = x^6 - 14x^4 - 82x^2 + 1 \tag{1}$$

If we load the database in Sage and let

$$f = t^6 - 14t^4 - 82t^2 + 1$$

then the command $\mathfrak{p}$=**ModPoint (f)** displays

$$\left(-1, -\frac{49281147}{5410276}, \frac{706232480445}{12584301976}, \frac{3071021069999403}{174296440211213762 56}\right)$$

If we ask if $\mathfrak{p} \in \mathcal{L}$, where $\mathcal{L}$ is the second dictionary from above, the answer will be **yes** and $\mathcal{L}(\mathfrak{p})$ will display

$$[82, 2^{14} \cdot 1163^5, 17^2 \cdot 12301^2, [4, 2], C, \mathbb{Q}, [[1, 0, -14, 0, -82, 0, 1]]]$$

which means that the minimal absolute height is $\mathrm{h} = 82$, automorphism group with GapId [4, 2] which is the Klein group V_4, minimal field of definition $\mathbb{Q}$, and minimal discriminant $\Delta = 17^2 \cdot 12301^2$. The moduli height is $\mathfrak{h} = 2^{14} \cdot 1163^5$. We chose not to display the conductor and all the twists.

In the Appendix A is given a list of all the functions used for the genus 2 curves package in Sage. In the next few sections we will go over the necessary definitions and procedures to construct such databases. The details will be explained in Section 9.

3. Heights of genus two curves

In this section we define heights on algebraic curves when such curves are given by some affine equation. Throughout this paper K denotes an algebraic number field and $\mathcal{O}_K$ its ring of integers.

Let $\mathcal{X}_g$ be an irreducible algebraic curve with affine equation $F(x, y) = 0$ for $F(x, y) \in K[x, y]$. We define the **height of the curve over** K to be

$$H_K(\mathcal{X}_g) := \min \{H_K(G) : H_K(G) \leq H_K(F)\}.$$

where the curve $G(x, y) = 0$ is isomorphic to $\mathcal{X}_g$ over K.

If we consider the equivalence over $\bar{K}$ then we get another height which we denote it as $\overline{H}_K(\mathcal{X}_g)$ and call it **the height over the algebraic closure**. Namely,

$$\overline{H}_K(\mathcal{X}_g) = \min\{H_K(G) : H_K(G) \leq H_K(F)\},$$

where the curve $G(x, y) = 0$ is isomorphic to $\mathcal{X}_g$ over $\overline{K}$.

In the case that $K = \mathbb{Q}$ we do not write the subscript K and use $H(\mathcal{X}_g)$ or $\overline{H}(\mathcal{X}_g)$. Obviously, for any algebraic curve $\mathcal{X}_g$ we have $\overline{H}_K(\mathcal{X}_g) \leq H_K(\mathcal{X}_g)$. In [**22**] is proved that given K a number field such that $[K : \mathbb{Q}] = d$, the height $H_K(\mathcal{X}_g)$ and $\overline{H}_K(\mathcal{X}_g)$ are well defined.

THEOREM 1 ([**22**]). *Let K be a number field such that $[K : \mathbb{Q}] \leq d$. Given a constant $\mathrm{h}_0 \geq 1$ there are only finitely many curves such that $H_K(\mathcal{X}_g) \leq \mathrm{h}_0$.*

As an immediate corollary we have the following

COROLLARY 1. *Let $\mathrm{h}_0 \geq 1$ be a fixed integer, K a number field, and $\mathcal{O}_K$ its ring of integers. For any genus $g \geq 2$ curve $\mathcal{X}_g$ defined over $\mathcal{O}_K$ with height $\mathrm{h}(\mathcal{X}_g) = \mathrm{h}_0$ there are only finitely many twists of $\mathcal{X}_g$ with height h_0.*

Given a genus two curve $\mathcal{X}_g$ the following algorithm computes a curve isomorphic over K to $\mathcal{X}_g$ of minimum height

ALGORITHM 1. ***Input:*** *an algebraic curve $\mathcal{X}_g : F(x, y) = 0$, where F has degree d and is defined over K*

Output: *an algebraic curve $\mathcal{X}'_g : G(x, y) = 0$ such that $\mathcal{X}'_g \cong_K \mathcal{X}_g$ and $\mathcal{X}'_g$ has minimum height.*

Step 1: *Compute $c_0 = H_K(F)$*

Step 2: *List all points $P \in \mathbb{P}^s(K)$ such that $H_K(P) \leq c_0$.*

Note: *s is the number of terms of F which is the number of monomials of degree d in n variables, and this is equal to $\binom{d+n-1}{d}$. From Thm. 1 there are only finitely many such points assume $P_1, \dots, P_r$.*

Step 3: *for* $i = 1$ *to* r *do*
Let $G_i(x, y) = p_i$;
if $g(G_i(x, y)) = g(\mathcal{X}_g)$ *then*
if $G_i(x, y) = 0 \cong_K F(x, y) = 0$
then add G_i *to the list* L
end if;
end if;

Step 4: *Return all entries of* L *of minimum height , L has curves isomorphic over* K *to* $\mathcal{X}_g$ *of minimum height.*

Note that this algorithm is not very efficient if we start with an algebraic curve of genus two and very big height. Hence, the question that can be raised at this point is: how can we reduce the height of the curve? This is done using reduction theory, see [**1**, **25**] and others for more details, and some elementary ways of reducing are given next. The following elementary lemma is useful.

LEMMA 1. *Let* $\mathcal{X}$ *be a superelliptic curve with Weierstrass equation* $y^m = \sum_{i=0}^d a_i x^i$, *defined over* $\mathbb{Z}$, *and height* $\mathrm{h}(\mathcal{X})$. *Let* p *be a prime such that* $p \,|\, a_0$ *and* $\mathfrak{v}_p(a_i) = \alpha_i$, *such that* $\alpha_0 \geq \alpha_i$, *for* $i = 0, \dots, d$. *Choose* m *to be the largest nonnegative integer which satisfies*

$$m \leq \frac{\alpha_0 - \alpha_i}{i}, \quad i = 1, \dots, d.$$

Then, there is a twist $\mathcal{X}'$ *of* $\mathcal{X}$ *such that*

$$\mathfrak{v}_p\left(\mathrm{h}(\mathcal{X}')\right) \leq \mathfrak{v}_p\left(\mathrm{h}(\mathcal{X})\right) - m.$$

PROOF. Let p be a prime such that

$$a_i = p^{\alpha_i} \cdot b_i, \quad \textit{such that } (p, b_i) = 1$$

In other words, $\mathfrak{v}_p(a_i) = \alpha_i$, for $i = 1, \dots, d$, as in the assumptions of the theorem. Hence, the equation of the curve is

$$y^m = \sum_{i=0}^d p^{\alpha_i} b_i \cdot x^i$$

Choose m as the largest nonnegative integer such that

$$m \leq \frac{\alpha_0 - \alpha_i}{i}, \quad i = 1, \dots, d.$$

If $m = 0$, then the curve can no further reduced by this method at the prime p. If $m > 0$, then we let

$$(x, y) \mapsto (p^m \cdot x, \, y)$$

which gives the curve

$$y^m = \sum_{i=0}^d a_i' x^i = \sum_{i=0}^d p^{\alpha_i + im} b_i x^i$$

Then,

$$\mathfrak{v}_p(a_i') = \alpha_i + i \cdot m.$$

Hence, to have reduction of $\mathfrak{v}_p\left(\mathrm{h}(\mathcal{X})\right)$ we must have $\alpha_i + im \leq \alpha_0$ for $i = 1, \dots, d$. Thus, $m \leq \frac{\alpha_0 - \alpha_i}{i}$, for $i = 1, \dots d$. Choosing the largest such m will result to the biggest possible reduction on p. Dividing both sides by the content of the polynomial, which has the maximum power of p as a factor, gives a twist $\mathcal{X}'$ of $\mathcal{X}$ with height which has valuation at p, $\mathfrak{v}_p(\mathrm{h}(\mathcal{X}')) \leq \mathfrak{v}_p(\mathrm{h}(\mathcal{X})) - m$. This completes the proof. □

REMARK 1. Notice that the curve $\mathcal{X}^\sigma$ could be in the same Γ-orbit of $\mathcal{X}$ or could be a twist of $\mathcal{X}$, depending on the values of m, where Γ is the modular group.

COROLLARY 2. *Let $\mathcal{X}$ be a curve with* $\mathrm{Aut}(\mathcal{X}) \cong D_6$ *and equation*

$$y^2 = x^6 + x^3 + s$$

where $s \in \mathbb{Z}$ such that it has a prime factorization

$$s = p^\alpha \cdot s', \quad \textit{where} \quad (s, s') = 1$$

Then, we can reduce the height by the transformation $x \mapsto p^m \cdot x$, where $m = \lfloor \frac{\alpha}{6} \rfloor$

EXAMPLE 2. Let us consider the curve from Ex. 4, namely

$$y^2 = x^6 + x^3 + 2^{33}$$

This curve has height $\mathrm{h} = 2^{33}$. Then m has to be the largest nonnegative integer such that it is $\leq$ to

$$\frac{33-0}{6}, \frac{33-0}{5},$$

which makes $m = 5$.

Consider the transformation $x \mapsto 2^5 \cdot x$. Then the curve becomes

$$y^2 = 2^{15} \cdot x^6 + x^3 + 2^{18}$$

which is with height $\mathrm{h} = 2^{18}$.

Next we will define the moduli height of genus g curves.

3.1. Moduli height of curves. Let g be an integer $g \geq 2$ and $\mathcal{M}_g$ denote the coarse moduli space of smooth, irreducible algebraic curves of genus g. It is known that $\mathcal{M}_g$ is a quasi projective variety of dimension $3g-3$. Hence, $\mathcal{M}_g$ is embedded in $\mathbb{P}^{3g-2}$. Let $\mathfrak{p} \in \mathcal{M}_g$. We call the moduli height $\mathfrak{h}(\mathfrak{p})$ the usual height $H(P)$ in the projective space $\mathbb{P}^{3g-2}$. Obviously, $\mathfrak{h}(\mathfrak{p})$ is an invariant of the curve. In [**22**] is proved the following result.

THEOREM 2. *For any constant $c \geq 1$, degree $d \geq 1$, and genus $g \geq 2$ there are finitely many superelliptic curves $\mathcal{X}_g$ defined over the ring of integers $\mathcal{O}_K$ of an algebraic number field K such that $[K : \mathbb{Q}] \leq d$ and $\mathfrak{h}(\mathcal{X}_g) \leq c$.*

While the above theorem shows that the number of curves with bounded moduli height is finite, determine this number seems to be a very difficult problem.

4. Genus 2 curves over $\mathbb{C}$

In this section we give a quick overview of the basic setup for genus two curves. The material is part of the folklore on the literature of genus 2 curves and we don't mention all the possible references. While the main definitions and results on what follows are valid for any $g \geq 2$ we only state them for the case $g = 2$. We mainly follow the approach of [**4, 7, 9–11, 16**].

4.1. Periods and invariants. Let $\mathcal{X}$ be a genus $g = 2$ algebraic curve. We choose a symplectic homology basis for $\mathcal{X}$, say $\{A_1, A_2, B_1, B_2\}$, such that the intersection products $A_i \cdot A_j = B_i \cdot B_j = 0$ and $A_i \cdot B_j = \delta_{ij}$, where δ_{ij} is the Kronecker delta. We choose a basis $\{w_i\}$ for the space of holomorphic 1-forms such that $\int_{A_i} w_j = \delta_{ij}$. The matrix

$$\Omega = \left[\int_{B_i} w_j \right]$$

is the *period matrix* of $\mathcal{X}$. The columns of the matrix $[I \, | \Omega]$ form a lattice Λ in $\mathbb{C}^g$ and the Jacobian of $\mathcal{X}$ is Jac $(\mathcal{X}) = \mathbb{C}^g / \Lambda$. Let $\mathbb{H}_g$ be the *Siegel upper-half space* and $Sp_4(\mathbb{Z})$ is the *symplectic group*. Then $\Omega \in \mathbb{H}_g$.

PROPOSITION 1. *Two period matrices Ω, Ω' define isomorphic principally polarized abelian varieties if and only if they are in the same orbit under the action of $Sp_4(\mathbb{Z})$ on $\mathbb{H}_g$.*

Hence, there is an injection

$$\mathcal{M}_2 \hookrightarrow \mathbb{H}_2 / Sp_4(\mathbb{Z}) =: \mathcal{A}_2$$

For any $z \in \mathbb{C}^2$ and $\tau \in \mathbb{H}_2$ *Riemann's theta function* is defined as

$$\theta(z, \tau) = \sum_{u \in \mathbb{Z}^2} e^{\pi i (u^t \tau u + 2 u^t z)}$$

where u and z are 2−dimensional column vectors and the products involved in the formula are matrix products. The fact that the imaginary part of τ is positive makes the series absolutely convergent over any compact sets. Therefore, the function is analytic. The theta function is holomorphic on $\mathbb{C}^2 \times \mathbb{H}_2$ and satisfies

$$\theta(z + \tau u) = \theta(z, \tau), \quad \theta(z + \tau, \tau u) = e^{-\pi i (u^t \tau u + 2 z^t u)} \cdot \theta(z, \tau),$$

where $u \in \mathbb{Z}^2$. Any point $e \in$ Jac $(\mathcal{X})$ can be written uniquely as $e = (b, a) \begin{pmatrix} 1_2 \\ \Omega \end{pmatrix}$, where $a, b \in \mathbb{R}^2$ are row vectors. We shall use the notation $[e] = \begin{bmatrix} a \\ b \end{bmatrix}$ for the characteristic of e. For any $a, b \in \mathbb{Q}^2$, the theta function with rational characteristics is defined as

$$\theta \begin{bmatrix} a \\ b \end{bmatrix} (z, \tau) = \sum_{u \in \mathbb{Z}^2} e^{\pi i ((u+a)^t \tau (u+a) + 2(u+a)^t (z+b))}.$$

When the entries of column vectors a^t and b^t are from the set $\{0, \frac{1}{2}\}$, then the characteristics $\begin{bmatrix} a \\ b \end{bmatrix}$ are called the *half-integer characteristics.* The corresponding theta functions with rational characteristics are called *theta characteristics.* A scalar obtained by evaluating a theta characteristic at $z = 0$ is called a *theta constant.* Any half-integer characteristic is given by

$$\mathfrak{m} = \frac{1}{2} m = \frac{1}{2} \begin{pmatrix} m_1 & m_2 \\ m_1' & m_2' \end{pmatrix}$$

where $m_i, m_i' \in \mathbb{Z}$. For $\gamma = \begin{bmatrix} \gamma' \\ \gamma'' \end{bmatrix} \in \frac{1}{2}\mathbb{Z}^4 / \mathbb{Z}^4$ we define

$$e_*(\gamma) = (-1)^{4(\gamma')^t \gamma''}.$$

Then,

$$\theta[\gamma](-z,\tau) = e_*(\gamma)\theta[\gamma](z,\tau).$$

We say that γ is an **even** (resp. **odd**) characteristic if $e_*(\gamma) = 1$ (resp. $e_*(\gamma) = -1$).

For any genus 2 curve we have six odd theta characteristics and ten even theta characteristics. The following are the sixteen theta characteristics, where the first ten are even and the last six are odd. For simplicity, we denote them by $\theta_i = \begin{bmatrix} a \\ b \end{bmatrix}$ instead of $\theta_i \begin{bmatrix} a \\ b \end{bmatrix} (z,\tau)$ where $i = 1, \ldots, 10$ for the even theta functions.

$$\theta_1 = \begin{bmatrix} 0 & 0 \\ 0 & 0 \end{bmatrix}, \theta_2 = \begin{bmatrix} 0 & 0 \\ \frac{1}{2} & \frac{1}{2} \end{bmatrix}, \theta_3 = \begin{bmatrix} 0 & 0 \\ \frac{1}{2} & 0 \end{bmatrix}, \theta_4 = \begin{bmatrix} 0 & 0 \\ 0 & \frac{1}{2} \end{bmatrix}, \theta_5 = \begin{bmatrix} \frac{1}{2} & 0 \\ 0 & 0 \end{bmatrix},$$

$$\theta_6 = \begin{bmatrix} \frac{1}{2} & 0 \\ 0 & \frac{1}{2} \end{bmatrix}, \theta_7 = \begin{bmatrix} 0 & \frac{1}{2} \\ 0 & 0 \end{bmatrix}, \theta_8 = \begin{bmatrix} \frac{1}{2} & \frac{1}{2} \\ 0 & 0 \end{bmatrix}, \theta_9 = \begin{bmatrix} 0 & \frac{1}{2} \\ \frac{1}{2} & 0 \end{bmatrix}, \theta_{10} = \begin{bmatrix} \frac{1}{2} & \frac{1}{2} \\ \frac{1}{2} & \frac{1}{2} \end{bmatrix},$$

and the odd theta functions correspond to the following characteristics

$$\begin{bmatrix} 0 & \frac{1}{2} \\ 0 & \frac{1}{2} \end{bmatrix}, \begin{bmatrix} 0 & \frac{1}{2} \\ \frac{1}{2} & \frac{1}{2} \end{bmatrix}, \begin{bmatrix} \frac{1}{2} & 0 \\ \frac{1}{2} & 0 \end{bmatrix}, \begin{bmatrix} \frac{1}{2} & \frac{1}{2} \\ \frac{1}{2} & 0 \end{bmatrix}, \begin{bmatrix} \frac{1}{2} & 0 \\ \frac{1}{2} & \frac{1}{2} \end{bmatrix}, \begin{bmatrix} \frac{1}{2} & \frac{1}{2} \\ 0 & \frac{1}{2} \end{bmatrix}$$

The complete set of thetanulls above are not independent, their relations are given via Frobenious relations. There are four theta constants which generate all the others, namely **fundamental theta constants** θ_1, θ_2, θ_3, θ_4; see [**24**] for details. The following is Igusa's result, which is valid for any $g \geq 2$. We only state it for $g = 2$.

THEOREM 3. *The complete set of theta constants uniquely determine the isomorphism class of a principally polarized abelian variety of dimension 2.*

For curves of genus 2 this can be made more precise. Let a genus 2 curve in Rosenheim form be given by

$$Y^2 = X(X-1)(X-\lambda)(X-\mu)(X-\nu). \tag{2}$$

By the sa called Picard's lemma λ, μ, ν can be written as follows:

$$\lambda = \frac{\theta_1^2\theta_3^2}{\theta_2^2\theta_4^2}, \quad \mu = \frac{\theta_3^2\theta_8^2}{\theta_4^2\theta_{10}^2}, \quad \nu = \frac{\theta_1^2\theta_8^2}{\theta_2^2\theta_{10}^2}. \tag{3}$$

We can determine an equation of the curve in terms of the fundamental thetas as follows:

PROPOSITION 2 ([**24**]). *Every genus two curve can be written in the form:*

$$y^2 = x\,(x-1)\left(x - \frac{\theta_1^2\theta_3^2}{\theta_2^2\theta_4^2}\right)\left(x^2 - \frac{\theta_2^2\,\theta_3^2 + \theta_1^2\,\theta_4^2}{\theta_2^2\,\theta_4^2}\cdot \alpha\, x + \frac{\theta_1^2\theta_3^2}{\theta_2^2\theta_4^2}\,\alpha^2\right), \tag{4}$$

where $\alpha = \frac{\theta_8^2}{\theta_{10}^2}$ *and in terms of* $\theta_1, \ldots, \theta_4$ *is given by*

$$\alpha^2 + \frac{\theta_1^4 + \theta_2^4 - \theta_3^4 - \theta_4^4}{\theta_1^2\theta_2^2 - \theta_3^2\theta_4^2}\,\alpha + 1 = 0$$

Furthermore, if $\alpha = \pm 1$ *then* $\mathcal{X}$ *has an extra involution.*

From the above we have that $\theta_8^4 = \theta_{10}^4$ implies that $\mathcal{X}$ has an extra involution. Hence, the Klein viergrouppe $V_4 \hookrightarrow Aut(\mathcal{X})$. The last part of the lemma above shows that if $\theta_8^4 = \theta_{10}^4$ then all coefficients of the genus 2 curve are given as rational functions of the 4 fundamental theta functions. Such fundamental theta functions determine the field of moduli of the given curve. Hence, the curve is defined over its field of moduli.

COROLLARY 3. *Let $\mathcal{X}$ be a genus 2 curve which has an extra involution. Then $\mathcal{X}$ is defined over its field of moduli.*

We will revisit the curves defined over their field of moduli again in the coming sections.

4.2. Siegel modular forms. Here we define Siegel modular forms ψ_4, ψ_6, χ_{10}, χ_{12}, of degree 4, 6, 10, and 12, as in [**11**, pg. 848].

$$\begin{aligned}
2^2 \cdot \psi_4 &= \sum (\theta_m)^8 \\
2^2 \cdot \psi_6 &= \sum_{syzygous} \pm (\theta_{m_1}\theta_{m_2}\theta_{m_3})^4 \\
-2^{14} \cdot \chi_{10} &= \prod (\theta_m)^2 \\
2^{17} \cdot 3 \cdot \chi_{12} &= \sum (\theta_{m_1}\theta_{m_2} \cdots \theta_{m_6})^4 \\
2^{39} \cdot 5^3 \cdot \chi_{35} &= \left(\prod \theta_m\right) \left(\sum_{azygous} \pm (\theta_{m_1}\theta_{m_2}\theta_{m_3})^{20} \right).
\end{aligned} \tag{5}$$

In definition of χ_{12} the summation is taken over all Göpel systems as explained in [**24**], where all the Göpel systems are displayed.

Theta constants provide a complete system of invariants for isomorphism classes of principally polarized varieties of dimension $g = 2$. But there are two main issues with this approach: First, these invariants are not independent. This can be fixed via the fundamental theta constants as in Prop. 2, however computationally things get difficult when we try to express all the results in terms of $\theta_1, \theta_2, \theta_3, \theta_4$. Secondly, and more importantly, they are defined analytically. Naturally, we would like to have algebraically defined invariants.

5. Algebraic invariants

Let $f(x, z)$ be a binary sextic defined over a field k, char $k = 0$, given by

$$\begin{aligned}
f(x, z) &= a_0 x^6 + a_1 x^5 z + \cdots + a_6 z^6 \\
&= (z_1 x - x_1 z)(z_2 x - x_2 z) \ldots (z_6 x - x_6 z)
\end{aligned} \tag{6}$$

A **covariant** I of $f(x, z)$ is a homogenous polynomial in x, z with coefficients in $k[a_0, \ldots, a_{2g+2}]$. The **order** of I is the degree of I as a polynomial in x, z and the **degree** of I is the degree of I as a polynomial in $k[a_0, \ldots, a_{2g+2}]$. An **invariant** is a covariant of order zero. The binary form $f(x, z)$ is a covariant of order $2g + 2$ and degree 1. Throughout this paper we will use as basic references [**7**], [**4**], and [**9**–**11**].

5.1. Invariants and covariants via transvections. For any two binary forms f and g the symbol $(f, g)_r$ denotes the r-transvection. Notice that the transvections are conveniently computed in terms of the coefficients of the binary forms.

Let $f(x,z)$ be a binary sextic as in Eq. (6) and consider the following covariants

$$\begin{aligned} \Delta &= ((f,f)_4,(f,f)_4)_2, & Y_1 &= (f,(f,f)_4)_4 \\ Y_2 &= ((f,f)_4,Y_1)_2, & Y_3 &= ((f,f)_4,Y_2)_2 \end{aligned} \tag{7}$$

The **Clebsch invariants** A, B, C, D are defined as follows

$$\begin{aligned} A &= (f,f)_6, & B &= ((f,f)_4,(f,f)_4)_4, \\ C &= ((f,f)_4,\Delta)_4, & D &= (Y_3,Y_1)_2 \end{aligned} \tag{8}$$

see Clebsch [**7**] or Bolza [**4**, Eq. (7), (8), pg. 51] for details.

We display the invariants A, B, C, D in terms of the coefficients in the Appendix. The following result is elementary but very important in our computations.

5.2. Root differences. Let $f(x,z)$ be a binary sextic as above and set $D_{ij} := \begin{pmatrix} x_i & x_j \\ z_i & z_j \end{pmatrix}$. For $\tau \in SL_2(k)$, we have

$$\tau(f) = (z_1^{'}x - x_1^{'}z)\dots(z_6^{'}x - x_6^{'}z), \qquad with \qquad \begin{pmatrix} x_i^{'} \\ z_i^{'} \end{pmatrix} = \tau^{-1}\begin{pmatrix} x_i \\ z_i \end{pmatrix}.$$

Clearly D_{ij} is invariant under this action of $SL_2(k)$ on $\mathbb{P}^1$. Let $\{i,j,k,l,m,n\} = \{1,2,3,4,5,6\}$. Treating a_i as variables, we construct the following elements in the ring of invariants $\mathcal{R}_6$

$$\mathfrak{A} = a_0^2 \prod_{fifteen} (12)^2(34)^2(56)^2 = \sum_{i<j,k<l,m<n} D_{ij}^2 D_{kl}^2 D_{mn}^2$$

$$\mathfrak{B} = a_0^4 \prod_{ten} (12)^2(23)^2(31)^2(45)^2(56)^2(64)^2 = \sum_{\substack{i<j,j<k,\\ l<m,m<n}} D_{ij}^2 D_{jk}^2 D_{ki}^2 D_{lm}^2 D_{mn}^2 D_{nl}^2$$

$$\begin{aligned} \mathfrak{C} &= a_0^6 \prod_{sixty} (12)^2(23)^2(31)^2(45)^2(56)^2(64)^2(14)^2(25)^2(36)^2 \\ &= \sum_{\substack{i<j,j<k,l<m,m<n\\ i<l',j<m',k<n'\\ l',m',n'\in\{l,m,n\}}} D_{ij}^2 D_{jk}^2 D_{ki}^2 D_{lm}^2 D_{mn}^2 D_{nl}^2 D_{il'}^2 D_{jm'}^2 D_{kn'}^2 \end{aligned} \tag{9}$$

$$\mathfrak{D} = a_0^{10} \prod_{i<j} (ij)^2$$

These invariants, sometimes called **integral invariants**, are defined in [**9**, pg. 620] where they are denoted by A, B, C, D. Incidentally even Clebsch invariants which are defined next are also denoted by A, B, C, D by many authors.

To quote Igusa *"if we restrict to integral invariants, the discussion will break down in characteristic 2 simply because Weierstrass points behave badly under reduction modulo 2"*; see [**9**, pg. 621]. Next we define invarints which will work in every characteristic.

5.3. Igusa invariants. In [**9**, pg. 622] Igusa defined what he called **basic arithmetic invariants**, which are now commonly known as **Igusa invariants**

$$J_2 = \frac{1}{2^3}\mathfrak{A}, \qquad J_4 = \frac{1}{2^5 \cdot 3}(4J_2^2 - \mathfrak{B}),$$
$$J_6 = \frac{1}{2^6 \cdot 3^2}(8J_2^3 - 160J_2J_4 - \mathfrak{C}), \qquad J_{10} = \frac{1}{2^{12}}\mathfrak{D}$$

While most of the current literature on genus 2 curves uses invariants $\mathfrak{A}, \mathfrak{B}, \mathfrak{C}, \mathfrak{D}$, which are now most commonly labeled as I_2, I_4, I_6, I_{10}, Igusa went to great lengths in [**9**] to define J_2, J_4, J_6, J_{10} and to show that they also work in characteristic 2.

LEMMA 2. *J_{2i} are homogeneous elements in $\mathcal{R}_6$ of degree $2i$, for i = 1,2,3,5.*

A degree $d \geq 2$ binary form $f(x, z)$ is called **semistable** if it has no root of multiplicity $> \frac{d}{2}$.

LEMMA 3. *A sextic has a root of multiplicity at least four if and only if the basic invariants vanish simultaneously.*

So a sextic for which all the basic invariants vanish simultaneously is not semistable.

Throughout this paper, we will use these invariants J_2, J_4, J_6, J_{10}. For a binary form $f(x, z) = \sum_{i=0}^{6} a_i \, x^i z^{6-i}$ as in Eq. (6) we display these invariants in terms of the coefficients $a_0, \dots, a_6$ in B.

Invariants $\{J_{2i}\}$ are homogeneous polynomials of degree $2i$ in $k[a_0, \dots, a_6]$, for $i = 1, 2, 3, 5$. These J_{2i} are invariant under the natural symmetries of the roots, namely the action of S_6 on the set of the roots. They are also natural on the action of $SL_2(k)$ on the sextic.

LEMMA 4. [**9**, Cor. pg. 632] *Two genus 2 curves C and C' are isomorphic if and only if there exists an $r \neq 0$ such that*

$$J_{2i}(C) = r^{2i} \cdot J_{2i}(C'), \qquad \textit{for} \quad i = 1, 2, 3, 5.$$

REMARK 2. There is a lot of confusion in the literature about the definitions of the generators of $\mathcal{R}_6$. Many authors have used different names and different notations for generators of $\mathcal{R}_6$. Moreover, many times the symbols $\mathfrak{A}, \mathfrak{B}, \mathfrak{C}, \mathfrak{D}$ are defined by scaling with different constants by many authors. Hence, sometimes equations in terms of these invariants might not be the same. To avoid any confusion, we display the expressions of j_2, J_4, J_6, j_{10} in terms of the coefficients of the binary sextic in Eq. (B).

Igusa invariants are expressed in terms of the Clebsch invariants as follows:

$$\begin{aligned}
J_2 &= \, - 2^3 \cdot 3 \cdot 5 \, A \, , \\
J_4 &= 2^3 \cdot 2^5 \, (75B - 8A^2) \\
J_6 &= 2^2 \cdot 3^3 \cdot 5 \, \left(16 \, A^3 - 200 \, A \, B + 375 \, C\right) \\
J_{10} &= 2^3 \cdot 4 \, \big(-384 \, A^5 + 6000 \, A^3 \, B + 10000 \, A^2 \, C - 18750 \, A \, B^2 \\
&\qquad -37500 \, B \, C - 28125 \, D\big)
\end{aligned} \tag{10}$$

Conversely, the invariants (A, B, C, D) are polynomial expressions in the Igusa invariants (J_2, J_4, J_6, J_{10}) with rational coefficients are displayed in Eq. B in the Appendix.

Thus, the invariants of a sextic define a point in a weighted projective space $[J_2 : J_4 : J_6 : J_{10}] \in \mathbb{WP}^3_{(2,4,6,10)}$. It was shown in [**12**] that points in the projective variety Proj $\mathbb{C}[J_2, J_4, J_6, J_{10}]$ which are not on $J_2 = 0$ form the variety $\mathcal{U}_6$ of moduli of sextics. Equivalently, points in the weighted projective space $\{[J_2 : J_4 : J_6 : J_{10}] \in \mathbb{WP}^3_{(2,4,6,10)} : J_{10} \neq 0\}$ are in one-to-one correspondence with isomorphism classes of sextics.

5.4. Absolute invariants. Dividing any $SL_2(k)$ invariant by another one of the same degree gives an invariant under $GL_2(k)$ action.

We follow [**10**, pg. 181] and define the following absolute invariants

$$i_1 := 144\frac{J_4}{J_2^2}, \quad i_2 := -1728\frac{J_2J_4 - 3J_6}{J_2^3}, \quad i_3 := 486\frac{J_{10}}{J_2^5} \tag{11}$$

for $J_2 \neq 0$. Notice that Igusa denotes them by x_1, x_2, x_3.

THEOREM 4. [**10**, Theorem 3] *The three absolute invariants can be expressed by the four modular forms in the form*

$$i_1 = \frac{\psi_4\chi_{10}^2}{\chi_{12}^2}, \quad i_2 = \frac{\psi_6\,\chi_{10}^3}{\chi_{12}^3}, \quad i_3 = \frac{\chi_{10}^6}{\chi_{12}^5} \tag{12}$$

Hence, the theorem says that every modular form, in the degree 2 case, is expressed in the Eisenstein series of weight four, six, ten, and twelve. It is the main result of [**10**].

There is another set of absolute invariants

$$j_1 := \frac{\mathfrak{A}^5}{\mathfrak{D}}, \quad j_2 := \frac{\mathfrak{B}\mathfrak{A}^3}{\mathfrak{D}}, \quad j_3 := \frac{\mathfrak{C}\mathfrak{A}^2}{\mathfrak{D}}, \tag{13}$$

The popularity of invariants j_1, j_2, j_3 is due to P. Van Wamelen who used them in [**27**] and later implemented them in Magma. They have no advantages over the invariants i_1, i_2, i_3 since both sets of invariants are not defined for $J_2 = 0$ or equivalently $\mathfrak{A} = 0$. Indeed, we could not find a direct proof that such invariants generate the field of invariants for $\mathcal{R}_6$, even though it is probably true. The results of other authors who have worked with j_1, j_2, j_3 can be converted into our results and vice-versa using the formulas

$$i_1 = 144\,\frac{j_2}{j_1}, \quad i_2 = -1728\,\frac{j_2 - 3j_3}{j_1}, \quad i_3 = 486\frac{1}{j_1}$$

and conversely

$$j_1 = 486\,\frac{1}{i_3}, \quad j_2 = \frac{27}{8}\frac{i_1}{i_3}, \quad j_3 = \frac{3}{32}\frac{i_2 + 12i_1}{i_3}$$

REMARK 3. It is to be noted that for computational purposes i_1, i_2, i_3 are much better than j_1, j_2, j_3 since they are of lower degrees. Especially, for our purposes it is very important that the moduli point $\mathfrak{p} = (r, i_1, i_2, i_3)$ (cf. next section) it is expressed in as small numbers as possible.

There is another set of invariants defined by Igusa in [**9**],

$$t_1 = \frac{J_2^5}{J_{10}}, \quad t_2 = \frac{J_4^5}{J_{10}^2}, \quad t_3 = \frac{J_6^5}{J_{10}^3}$$

which are defined everywhere in the moduli space. Due to their high degrees, they become difficult to use especially when someone wants to do symbolic computations.

In the case $J_2 = 0$ we define

$$a_1 := \frac{J_4 \cdot J_6}{J_{10}}, \quad a_2 := \frac{J_6 \cdot J_{10}}{J_4^4} \tag{14}$$

to determine genus two fields with $J_2 = 0$, $J_4 \neq 0$, and $J_6 \neq 0$ up to isomorphism. They were used in [**8, 16, 19, 23**] and others. Moreover, when $J_2 = 0, J_4 = 0, J_6 \neq 0$ we have $\frac{J_6^5}{J_{10}^3}$ as an invariant and when $J_2 = 0, J_6 = 0, J_4 \neq 0$ we have $\frac{J_4^5}{J_{10}^2}$ as an invariant.

All our computations are made in terms of absolute invariants i_1, i_2, i_3 or the Igusa invariants J_2, J_4, J_6, J_{10}. In the Appendix, we provide formulas how to convert back and forth among all sets of invariants. When $J_2 = 0$ we will use invariants t_1, t_2, t_3 instead, in this case $t_1 = 0$, so this locus is determined by t_2 and t_3.

5.5. Representing a point in the moduli space. In creating a large database of genus two curves we need a way to identify uniquely a point $\mathfrak{p} \in \mathcal{M}_2$. Such point would ideally be defined everywhere, computationally feasable, and should not involve large integers. Of course, the representation $\mathfrak{p} = (t_1, t_2, t_3)$ is unique, but these invariants are very large and extremely difficult to handle in a large database. The set of invariants (i_1, i_2, i_3) is nicer, but they are not defined for $J_2 = 0$.

To find a compromise and simplify the implementation in [**3**], for a given genus 2 curve $\mathcal{X}$ we define the corresponding **moduli point** $\mathfrak{p} = [\mathcal{X}]$ to be

$$\mathfrak{p} = \begin{cases} (i_1, i_2, i_3) \ \ if \ J_2 \neq 0 \\ (t_1, t_2, t_3) \ \ if \ J_2 = 0 \end{cases} \tag{15}$$

This moduli point $\mathfrak{p}$ together with J_2 will be the *key* in our dictionary of genus 2 curves. It will be stored as a 4-tuple $(r, \mathfrak{p})$ as explained before.

For $I_2 \neq 0$ we use the variables i_1, i_2, i_3 to write

$$\left[J_2 : J_4 : J_6 : J_{10}\right] = \left[1 : \frac{1}{2^4\, 3^2}\, i_1 : \frac{1}{2^6\, 3^4}\, i_2 + \frac{1}{2^4\, 3^3}\, i_1 : \frac{1}{2 \cdot 3^5}\, i_3\right] \in \mathbb{WP}^3_{(2,4,6,10)},$$

see [**15**] *pg. 13 for details. Notice that i_1, i_2, i_3 are denoted by $\mathbf{x}_1, \mathbf{x}_2, \mathbf{x}_3$ and J_2, J_4, J_6, J_{10} respectively by I_2, I_4, I_6, I_{10} in [**15**].

5.6. Relations among different sets of invariants. There are three sets of $SL_2(k)$ invariants most commonly used, namely (A, B, C, D), $(\mathfrak{A}, \mathfrak{B}, \mathfrak{C}, \mathfrak{D})$, and (J_2, J_4, J_6, J_{10}). Many authors will use the notation (I_2, I_4, I_6, I_{10}) for $(\mathfrak{A}, \mathfrak{B}, \mathfrak{C}, \mathfrak{D})$ due to the Magma notation (they are called *Igusa Clebsch invariants* in Magma) or some scaling of them.

Notice that

$$J_{2i} = \frac{1}{2^{4i}}\, I_{2i}, \qquad i = 1, 2, 3, 5. \tag{16}$$

For a given genus two curve $\mathcal{X}$ with equation $y^2 = f(x)$, the relations between invariants $\mathfrak{A}, \mathfrak{B}, \mathfrak{C}, \mathfrak{D}$ and even Siegel modular forms $\psi_4, \psi_6, \chi_{10}, \chi_{12}, \chi_{35}$ of $\mathcal{A}_2$ are given by Igusa in [**11**, p. 848]:

$$\begin{aligned} \mathfrak{A}(f) &= -2^3 \cdot 3\, \frac{\chi_{12}(\tau)}{\chi_{10}(\tau)}, & \mathfrak{B}(f) &= \ 2^2\, \psi_4(\tau), \\ \mathfrak{C}(f) &= -\frac{2^3}{3}\, \psi_6(\tau) - 2^5\, \frac{\psi_4(\tau)\, \chi_{12}(\tau)}{\chi_{10}(\tau)}, & \mathfrak{D}(f) &= -2^{14}\, \chi_{10}(\tau). \end{aligned} \tag{17}$$

In the Appendix are given the relations between A, B, C, D and J_2, J_4, J_6, J_{10}. Using such relations we get the absolute invariants i_1, i_2, i_3 in terms of A, B, C, D are as follows:

$$
\begin{aligned}
i_1 &= \frac{9}{10} \cdot \frac{75B - 8A^2}{A^2} \\
i_2 &= \frac{27}{50} \cdot \frac{112A^3 - 900AB - 1125C}{A^3} \\
i_3 &= \frac{81}{2^{13} \cdot 5^5} \cdot \frac{384A^5 - 6000A^3B - 10000A^2C + 18750AB^2 + 37500BC + 28125CD}{A^5}
\end{aligned}
$$

Next we see how to construct the equation of a curve starting from a moduli point $\mathfrak{p} \in \mathcal{M}_2$.

5.7. Recovering an equation of the curve from the moduli point. Some other invariants are as follows

$$
\begin{aligned}
A_{ij} &= (Y_i Y_j)_2, & (1 \leq i, j \leq 3) \\
H_{ijk} &= (fY_i)_2 (fY_j)_2 (fY_k)_2 & (1 \leq i, j, k \leq 3) \\
R &= -(Y_1 Y_2)(Y_2 Y_3)(Y_3 Y_1)
\end{aligned}
$$

Note that A_{ij} and H_{ijk} can be expressed in terms of the Clebsch invariants, see [**16**, pg. 318]. Moreover, $R^2 = \frac{1}{2} \det M$ where M is the **Clebsch matrix** defined as follows.

$$
M = \begin{bmatrix} A_{11} & A_{12} & A_{13} \\ A_{12} & A_{22} & A_{23} \\ A_{13} & A_{23} & A_{33} \end{bmatrix} \tag{18}
$$

The following is an immediate consequence of classical invariant theory

Let J_{30} denote the degree 30 invariant

$$
J_{30} := \det M
$$

An expression of J_{30} in terms of J_2, J_4, J_6, J_{10} is given in [**23**, Theorem 3], where it is also expressed in terms of the roots of the sextic and its coefficients.

Let us now compute the determinant of the minor $S = \begin{pmatrix} A_{1,1} & A_{1,2} \\ A_{2,1} & A_{2,2} \end{pmatrix}$ from the matrix $M = [A_{i,j}]$.

$$
\begin{aligned}
\det S &= A_{1,1}A_{2,2} - A_{1,2}^2 \\
&= \frac{2^{16}}{3^6 \cdot 5^4 J_2^8} \big(15\, J_2^3 J_4\, J_6 - 4\, J_2^4 J_4^2 - 175\, J_2^2 J_4^3 + 2430\, J_{10}\, J_2^3 - 9\, J_2^2 J_6^2 \\
&\quad + 1488\, J_2\, J_4^2 J_6 - 64\, J_4^4 + 113400\, J_{10}\, J_2\, J_4 - 2880\, J_4\, J_6^2 - 648000\, J_{10}\, J_6 \big)
\end{aligned}
$$

We define the new invariant

$$
\begin{aligned}
J_{16} := & 15\, J_2^3 J_4\, J_6 - 4\, J_2^4 J_4^2 - 175\, J_2^2 J_4^3 + 2430\, J_{10}\, J_2^3 - 9\, J_2^2 J_6^2 + 1488\, J_2\, J_4^2 J_6 \\
& - 64\, J_4^4 + 113400\, J_{10}\, J_2\, J_4 - 2880\, J_4\, J_6^2 - 648000\, J_{10}\, J_6
\end{aligned}
$$

which is a degree 16 invariant.

6. Automorphisms

Let $\mathcal{X}$ be a genus 2 curve defined over an algebraically closed field k. Let K be the function field of $\mathcal{X}$. Then K has exactly one genus 0 subfield of degree 2, denote it $k(x)$. Since $k(x)$ is the only genus 0 subfield of degree 2 of K, then $G = \mathrm{Aut}(K/k)$ (or equivalently $\mathrm{Aut}(\mathcal{X})$) fixes $k(x)$. It is the fixed field of the hyperelliptic involution τ in $\mathrm{Aut}(K)$. Thus τ is in the center of $\mathrm{Aut}(K)$ and $\langle\tau\rangle \triangleleft \mathrm{Aut}(K)$. The quotient group $\overline{\mathrm{Aut}}(K) = \mathrm{Aut}(K)/\langle\tau\rangle$ is called the **reduced automorphism group** of $\mathcal{X}$ (or equivalently of K).

The reduced automorphism group embeds as a finite subgroup of $PGL(2,\mathbb{C})$, therefore it is isomorphic to C_n, D_n, A_4, S_4, A_5. Hence, the full automorphism group $\mathrm{Aut}(\mathcal{X})$ is a degree 2 central extension of $\overline{\mathrm{Aut}}(\mathcal{X})$.

The extension $K/k(x)$ is ramified at exactly six distinct points, namely the points $P = \{\omega_1, \cdots, \omega_6\}$ in $\mathbb{P}^1$. The corresponding places of this points in K are called Weierstrass points of K. The group $\mathrm{Aut}(K)$ permutes the 6 Weierstrass points and the group $\overline{\mathrm{Aut}}(K)$ permutes accordingly $\{\omega_1, \cdots, \omega_6\}$ in its action on $\mathbb{P}^1$ as subgroup of $PLG(2,\mathbb{C})$. This yields an embedding $\overline{\mathrm{Aut}}(K) \hookrightarrow S_6$, so all elements of $\overline{\mathrm{Aut}}(K)$ have order ≤ 6.

In any characteristic different from 2, the automorphism group $\mathrm{Aut}(\mathcal{X})$ is isomorphic to one of the groups given by the following theorem from [**23**].

THEOREM 5 ([**23**]). *The automorphism group G of a genus 2 curve $\mathcal{X}$ in characteristic $\neq 2$ is isomorphic to $\mathbb{Z}_2$, $\mathbb{Z}_{10}$, V_4, D_8, D_{12}, $SL_2(3)$, $GL_2(3)$, or 2^+S_5. The case when $G \cong 2^+S_5$ occurs only in characteristic 5. If $G \cong SL_2(3)$ (resp., $GL_2(3)$) then $\mathcal{X}$ has equation $Y^2 = X^6 - 1$ (resp., $Y^2 = X(X^4 - 1)$). If $G \cong \mathbb{Z}_{10}$ then $\mathcal{X}$ has equation $Y^2 = X^6 - X$.*

An **elliptic involution** of K is an involution in G which is different from z_0 (the hyperelliptic involution). Thus the elliptic involutions of G are in 1-1 correspondence with the elliptic subfields of K of degree 2 (by the Riemann-Hurwitz formula). For the number of elliptic involutions of a genus 2 curve see [**23**].

6.1. Automorphisms in terms of theta functions.

The locus $\mathcal{L}_2$ of genus 2 curves $\mathcal{X}$ which have an elliptic involution is a closed subvariety of $\mathcal{M}_2$.

LEMMA 5. *Let $\mathcal{X}$ be a genus 2 curve. Then $Aut(\mathcal{X}) \cong V_4$ if and only if the theta functions of $\mathcal{X}$ satisfy*

(19)
$$
\begin{aligned}
&(\theta_1^4-\theta_2^4)(\theta_3^4-\theta_4^4)(\theta_8^4-\theta_{10}^4)(-\theta_1^2\theta_3^2\theta_8^2\theta_2^2-\theta_1^2\theta_2^2\theta_4^2\theta_{10}^2+\theta_1^4\theta_3^2\theta_{10}^2+\theta_3^2\theta_2^4\theta_{10}^2)\\
&(\theta_3^2\theta_8^2\theta_2^2\theta_4^2-\theta_2^2\theta_4^4\theta_{10}^2+\theta_1^2\theta_3^2\theta_4^2\theta_{10}^2-\theta_3^4\theta_2^2\theta_{10}^2)(-\theta_8^4\theta_3^2\theta_2^2+\theta_8^2\theta_2^2\theta_{10}^2\theta_4^2+\theta_1^2\theta_3^2\theta_8^2\theta_{10}^2-\theta_3^2\theta_2^2\theta_{10}^4)\\
&(-\theta_1^2\theta_8^4\theta_4^2-\theta_1^2\theta_{10}^4\theta_4^2+\theta_8^2\theta_2^2\theta_{10}^2\theta_4^2+\theta_1^2\theta_3^2\theta_8^2\theta_{10}^2)(-\theta_1^2\theta_8^2\theta_3^2\theta_4^2+\theta_1^2\theta_{10}^2\theta_4^4+\theta_1^2\theta_3^4\theta_{10}^2-\theta_3^2\theta_2^2\theta_{10}^2\theta_4^2)\\
&(-\theta_1^2\theta_8^2\theta_2^2\theta_4^2+\theta_1^4\theta_{10}^2\theta_4^2-\theta_1^2\theta_3^2\theta_2^2\theta_{10}^2+\theta_2^4\theta_4^2\theta_{10}^2)(-\theta_8^4\theta_2^2\theta_4^2+\theta_1^2\theta_8^2\theta_{10}^2\theta_4^2-\theta_2^2\theta_{10}^4\theta_4^2+\theta_3^2\theta_8^2\theta_2^2\theta_{10}^2)\\
&(\theta_1^4\theta_8^2\theta_4^2-\theta_1^2\theta_2^2\theta_4^2\theta_{10}^2-\theta_1^2\theta_3^2\theta_8^2\theta_2^2+\theta_8^2\theta_2^4\theta_4^2)(\theta_1^4\theta_3^2\theta_8^2-\theta_1^2\theta_8^2\theta_2^2\theta_4^2-\theta_1^2\theta_3^2\theta_2^2\theta_{10}^2+\theta_3^2\theta_8^2\theta_2^4)\\
&(\theta_1^2\theta_8^4\theta_3^2-\theta_1^2\theta_8^2\theta_{10}^2\theta_4^2+\theta_1^2\theta_3^2\theta_{10}^4-\theta_3^2\theta_8^2\theta_2^2\theta_{10}^2)(\theta_1^2\theta_8^2\theta_4^4-\theta_1^2\theta_3^2\theta_4^2\theta_{10}^2+\theta_1^2\theta_3^4\theta_8^2-\theta_3^2\theta_8^2\theta_2^2\theta_4^2)=0
\end{aligned}
$$

This was done in [**24**]. We would like to express the conditions of the previous lemma in terms of the fundamental theta constants only.

LEMMA 6. *Let $\mathcal{X}$ be a genus 2 curve. Then we have the following:*

i: $V_4 \hookrightarrow Aut(\mathcal{X})$ *if and only if the fundamental theta constants of $\mathcal{X}$ satisfy*

$$
\begin{aligned}
&(\theta_3^4-\theta_4^4)(\theta_1^4-\theta_3^4)(\theta_2^4-\theta_4^4)(\theta_1^4-\theta_4^4)(\theta_3^4-\theta_2^4)(\theta_1^4-\theta_2^4)\\
&(-\theta_4^2+\theta_3^2+\theta_1^2-\theta_2^2)(\theta_4^2-\theta_3^2+\theta_1^2-\theta_2^2)(-\theta_4^2-\theta_3^2+\theta_2^2+\theta_1^2)(\theta_4^2+\theta_3^2+\theta_2^2+\theta_1^2)\\
&({\theta_1}^4{\theta_2}^4+{\theta_3}^4{\theta_2}^4+{\theta_1}^4{\theta_3}^4-2\,\theta_1^2\theta_2^2\theta_3^2\theta_4^2)(-{\theta_3}^4{\theta_2}^4-{\theta_2}^4{\theta_4}^4-{\theta_3}^4{\theta_4}^4+2\,\theta_1^2\theta_2^2\theta_3^2\theta_4^2)\\
&({\theta_2}^4{\theta_4}^4+{\theta_1}^4{\theta_2}^4+{\theta_1}^4{\theta_4}^4-2\,\theta_1^2\theta_2^2\theta_3^2\theta_4^2)({\theta_1}^4{\theta_4}^4+{\theta_3}^4{\theta_4}^4+{\theta_1}^4{\theta_3}^4-2\,\theta_1^2\theta_2^2\theta_3^2\theta_4^2)=0
\end{aligned}
\tag{20}
$$

ii: $D_8 \hookrightarrow Aut(\mathcal{X})$ *if and only if the fundamental theta constants of $\mathcal{X}$ satisfy Eq. (3) in* [**24**]

iii: $D_6 \hookrightarrow Aut(\mathcal{X})$ *if and only if the fundamental theta constants of $\mathcal{X}$ satisfy Eq. (4) in* [**24**]

Hence, one can determine very easily the automorphism group of the curve once its thetanulls are known. However, this is not quite easy since thetanulls are defined in terms of the complex integrals. However, since the Siegel modular forms are expressed in terms of the thetanulls as in Eq. (5), we should be able to express such loci in terms of the modular forms.

Instead, it is much easier to determine the automorphism groups in terms of the algebraic invariants which are given in terms of the coefficients of the curve, but are invariant of the equation of the curve. In the next section we will express such loci in terms of the absolute invariants i_1, i_2, i_3. Then, using formulas in Eq.(12) we can always express these equations in terms of the modular forms.

6.2. Automorphisms in terms of algebraic invariants. The locus of genus 2 curves with an extra involution is computed in [**23**, Theorem 3] in terms of J_2, J_4, J_6, J_{10}. It corresponds precisely to the locus $\det M = 0$.

PROPOSITION 3. *Let $\mathfrak{p} \in \mathcal{M}_2$. The following hold:*

(1) *If $J_{30}(\mathfrak{p}) \neq 0$, then $|\operatorname{Aut}(\mathfrak{p})| = 2$ or $\mathfrak{p}$ correspond to the curve $y^2 = x(x^5-1)$. Moreover, $J_{30}(\mathfrak{p}) = 0$ if and only if $V_4 \hookrightarrow \operatorname{Aut}(\mathfrak{p})$.*

(2) *If $J_{30}(\mathfrak{p}) = 0$ and $J_{16}(\mathfrak{p}) = 0$, then $V_4 \hookrightarrow \overline{\operatorname{Aut}}(\mathfrak{p})$.*

Part i) was known to Clebsch and proved independently in [**23**, Theorem 3]. The proof of 2) can be found in [**23**, Lemma 3]. In the following proposition we give conditions on the absolute invariants i_1, i_2 and i_3 for each automorphism to happen.

PROPOSITION 4. *Let $\mathfrak{p} = (i_1, i_2, i_3) \in \mathcal{M}_2 \setminus \{J_2 = 0\}$. Then the following hold:*

i) *If $\mathfrak{p} = (0,0,0)$ then* $\operatorname{Aut}(\mathfrak{p}) \cong C_{10}$

ii) *If $\mathfrak{p} = (\frac{81}{20}, -\frac{729}{200}, \frac{729}{25600000})$, then* $\operatorname{Aut}(\mathfrak{p}) \cong SL_2(3)$

iii) *If $\mathfrak{p} = (-\frac{36}{5}, \frac{1512}{25}, \frac{243}{200000})$, then* $\operatorname{Aut}(\mathfrak{p}) \cong GL_2(3)$

iv) *If the following conditions are satisfied*

$$
\begin{aligned}
&27i_2^4-9i_2^4i_1+9459597312000i_3^2i_1^2-111451255603200i_3^2i_1+55240704i_3i_1^4\\
&\quad-161243136i_3i_1^3+27i_1^6+2407347121029 12i_3^2+2641807540224000000i_3^3\\
&-9i_1^7+18i_2^2i_1^4-54i_1^3i_2^2-161243136i_3i_2^2+12441600i_3i_2^3-107495424i_3i_2i_1^2\\
&\quad+52254720i_3i_2^2i_1-2i_2i_1^6+4i_2^3i_1^3-20639121408000i_3^2i_2+8294400i_3i_2^2i_1^2\\
&\qquad-331776i_3i_1^5+2866544640000i_3^2i_1i_2+47278080i_3i_1^3i_2-2i_2^5=0\\
&\qquad\qquad-243i_1^2+80i_1^3-1458i_2+540i_2i_1+100i_2^2=0
\end{aligned}
\tag{21}
$$

and $\mathfrak{p}$ is not one of the cases ii) and iii), then $\operatorname{Aut}(\mathfrak{p}) \cong D_4$

v) *If the following conditions are satisfied*

$$
\begin{aligned}
&3888i_1 + 432i_2 - 1188i_1^2 + 5i_1^3 - 25i_2^2 - 360i_2i_1 = 0 \\
&26873856i_3 + 5184000i_3i_1^2 - 9331200i_3i_1 - 149299200000i_3^2 - 729i_1^2 - \\
&\qquad 27i_1^4 + 243i_1^3 + i_1^5 = 0
\end{aligned} \tag{22}
$$

and $\mathfrak{p}$ *is not one of the cases ii) and iii), then* $\mathrm{Aut}(\mathfrak{p}) \cong D_6$

vi) *If the following is satisfied*

$$
\begin{aligned}
&-27i_1^6 - 9459597312000i_3^2i_1^2 + 20639121408000i_3^2i_2 + 111451255603200i_3^2i_1 \\
&-240734712102912i_3^2 - 55240704i_3i_1^4 - 18i_1^4i_2^2 - 8294400i_3i_2^2i_1^2 - 47278080i_3i_2i_1^3 \\
&-2641807540224000000i_3^3 - 2866544640000i_3^2i_2i_1 + 2i_1^6i_2 - 4i_1^3i_2^3 + 9i_1^7 + 331776i_3i_1^5 \\
&+107495424i_3i_2i_1^2 - 27i_2^4 + 9i_1i_2^4 - 52254720i_3i_2^2i_1 + 2i_2^5 + 161243136i_3i_2^2 \\
&+161243136i_3i_1^3 - 12441600i_3i_2^3 + 54i_1^3i_2^2 = 0
\end{aligned} \tag{23}
$$

and $\mathfrak{p}$ *is not one of the above cases, then* $\mathrm{Aut}(\mathfrak{p}) \cong V_4$.

PROOF. The proof is computational and straightforward. The reader can check [**18**] or [**23**] for details. □

If $\mathcal{X}$ has extra involutions (i.e. it has automorphisms other then the hyperelliptic involution), then X is isomorphic to a curve given by

$$y^2 = x^6 + ax^4 + bx^2 + 1 \tag{24}$$

for appropriate values of a, b (i.e. the discriminant is nonzero). Such form of genus 2 curves with automorphisms was know to XIX century mathematicians, but it has been used in the last decade quite often mostly due to Shaska and Völklein paper [**23**] which first appeared in 2000.

In [**23**] for curves with automorphism were defined the dihedral invariants

$$u := ab, \qquad v := a^3 + b^3 \tag{25}$$

which give a birational parametrization of this locus $\mathcal{L}_2$ which is a 2-dimensional subvariety of $\mathcal{M}_2$. It was the first time that such parametrization was discovered and since it has become the common way of computing with genus 2 curves with automorphisms. We can express u, and v in terms of the absolute invariants i_1, i_2, i_3 as in [**23**]. Moreover, the invariant J_{16} is expressed in terms of u, and v as follows

$$J_{16} = \left(4u^3 - v^2\right)\left(u^2 - 110\,u - 4\,v + 1125\right)^2$$

We have the following:

PROPOSITION 5. *Let* $\mathfrak{p}$ *be a genus 2 curve such that* $G := \mathrm{Aut}(\mathfrak{p})$ *has an elliptic involution. Then,*

a) $G \cong SL_2(3)$ *if and only if* $(u, v) = (0, 0)$ *or* $(u, v) = (225, 6750)$.

b) $G \cong GL_2(3)$ *if and only if* $u = 25$ *and* $v = -250$.

c) $G \cong D_6$ *if and only if* $4v - u^2 + 110u - 1125 = 0$, *for* $u \neq 9, 70 + 30\sqrt{5}, 25$.

d) $G \cong D_4$ *if and only if* $v^2 - 4u^3 = 0$, *for* $u \neq 1, 9, 0, 25, 225$. *Cases* $u = 0, 225$ *and* $u = 25$ *are reduced to cases a),and b) respectively.*

The V_4-locus (i.e., the locus $J_{30} = 0$) is birationally parametrized by dihedral invariants u, v as in [**23**].

6.3. The locus $J_2 = 0$. If a point $p \in \mathcal{M}_2$ is in the locus $J_2 = 0$ then the following result holds true:

PROPOSITION 6. *Let $\mathfrak{p} = (0, t_1, t_2, t_3) \in \mathcal{M}_2$. Then $|\operatorname{Aut}(\mathfrak{p})| > 2$ if and only if the invariants t_2, t_3 satisfy the Eq.* (34) *in the Appendix. Moreover,* $\operatorname{Aut}(\mathfrak{p}) \cong D_4$ *if and only if*

$$y^2 = x^5 + x^3 - \frac{3}{20}x$$

and $\operatorname{Aut}(\mathfrak{p}) \cong D_6$ *if and only if*

$$y^2 = x^6 + x^3 + -\frac{1}{40}$$

PROOF. We let $\mathfrak{p}$ be a curve with extra automorphisms. Then its equation can be written as

$$y^2 = x^6 + ax^4 + bx^2 + 1$$

Compute t_2, t_3 in terms of the dihedral invariants u, v. Since $J_2 = 0$, then $u = 0$. Hence, t_2 and t_3 are expressed only in terms of v. Eliminate v and we get the equation in Eq. (34).

Assume that $\operatorname{Aut}(\mathfrak{p}) \cong D_4$. Then, the curve has equation

$$y^2 = x^5 + x^3 + sx,$$

cf. section 8.2. We then have

$$J_2 = 6 + 40s = 0$$

Hence, $s = -\frac{3}{20}$.

If $\operatorname{Aut}(\mathfrak{p}) \cong D_4$, then the equation of the curve is

$$y^2 = x^6 + x^3 + w,$$

cf. section 8.3. We then have

$$J_2 = 6 - 240w = 0$$

Hence, $w = -\frac{1}{40}$. ∎

In the next section we will focus on genus two curves over $\mathbb{Q}$.

7. Genus 2 curves defined over $\mathbb{Q}$

Let $\mathcal{X}$ be a genus two curve defined over $\mathbb{Q}$. The moduli point in $\mathcal{M}_2$ corresponding to $\mathcal{X}$ is given by $\mathfrak{p} = (i_1, i_2, i_3)$. Since i_1, i_2, i_3 are rational functions in terms of the coeffiecients of $\mathcal{X}$, then $i_1, i_2, i_3 \in \mathbb{Q}$.

The converse isn't necessarily true. Let $\mathfrak{p} = (i_1, i_2, i_3) \in \mathcal{M}_2(\mathbb{Q})$. The universal equation of a genus 2 curve corresponding to $\mathfrak{p}$ is determined in [**15**], which is defined over a quadratic number field K. The main questions we want to consider is what percentage of the rational moduli points are defined over $\mathbb{Q}$? How can we determine a minimal equation for such curves?

The quotient space $C := \mathcal{X}/\operatorname{Aut}(\mathcal{X})$ is a genus zero curve, i.e. is isomorphic to a conic which is determined as follows. Let $\mathbf{x} = (x_1, x_2, x_3)$ be a coordinate in $\mathbb{P}^2$ and

$$C: \ \psi(x_1, x_2, x_3) := \mathbf{x}^t M \mathbf{x} \tag{26}$$

be the corresponding conic for the symmetric matrix M given as in Eq. (18).

There is also a cubic L given by the equation

$$L: \sum_{1\leq j,k,l\leq 3} a_{jkl} x_j x_k x_l = 0 \tag{27}$$

where the coefficients a_{jkl} are given explicitly in terms of the invariants in [**15**].

LEMMA 7 ([**16**]). *The genus 2 curve $\mathcal{X}$ is the intersection of the conic C and the cubic L.*

Mestre's algorithm has been implemented in Magma, Maple, Sage. For an implementation in Sage see Bouyer/Streng and their paper [**6**]. However, all these implementations have had issues. For example, the implementation in Sage will not work for curves with $J_2 = 0$.

EXAMPLE 3. Let $\mathcal{X}$ be the genus 2 curve given by the equation

$$y^2 = x^6 + x^5 + x + 1/6.$$

One can check that for this curve the Igusa-Clebsch invariants are

$$[\,0,\ -32000,\ 5120000/3,\ 295116800000/81\,]$$

and the algorithm fails in this case.

THEOREM 6. [**15**, Thm. 2] *For every point $\mathfrak{p} \in \mathcal{M}_2 \setminus \{D = 0\}$ such that $\mathfrak{p} \in \mathcal{M}_2(k)$, for some number field K, there is a pair of genus-two curves $\mathcal{C}^{\pm}$ given by*

$$\mathcal{C}^{\pm}: \quad y^2 = \sum_{i=0}^{6} a_{6-i}^{\pm}\, x^i\,,$$

corresponding to $\mathfrak{p}$, such that $a_i^{\pm} \in K(d)$, $i = 0, \dots, 6$ as given explicitly in [**15**, Eq. 41].

PROOF. The proof is exactly the same as [**15**, Thm. 4], other then the fact that the coefficients are expressed in terms of i_1, i_2, i_3. We can take $J_2, \dots, J_{10}$ as in Eq. (5.5). Substituting in formulas Eq. (B) we get,

(28)

$$\begin{aligned}
A &= -\frac{1}{120}\\
B &= \frac{1}{2^5\cdot 3^5\cdot 5^4}\,(36+5i_1)\\
C &= \frac{1}{2^8\cdot 3^8\cdot 5^6}\,(25i_2+180i_1-216)\\
D &= \frac{1}{2^{12}\cdot 3^{14}\cdot 5^{10}}\,(2700i_2-675i_1^2-250i_1i_2-86400000i_3+13500i_1-34992)
\end{aligned}$$

Then d^2 in [**15**] becomes

(29)

$$
\begin{aligned}
d^2 = & -\frac{1}{2^{50}\cdot 3^{56}\cdot 5^{30}}\,(9i_1^7 + 2i_1^6 i_2 - 27i_1^6 - 18i_1^4 i_2^2 - 4i_1^3 i_2^3 + 331776 i_1^5 i_3 + 54 i_1^3 i_2^2 + 9 i_1 i_2^4 \\
& + 2i_2^5 - 55240704 i_1^4 i_3 - 47278080 i_1^3 i_2 i_3 - 8294400 i_1^2 i_2^2 i_3 - 27 i_2^4 + 161243136 i_1^3 i_3 \\
& + 107495424 i_1^2 i_2 i_3 - 52254720 i_1 i_2^2 i_3 - 12441600 i_2^3 i_3 - 9459597312000 i_1^2 i_3^2 \\
& - 2866544640000 i_1 i_2 i_3^2 + 161243136 i_2^2 i_3 + 111451255603200 i_1 i_3^2 \\
& + 20639121408000 i_2 i_3^2 - 2641807540224000000 i_3^3 - 2407347121029 12 i_3^2)\cdot \\
& \cdot (675 i_1^2 + 250 i_1 i_2 - 13500 i_1 - 2700 i_2 + 86400000 i_3 + 34992)
\end{aligned}
$$

Notice that d^2 has two significant factors: one is J_{30} which correspond exactly to the locus of the curves with extra involutions, and the other one is the Clebsch invariant D expressed in terms of i_1, i_2, i_3. Substituting A, B, C, D for the expressions in Eq. (28) we get $a_0, \dots, a_6$ in terms of i_1, i_2, i_3 as in the Appendix. $\square$

REMARK 4. Notice that the above theorem does not provide an equation for the curve if the Clebsch invariant $D = 0$.

7.1. Genus 2 curve with $\mathrm{Aut}(\mathcal{X}) \cong V_4$**.** The above theorem is valid for any point $\mathfrak{p} \in \mathcal{M}_2$, including the points $\mathfrak{p} \in \mathcal{M}_2$ with $V_4 \hookrightarrow \mathrm{Aut}(\mathfrak{p})$. On contrary, Mestre's approach which has been implemented in many computer algebra packages is valid only for $\mathfrak{p} \in \mathcal{M}_2$ with $\mathrm{Aut}(\mathfrak{p}) \cong C_2$.

Let $\mathcal{X}$ be a genus 2 curve with $\mathrm{Aut}(\mathcal{X}) \cong V_4$. The V_4 locus has dimension two and is parametrized by the parameters u, and v as explained above. The rational model of a genus 2 curve with automorphism group V_4 given in terms of this parameters u, and v has equation given as follows

(30)

$$
\begin{aligned}
f(x,z) = & \left(v^2 + u^2 v - 2u^3\right) x^6 + 2\left(u^2 + 3v\right)\left(v^2 - 4u^3\right) x^5 z + \\
& + \left(15v^2 - u^2 v - 30u^3\right)\left(v^2 - 4u^3\right) x^4 z^2 + 4\left(5v - u^2\right)\left(v^2 - 4u^3\right)^2 x^3 z^3 + \\
& + \left(v^2 - 4u^3\right)^2 \left(15v^2 - u^2 v - 30u^3\right) x^2 z^4 + 2\left(v^2 - 4u^3\right)^3 \left(u^2 + 3v\right) x z^5 + \\
& + \left(v^2 - 4u^3\right)^3 \left(v^2 + u^2 v - 2u^3\right) z^6
\end{aligned}
$$

where u, and v are expressed in terms of the absolute invariants (i_1, i_2, i_3) as in [**23**]. The discriminant of the form is

$$
\Delta_f = 2^{36}\,\left(u^2 + 18u - 4v - 27\right)^2 \left(4u^3 - v^2\right)^{15} u^{30}.
$$

The case $\Delta_f = 0$ correspond exactly to the cases when $|\mathrm{Aut}(\mathfrak{p})| > 4$ which is treated below. Notice that the factors 2 and u have exponents ≥ 30.

7.2. Genus 2 curve with $\mathrm{Aut}(\mathcal{X}) \cong D_4$**.** Let $\mathcal{X}$ be a genus 2 curve with $\mathrm{Aut}(\mathcal{X}) \cong D_4$. The D_4 locus is a dimension one locus parametrized by the parameter s, as proved in [**19**, Lemma 3]ants. The rational model of a genus 2 curve with automorphism group D_4 given in terms of this parameter s is as follows

$$
y^2 = x^5 + x^3 + s\,x
$$

where s can be expressed in terms of the absolute invariants (i_1, i_2, i_3) as follows

$$
s = -\frac{3}{4}\frac{345\,i_1^2 + 50\,i_1 i_2 - 1296 i_1 - 90 i_2}{2925 i_1^2 + 250\,i_1 i_2 - 54000 i_1 - 9450 i_2 + 139968}.
$$

The discriminant is $\Delta_f = 2^4 \cdot s^3\,(4s - 1)^2$.

7.3. Genus 2 curve with $\mathrm{Aut}(\mathcal{X}) \cong D_6$. Now let us consider the case when $\mathcal{X}$ is a genus 2 curve with $\mathrm{Aut}(\mathcal{X}) \cong D_6$. The D_6 locus is a dimension one locus parametrized by the parameter w as proved in [**19**, Lemma 3]. The rational model of a genus 2 curve with automorphism group D_6 given in terms of this parameter w is as follows

$$y^2 = x^6 + x^3 + w$$

where the parameter w given in terms of the absolute invariants (i_1, i_2, i_3) is

$$w = \frac{1}{4} \frac{540 i_1^2 + 100 i_1 i_2 - 1728 i_1 + 45 i_2}{2700 i_1^2 + 1000 i_1 i_2 + 204525 i_1 + 40950 i_2 - 708588}$$

The discriminant of the form is $\Delta_f = -3^6 \cdot w^2 \, (4\,w - 1)^3$.

Hence, for every point $\mathfrak{p} \in \mathcal{M}_2$ we can find an equation of a curve C corresponding to $\mathfrak{p}$. Below we give an example to illustrate how this work in our code in Sage.

Example 4. Consider the curve

$$y^2 = 4294967297\,t^6 + 77309411328\,t^5 + 579820584969\,t^4 + 2319282339816\,t^3 + 5218385264643\,t^2 + 6262062317592\,t + 3131031158771$$

By using the functions above we find that this curve has automorphism group D_6. Hence, we compute $w = 2^{33}$. The equation of a curve over $\mathbb{Q}$ isomorphic (over $\mathbb{C}$) to our curve is

$$y^2 = x^6 + x^3 + 2^{33}$$

In the next section we will see if we can transform this curve over $\mathbb{Q}$ to another curve with smaller coefficients.

In the next section we will see how to minimize the discriminant of a genus two curve when its equation is given in Weierstrass form.

8. Minimal discriminant for Weierstrass equations

Let K be a field with a discrete valuation $\mathfrak{v}$ and ring of integers $\mathcal{O}_K$ and C an irreducible, smooth, algebraic curve of genus $g \geq 1$ defined over K and function field $K(C)$. The discriminant $\mathfrak{D}_{C/K}$ is an important invariant of the function field of the curve and therefore of the curve. Since the discriminant is a polynomial given in terms of the coefficients of the curve, then it is an ideal in the ring of integers $\mathcal{O}_K$ of K. The valuation of this ideal is a positive integer. A classical question is to find an equation of the curve such that this valuation is minimal, in other words the discriminant is minimal.

When $g = 1$, so that C is an elliptic curve, there is an extensive theory of the minimal discriminant ideal $\mathfrak{D}_{C/K}$. Tate [**26**] devised an algorithm how to determine the Weierstrass equation of an elliptic curve with minimal discriminant as part of his larger project of determining Neron models for elliptic curves. Such ideas were extended for genus 2 in [**13**]. Here we mostly follow [**17**].

For a binary form $f(x, z)$ and a matrix $M = \begin{bmatrix} a, b \\ c, d \end{bmatrix}$, such that $M \in GL_2(k)$, we have that $f^M := f(aX + bZ, cX + dZ)$ has discriminant

$$\Delta(f^M) = (\det M)^{\frac{d(d-1)}{2}} \cdot \Delta(f)$$

This property of the discriminant is crucial in our algorithm which is explained later.

8.1. Discriminant of a curve. The concept of a minimal discriminant for elliptic curves was defined by Tate and others in the 1970-s; see [**26**]. Such definitions and results were generalized by Lockhart in [**14**] for hyperelliptic curves.

Let us assume now that K is an algebraic number field with field of integers $\mathcal{O}_K$. Let M_K be the set of all inequivalent absolute values on K and M_K^0 the set of all non-archimedean absolute values in M_K. We denote by $K_\mathfrak{v}$ the completion of K for each $\mathfrak{v} \in M_K^0$ and by $\mathcal{O}_\mathfrak{v}$ the valuation ring in $K_\mathfrak{v}$. Let $\mathfrak{p}_v$ be the prime ideal in $\mathcal{O}_K$ and $\mathfrak{m}_v$ the corresponding maximal ideal in $K_\mathfrak{v}$. Let $(\mathcal{X}, P)$ be a superelliptic curve of genus $g \geq 2$ over K.

If $\mathfrak{v} \in M_K^0$ we say that $\mathcal{X}$ is **integral at** $\mathfrak{v}$ if $\mathcal{X}$ is integral when viewed as a curve over $K_\mathfrak{v}$. We say that $\mathcal{X}$ is **minimal at** $\mathfrak{v}$ when it is minimal over $K_\mathfrak{v}$.

An equation of $\mathcal{X}$ over K is called **integral** (resp. **minimal**) over K if it is integral (resp. minimal) over $K_\mathfrak{v}$, for each $\mathfrak{v} \in M_K^0$.

Next we will define the minimal discriminant over K to be the product of all the local minimal discriminants. For each $\mathfrak{v} \in M_K^0$ we denote by $\Delta_\mathfrak{v}$ the minimal discriminant for $(\mathcal{X}, P)$ over $K_\mathfrak{v}$. The **minimal discriminant** of $(\mathcal{X}, P)$ over K is the ideal

$$\Delta_{\mathcal{X}/K} = \prod_{\mathfrak{v} \in M_K^0} \mathfrak{m}_\mathfrak{v}^{\mathfrak{v}(\Delta_\mathfrak{v})}$$

We denote by $\mathfrak{a}_\mathcal{X}$ the ideal $\mathfrak{a}_\mathcal{X} = \prod_{\mathfrak{v} \in M_K^0} \mathfrak{p}_\mathfrak{v}^{\mathfrak{v}(\Delta_\mathfrak{v})}$.

Let $\mathcal{X}$ be a genus two curve with equation $y^2 = f(x)$. The discriminant of $\mathcal{X}$ is the discriminant of $f(x)$, hence $J_{10}(f)$, a degree 10 polynomial in terms of the coefficients of $f(x)$.

Let $M \in GL_2(K)$ such that $\det M = \lambda$. Then, $\Delta(f^M) = \lambda^{30} \Delta(f)$. Assume $\Delta(f) = p^\alpha \cdot N$, where $\alpha > 30$, for some prime p and some integer N such that $(p, N) = 1$. We perform the coordinate change

$$x \to \frac{1}{p} x$$

on $f(x)$. Then, the new discriminant is $\Delta' = \frac{1}{p^{30}} \cdot p^\alpha \cdot N = p^{\alpha-30} \cdot N$. Hence, we have the following

LEMMA 8. *A genus 2 curve $\mathcal{X}_g$ with integral equation*

$$y^2 = a_6 x^6 + \cdots a_1 x + a_0$$

has minimal discriminant if $\mathfrak{v}(\Delta) < 30$.

Thus, we factor Δ as a product of primes, say $\Delta = p_1^{\alpha_1} \cdots p_r^{\alpha_r}$, and take u to be the product of those powers of primes with exponents $\alpha_i \geq 30$. Then, the transformation $x \mapsto \frac{1}{p^u} \cdot x$ will reduce the discriminant.

8.2. Twists. Of course we can make the discriminant as small as we want it if we allow twists.

Let $\mathcal{X}$ be hyperelliptic curve with an extra automorphism of order $n \geq 2$. Then, from [**20**] we know that the equation of $\mathcal{X}$ can be written as

$$y^2 = f(x^n)$$

If $\mathcal{X}$ is given with such equation over $\mathbb{Q}$ and discriminant

$$\Delta = u\, N, \quad \textit{such that} \quad (u, N) = 1$$

then for any transformation $\tau : x \mapsto u^{-\frac{n}{30}} \cdot x$ would lead to $\mathcal{X}^\tau$ defined over $\mathbb{Q}$ and isomorphic to $\mathcal{X}$ over $\mathbb{C}$. Hence, $\mathcal{X}^\tau$ is a twist of $\mathcal{X}$ with discriminant

$$\Delta' = \frac{1}{u} \cdot u N = N,$$

Hence, in this case we can reduce the discriminant even further.

LEMMA 9. *Let $\mathcal{X}$ be a genus 2 curve with $\mathrm{Aut}(\mathcal{X}) \cong V_4$. Assume that $\mathcal{X}$ has equation over $\mathbb{Q}$ as $y^2 = f(x^2)$. Then, $\mathcal{X}$ has minimal discriminant Δ if $\mathfrak{v}(\Delta) < 15$.*
If $\mathrm{Aut}(\mathcal{X}) \cong D_6$ and $\mathcal{X}$ has equation over $\mathbb{Q}$ as $y^2 = f(x^3)$. Then, $\mathcal{X}$ has minimal discriminant Δ if $\mathfrak{v}(\Delta) < 10$.

PROOF. The proof is an immediate consequence of the above remarks. □

The following gives a universal curve with minimal discriminant for the V_4-locus.

LEMMA 10. *Let $\mathcal{X}$ be a genus 2 curve such that $\mathrm{Aut}(\mathcal{X}) \cong V_4$ and (u, v) the corresponding dihedral invariants. Then, $\mathcal{X}$ has minimal discriminant and is defined over its field of moduli when given by the equation*

$$y^2 = b_6 x^6 + \cdots + b_1 x + b_0 \tag{31}$$

where

$$\begin{aligned}
b_6 &= -\frac{(2u^3 - u^2 v - v^2)}{2^6 u^6} \\
b_5 &= -2\frac{(u^2 + 3v)(4u^3 - v^2)}{2^5 u^5} \\
b_4 &= \frac{(30u^3 + u^2 v - 15v^2)(4u^3 - v^2)}{2^4 u^4} \\
b_3 &= -4\frac{(u^2 - 5v)(4u^3 - v^2)^2}{2^3 u^3} \\
b_2 &= -\frac{(30u^3 + u^2 v - 15v^2)(4u^3 - v^2)^2}{2^2 u^2} \\
b_1 &= -\frac{(u^2 + 3v)(4u^3 - v^2)^3}{u} \\
b_0 &= (2u^3 - u^2 v - v^2)(4u^3 - v^2)^3
\end{aligned}$$

PROOF. We know that the equation of the curve in Eq. (30) is defined over its field of moduli. Let

$$\sigma : (x, y) \mapsto \left(\frac{1}{2u} x, \frac{1}{(2u)^6} y \right).$$

Then the discriminant of the new curve $\mathcal{X}^\sigma$ is

$$\Delta_f = 2^6 \, (u^2 + 18u - 4v - 27)^2 \, (4u^3 - v^2)^{15}.$$

Since all exponents are < 15, this discriminant can not be further reduced over $\mathbb{Q}$. The equation of the curve in this case becomes as in Eq. (31). □

Notice that for any curve $\mathcal{X}$ with $\mathrm{Aut}(\mathcal{X}) \cong V_4$, we have $u \neq 0$, so the curve in Eq. (31) is defined everywhere.

In the next section we will see that smaller discriminant doesn't necessarily mean small coefficients. We illustrate below with an example.

EXAMPLE 5. Let $u = 35$ and $v = 6$. Using the function

```
RatModSha_uv (35, 6)
```

we get

(-4)*(19591*t^6+106564876*t^5-55428309960*t^4+35132884438720*t^3
+9503959738981440*t^2+3132997049150231296*t-98758721142240034304)

Notice that this curve has large coefficients. Especially, when we have that for $u = 35$ and $v = 6$ we get corresponding $(a, b) = (2, 3)$ and therefore a curve

$$y^2 = x^6 + 2x^4 + 3x^3 + 1$$

As it is shown in [**22**], it is almost always true that the curve $y^2 = x^6 + ax^4 + bx^2 + 1$ has smaller coefficients when such $a, b \in \mathbb{Q}$ do exist.

9. Constructing the databases

Now that we have all the necessary results we are ready to construct all three dictionaries.

9.1. Curves with height ≤ 10: the dictionary $\mathcal{L}_1$. In order to create the list $\mathcal{L}_1$ we first compile a list of all 7-tuples $(a_0, \dots, a_6)$ and from this list eliminate all the tuples with $J_{10} = 0$. Then, for each tuple we compute the moduli point $\mathfrak{p} = (i_1, i_2, i_3)$ and all the other invariants as described in the previous sections.

TABLE 1. The number of curves for a given height

h	$\mathbb{P}^6$	# of curves	V_4	D_4	D_6
1	1 093	230	28	11	2
2	37 969	8 593	230	40	7
3	409 585	88 836	1054	112	26

REMARK 5. The bound for the number of genus 2 curves of height h is $< (2h+1)^7$. For example, for h $= 1$ this bound is $< 3^7 = 2187$. In fact, as shown in Table 1 the number of such curves is 230, all of which are listed in [**22**, Table 2, 3].

In the Table 1 we display the number of curves with extra automorphism for a given height h.

9.2. Curve with extra involutions: the dictionary $\mathcal{L}_2$. The list $\mathcal{L}_2$ was computed in [**22**] for $1 \leq$ h ≤ 100. Since every genus 2 curve with extra automorphisms can be written (over $\mathbb{C}$) as

$$y^2 = x^6 + ax^4 + bx^2 + 1 \tag{32}$$

we create the list of tuples $(1, 0, b, 0, a, 0, 1)$ for $a, b \leq$ h. We go through the lists and delete the ones which have $J_{10} = 0$. This number is displayed in [**22**, Table 1] for $1 \leq \mathfrak{h} \leq 100$, including information on how many points from $\mathcal{L}_2$ are already in $\mathcal{L}_1$, how many of these curves have automorphism group D_4 and how many have automorphism group D_6.

TABLE 2. Curves with extra involutions and height $\mathrm{h} \leq 101$

h	$J_{10} \neq 0$	new pts. in $\mathcal{M}_2$	D_4	D_6	Total pts
1	8	4	1	0	5
2	24	9	3	0	14
3	47	12	4	0	26
4	79	17	6	0	43
5	119	20	7	0	63
6	167	25	9	0	88
7	223	28	11	0	116
8	287	33	13	0	149
9	359	36	15	0	185
10	439	41	17	0	226

In [**22**] it is discussed when such curves have minimal height and how a reduction as in [**1**] is easier to perform in this case. In Table 2 we display only the first ten rows of Table 1 from [**22**].

There are 20 697 curves in $\mathcal{L}_2$, such that for each h we have roughly 4 h curves. So it is expected that the number of curves of height $\leq h$, with equation Eq. (32), defined over $\mathbb{Q}$ is $\leq 4\frac{\mathrm{h}(\mathrm{h}+1)}{2}$; see Beshaj [**22**] for more details.

The first curve with automorphism group isomorphic to D_4 occurs for $\mathrm{h} = 1$. It is only one such curve, namely the curve with equation

$$y^2 = x^6 + x^4 + x^2 + 1.$$

The are only two curves with automorphism group isomorphic to D_6, which occur for $\mathrm{h} = 79$ and $\mathrm{h} = 83$. namely the curve

$$y^2 = x^6 + 79x^4 - 17x^2 + 1$$

and the curve

$$y^2 = x^6 + 83x^6 + 19x^2 + 1.$$

There are 195 curves with automorphism group isomorphic to D_4. For the largest height $\mathrm{h} = 101$ there are 405 curves two of which have automorphism group isomorphic to D_4.

In [**1**] is proved the following theorem:

THEOREM 7 ([**1**]). *Let $\mathfrak{p} \in \mathcal{M}_2(\mathbb{Q})$ be such that $\mathrm{Aut}(\mathfrak{p}) \cong V_4$. There is a genus 2 curve $\mathcal{X}$ corresponding to $\mathfrak{p}$ with equation $y^2 z^4 = f(x^2, z^2)$, where*

$$f(x, z) = x^6 - s_1 x^4 z^2 + s_2 x^2 z^4 - z^6. \tag{33}$$

If $f \in \mathbb{Z}[x, z]$, then $f(x, z)$ or $f(-z, x)$ is a reduced binary form.

It is interesting to not that from 20 292 such curves we found only 57 which do not have minimal absolute height. For more details see the discussion in the last section of [**1**].

9.3. Curves with small moduli height: the dictionary $\mathcal{L}_3$. For the last dictionary $\mathcal{L}_3$ we create a list of all points $[x_0 : x_1 : x_2 : x_3]$ of projective height $\leq \mathfrak{h}$ in $\mathbb{P}^3(\mathbb{Q})$, for some integer $\mathfrak{h} \geq 1$. Each such point correspond to the point $[J_2^5 : J_4 J_2^3 : J_6 J_2^2 : J_{10}]$. The number of such tuples (up to equivalence in $\mathbb{P}^3$) is given in column 2 of Table 3. Not all such points correspond to a genus 2 curve. We delete all the points such that $x_3 = 0$ since they correspond to the cases when $J_{10} = 0$. What is left on the list is the number of points in the moduli space $\mathcal{M}_2$ with moduli height $\leq \mathfrak{h}$. This number is given in the third column in the Table 3.

For each given point $\mathfrak{p} = (r, i_1, i_2, i_3)$ in this list we find the equation of the curves as described above. As we already know we don't get a curve defined over $\mathbb{Q}$ in each case. The number of points which are defined over $\mathbb{Q}$ is given in column four. From those points $\mathfrak{p} \in \mathcal{M}_2$ such that the field of definition is $\mathbb{Q}$, the number of points with automorphisms is given in column five.

TABLE 3. Genus 2 curves with bounded moduli height

$\mathfrak{h}$	$\mathbb{P}^3(\mathbb{Q})$ n_1	$\mathfrak{p} \in \mathcal{M}_2(\mathbb{Q})$ n_2	$M_\mathfrak{p} = F_\mathfrak{p}$ n_3	$\lvert \mathrm{Aut}(\mathfrak{p}) \rvert > 2$ n_4	ratio
1	40	27	20	15	0.25
2	272	223	124	75	0.4
3	1120	975	514	243	0.53
4	2928	2639	1311	507	0.61
5	6928	6351	3056	1035	0.66
6	12768	11903	5561	1587	0.71
7	23760	22319	9963	2667	0.73
8	38128	36111	15648	3771	0.76
9	60640	57759	24214	5427	0.78
10	88448	84703	34936	7107	0.8
11	131088	125903	50630	9867	0.81
12	177712	171375	68046	12123	0.82
13	248080	239727	93229	16011	0.83
14	324736	314655	120358	19395	0.84
15	427968	415583	157569	23907	0.85
16	542720	528031	199136	28419	0.86
17	700032	681887	252947	35139	0.86
18	857328	836591	307964	40251	0.87
19	1076928	1051871	381219	48675	0.87

The following question is natural: What percentage of rational points $\mathfrak{p} \in \mathcal{M}_2(\mathbb{Q})$ with a fixed moduli height $\mathfrak{h}$ have $\mathbb{Q}$ as a field of definition, when $\mathfrak{h}$ becomes arbitrarily large? In other words, what is the limit

$$\lim_{\mathfrak{h} \to \infty} \frac{n_3}{n_2}?$$

LEMMA 11. *For large enough moduli height $\mathfrak{h} \in \mathcal{M}_2(\mathbb{Q})$, the majority of genus 2 curves with moduli height $\mathfrak{h}$ are not defined over $\mathbb{Q}$.*

PROOF. Let $\mathfrak{p} \in \mathcal{M}_2(\mathbb{Q})$ with moduli height $\mathfrak{h}_0$. Then, $\mathfrak{p} = (i_1, i_2, i_3)$ for $i_1, i_2, i_3 \in \mathbb{Q}$. The equation of a curve $\mathcal{X}$ corresponding to $\mathfrak{p}$ is the intersection of the conic $C : \psi(x_1, x_2, x_3) = 0$ and the cubic $L : \phi(x_1, x_2, x_3) = 0$ which are both defined over $\mathbb{Q}$, since their equations are given in terms of i_1, i_2, i_3. The intersection $C \cap L$ is obtained as the resultant of the corresponding equations with respect to one of the variables x_1, x_2, x_3. This resultant contains a square root which is given in terms of i_1, i_2, i_3; see [**15**]. $\mathcal{X}$ is defined over $\mathbb{Q}$ if and only if the expression inside the square root is a complete square. Since this is a surface in $\mathbb{Q}^3$, the set of points of fixed projective height which make it a complete square has measure zero. This completes the proof. □

Next we will consider a similar question. What percentage of curves defined over $\mathbb{Q}$ don't have extra automorphisms when $\mathfrak{h}$ becomes arbitrary large? In other words, what it the limit

$$\lim_{\mathfrak{h}\to\infty} \frac{n_3 - n_4}{n_3}?$$

Thus, we want to determine the ratio of points of height $\mathfrak{h}$ in $\mathcal{M}_2(\mathbb{Q})$ for which $M_\mathfrak{p} \neq F_\mathfrak{p}$ over the total number of points of height $\mathfrak{h}$ in $\mathcal{M}_2(\mathbb{Q})$ as $\mathfrak{h} \to \infty$.

LEMMA 12. *For large enough moduli height* $\mathfrak{h}$*, the majority of genus 2 curves with moduli height* $\mathfrak{h}$ *don't have extra automorphisms.*

PROOF. The number of points with height $\mathfrak{h}$ in the projective space increases as $\mathfrak{h}$ increases. However, such points $\mathfrak{p} = (i_1, i_2, i_3)$ have to satisfy the equation $J_{30} = 0$ when they have automorphisms. □

There is another database of genus 2 curves described in [**5**] which is has all the curves with discriminant < 1000. We checked our database to see how many of our curves would be with small discriminant. We have to warn the reader that our definition of the discriminant is just J_{10} and slightly different from what is used in [**5**]. The majority of curves in the third dictionary have small discriminant, but this is not a surprise since small moduli height forces J_{10} to be small. Only one curve in the second dictionary has $|J_{10}| < 1000$. Six curves in the first dictionary have $|J_{10}| < 10000$. It would be interesting to see exactly how the database in [**3**] and [**5**] intersect.

Appendix A. Functions of the genus 2 package

Genus 2 package

Input: a sextic polynomial $f(t)$

J2	Igusa invariant J_2
J4	Igusa invariant J_4
J6	Igusa invariant j_6
J10	Igusa invariant J_{10}
J30	J_{30}: the V_4-locus
Igusa	Igusa invariants $[J_2, J_4, J_6, J_{10}]$
Clebsch	Invariants [A, B, C, D]
i_1	absolute invariant i_1
i_2	absolute invariant i_2
i_3	absolute invariant i_3
j1	Igusa function j_1
j2	Igusa function j_2
j3	Igusa function j_3
RatMod	Rational model of the curve over when such model exists.
RatModSha	Rational model of the curve over its minimal field of definition as in Shaska **[23]**
RatModMe	Rational model over $\mathbb{Q}$, when such model exists, as in Mestre **[16]**
height	Height of the sextic
EquivBin	Checks if sextics are equivalent
RatModTable	Rational Model from the Table of minimal models
MinField	Minimal field of definition
Info	Displays information about the curve $y^2 = f(t)$
RatForm	Rational Model from Malmendier/Shaska **[15]**

Input: the moduli point (i_1, i_2, i_3)

J30_j	J_{30} in terms of i_1, i_2, i_3
Igusa_i	$J_2, \ldots, J_{10}$
Clebsch_i	Invariants [A, B, C, D]
ClebschMatrix	ClebschMatrix in terms of invariants i_1, i_2, i_3
Cubic	Cubic defined in Eq. (27)
Conic	Conic defined in Eq. (26)
L_D4	Locus of curves with group D_4
L_D6	Locus of curves with group D_6
AutGroup	Automorphism group of the curve
Sh_u_v	Shaska invariants u, v
ModHeight	Modular height

Moduli Space

curves_moduli	Computes the number of rational points of height $\mathfrak{h}$ in the moduli space and how many of those have a rational model
NumbCurvMod	This function computes the total number of rational points of moduli height $\mathfrak{h}$, how many of them have a rational model over $\mathbb{Q}$, how many of them have automorphisms and the ratio=obstruction/ fine points
moduli_points	Computes the number of rational points of height $\mathfrak{h}$ in the moduli space
MoPtsCurvAut	Moduli points with automorphisms

Creating the databases

Curves(h, L)	Creates the dictionary $\mathcal{L}_1$ of curves with height h
CurvesAut(h, L)	Creates the dictionary $\mathcal{L}_2$ of curves with automorphisms
CurvHe	Number of curves with height h
SelfRec(f)	Checks if a sextic is self reciprocal
CurvHeW(h, w)	Number of curves with height h and w
NCWT(h, w)	Number of curves with height h and twists w
CurvesTabOverQ(h, w)	Counts the number of curves over $\mathbb{Q}$, including twist, for given height.

Appendix B. Basic Invariants and relations among them

Clebsch invariants A, B, C are in terms of the coefficients

$$A = 2\, a_6\, a_0 - \frac{1}{3}\, a_5\, a_1 + \frac{2}{15}\, a_4\, a_2 - \frac{1}{20}\, a_3^2$$

$$\begin{aligned}
B = - \Big(& \frac{4}{225}\, a_6\, a_2\, a_4\, a_0 - \frac{2}{75}\, a_6\, a_2\, a_3\, a_1 + \frac{8}{1125}\, a_6\, a_2^3 - \frac{2}{75}\, a_5\, a_3\, a_4\, a_0 \frac{2}{225}\, a_5 a_3^2\, a_1 \\
& - \frac{2}{1125}\, a_5\, a_3\, a_2^2 + \frac{8}{1125}\, a_4^3\, a_0 - \frac{2}{1125}\, a_4^2\, a_3\, a_1 + \frac{14}{5625}\, a_4^2\, a_2^2 - \frac{2}{225}\, a_5\, a_2\, a_4\, a_1 \\
& + \frac{2}{45}\, a_5^2\, a_2\, a_0 + \frac{2}{45}\, a_6\, a_1^2\, a_4 - \frac{2}{9}\, a_6\, a_1\, a_5\, a_0 - \frac{8}{5625}\, a_4\, a_3^2\, a_2 + \frac{2}{3}\, a_6^2\, a_0^2 \\
& + \frac{2}{75}\, a_6\, a_0\, a_3^2 + \frac{1}{3750}\, a_3^4 \Big)
\end{aligned}$$

$$\begin{aligned}
C = & - \frac{2}{46875}\, a_4^3\, a_2^3 - \frac{1}{5625}\, a_4^4\, a_1^2 - \frac{1}{5625}\, a_5^2\, a_2^4 - \frac{2}{9}\, a_6^3\, a_0^3 + \frac{11}{5625}\, a_6\, a_2\, a_3^2\, a_4\, a_0 \\
& - \frac{7}{1125}\, a_6\, a_2^2\, a_5\, a_0\, a_3 + \frac{4}{225}\, a_6\, a_2\, a_4\, a_1\, a_5\, a_0 - \frac{1}{562500}\, a_3^6 + \frac{8}{1125}\, a_6^2\, a_2^3\, a_0 \\
& - \frac{8}{28125}\, a_6\, a_2^4\, a_4 + \frac{1}{9375}\, a_6\, a_2^3\, a_3^2 + \frac{1}{5625}\, a_5\, a_3^4\, a_1 + \frac{1}{90}\, a_5^3\, a_3\, a_0^2 - \frac{2}{28125}\, a_5\, a_3^3\, a_2^2 \\
& - \frac{8}{28125}\, a_4^4\, a_2\, a_0 + \frac{8}{1125}\, a_4^3\, a_6\, a_0^2 + \frac{1}{9375}\, a_4^3\, a_3^2\, a_0 - \frac{2}{28125}\, a_4^2\, a_3^3\, a_1 - \frac{1}{225}\, a_4^2\, a_5^2\, a_0^2 \\
& - \frac{2}{140625}\, a_4^2\, a_3^2\, a_2^2 + \frac{1}{90}\, a_6^2\, a_1^3\, a_3 - \frac{1}{225}\, a_6^2\, a_1^2\, a_2^2 + \frac{2}{140625}\, a_4\, a_3^4\, a_2 - \frac{1}{3750}\, a_6\, a_0\, a_3^4 \\
& - \frac{1}{75}\, a_6^2\, a_0^2\, a_3^2 + \frac{14}{225}\, a_6^2\, a_2\, a_0^2\, a_4 - \frac{1}{3750}\, a_6\, a_2\, a_3^3\, a_1 - \frac{2}{45}\, a_6\, a_2\, a_5^2\, a_0^2 \\
& - \frac{1}{3750}\, a_5\, a_3^3\, a_4\, a_0 \\
& + \frac{1}{2250}\, a_5^2\, a_3^2\, a_0\, a_2 + \frac{7}{28125}\, a_5\, a_3\, a_4\, a_2^3 + \frac{1}{2250}\, a_5\, a_3\, a_4^2\, a_1^2 + \frac{7}{28125}\, a_4^3\, a_2\, a_3\, a_1 \\
& + \frac{2}{1125}\, a_4^3\, a_1\, a_5\, a_0 - \frac{1}{5625}\, a_5\, a_2^2\, a_4^2\, a_1 + \frac{2}{1125}\, a_5\, a_2^3\, a_6\, a_1 + \frac{1}{2250}\, a_5^2\, a_2^2\, a_3\, a_1 \\
& + \frac{1}{9}\, a_6^2\, a_1\, a_0^2\, a_5 + \frac{1}{2250}\, a_6\, a_1^2\, a_3^2\, a_4 - \frac{2}{45}\, a_6^2\, a_1^2\, a_0\, a_4 - \frac{2}{1875}\, a_6\, a_2^2\, a_4^2\, a_0 \\
& + \frac{1}{1875}\, a_6\, a_2^2\, a_4\, a_3\, a_1 - \frac{1}{75}\, a_6^2\, a_2\, a_0\, a_3\, a_1 + \frac{1}{1875}\, a_5\, a_3\, a_4^2\, a_2\, a_0 - \frac{1}{75}\, a_5\, a_3\, a_6\, a_0^2\, a_4 \\
& - \frac{7}{11250}\, a_5\, a_3^2\, a_4\, a_2\, a_1 + \frac{1}{90}\, a_5\, a_3^2\, a_6\, a_0\, a_1 - \frac{1}{225}\, a_5^2\, a_3\, a_4\, a_1\, a_0 \\
& - \frac{7}{1125}\, a_4^2\, a_6\, a_0\, a_3\, a_1 - \frac{1}{225}\, a_5\, a_2\, a_6\, a_1^2\, a_3
\end{aligned}$$

Clebsch invariants A, B, C, D in terms of the Igusa invariants J_2, J_4, J_6, J_{10}

$$\begin{aligned}
A &= -\frac{1}{2^3 3\cdot 5}\, J_2\,, \\
B &= \frac{1}{2^3 3^3 5^4}\left(J_2^2 + 20\,J_4\right)\,, \\
C &= -\frac{1}{2^5 3^5 5^6}\left(J_2^3 + 80\,J_2 J_4 - 600\,J_6\right)\,, \\
D &= -\frac{1}{2^8 3^9 5^{10}}\left(9\,J_2^5 + 700\,J_2^3 J_4 - 3600\,J_2^2 J_6 - 12400\,J_2 J_4^2 + 48000\,J_4\,J_6 \right. \\
&\quad \left. +10800000\,J_{10}\right)
\end{aligned}$$

The entries of the Clebsch matrix M.

$$\begin{aligned}
A_{11} &= 2\,C + \frac{1}{3}\,A\,B\,, \\
A_{22} &= A_{13} = \ D\,, \\
A_{33} &= \frac{1}{2}\,B\,D + \frac{2}{9}\,C\,(B^2 + A\,C)\,, \\
A_{23} &= \frac{1}{3}\,B\,(B^2 + A\,C) + \frac{1}{3}\,C\,(2\,C + \frac{1}{3}\,A\,B)\,, \\
A_{12} &= \frac{2}{3}\,(B^2 + A\,C)
\end{aligned}$$

Igusa invariants J_2, J_4, J_6, J_{10} we display them below in terms of the coefficients

$$\begin{aligned}
J_2 &= 6a_3^2 - 240a_0a_6 + 40a_1a_5 - 16a_2a_4 \\
J_4 &= 48a_0a_4^3 + 48a_2^3a_6 + 4a_2^2a_4^2 + 1620a_0^2a_6^2 + 36a_1a_3^2a_5 - 12a_1a_3a_4^2 - 12a_2^2a_3a_5 \\
&\quad + 300a_1^2a_4a_6 + 300a_0a_5^2a_2 + 324a_0a_6a_3^2 - 504a_0a_4a_2a_6 - 180a_0a_4a_3a_5 \\
&\quad - 180a_1a_3a_2a_6 + 4a_1a_4a_2a_5 - 540a_0a_5a_1a_6 - 80a_1^2a_5^2 \\
J_6 &= 176a_1^2a_5^2a_3^2 + 64a_1^2a_5^2a_4a_2 + 1600a_1^3a_5a_4a_6 + 1600a_1a_5^3a_0a_2 - 160a_0a_4^4a_2 \\
&\quad - 96a_0^2a_4^3a_6 + 60a_0a_4^3a_3^2 + 72a_1a_3^4a_5 - 24a_1a_3^3a_4^2 - 119880a_0^3a_6^3 - 320a_1^3a_5^3 \\
&\quad - 160a_2^4a_4a_6 - 96a_2^3a_0a_6^2 + 60a_2^3a_3^2a_6 - 24a_2^2a_3^3a_5 + 8a_2^2a_3^2a_4^2 - 900a_2^2a_1^2a_6^2 \\
&\quad - 24a_2^3a_4^3 - 36a_2^4a_5^2 - 36a_1^2a_4^4 + 424a_0a_4^2a_2^2a_6 - 2240a_1^2a_5^2a_0a_6 + 2250a_1^3a_3a_6^2 \\
&\quad + 492a_0a_4^2a_2a_3a_5 + 20664a_0^2a_4a_6^2a_2 + 3060a_0^2a_4a_6a_3a_5 - 468a_0a_4a_3^2a_2a_6 \\
&\quad - 198a_0a_4a_3^3a_5 - 640a_0a_4a_2^2a_5^2 + 3472a_0a_4a_2a_5a_1a_6 - 18600a_0a_4a_1^2a_6^2 \\
&\quad - 876a_0a_4^2a_1a_6a_3 + 492a_1a_3a_2^2a_4a_6 - 238a_1a_3^2a_2a_4a_5 + 76a_1a_3a_2a_4^3 \\
&\quad + 3060a_1a_3a_0a_6^2a_2 + 1818a_1a_3^2a_0a_6a_5 + 26a_1a_3a_2^2a_5^2 - 1860a_1^2a_3a_2a_5a_6 \\
&\quad + 330a_1^2a_3^2a_6a_4 + 76a_2^3a_4a_3a_5 - 876a_2^2a_0a_6a_3a_5 + 616a_2^3a_5a_1a_6 + 2250a_0^2a_5^3a_3 \\
&\quad - 10044a_0^2a_6^2a_3^2 + 28a_1a_4^2a_2^2a_5 - 640a_1^2a_4^2a_2a_6 + 26a_1^2a_4^2a_3a_5 - 1860a_1a_4a_0a_5^2a_3 \\
&\quad + 616a_1a_4^3a_0a_5 - 18600a_0^2a_5^2a_6a_2 + 59940a_0^2a_5a_6^2a_1 + 330a_0a_5^2a_3^2a_2 + 162a_0a_6a_3^4 \\
&\quad - 900a_0^2a_5^2a_4^2 - 198a_1a_3^3a_2a_6 \\
J_{10} &= a_6^{-1} Res_X\left(f, \frac{\partial f}{\partial X}\right)
\end{aligned}$$

$$
\begin{aligned}
&t_2^6t_3 - 15265260t_2^5t_3 - 27949860t_2^4t_3^2 - 118098t_2^3t_3^3 + 14693280768t_2^5\\
&+ 93437786558880t_2^4t_3 - 878290475269680t_2^3t_3^2 + 85811055510240t_2^2t_3^3\\
&- 1139016237660t_2t_3^4 + 3486784401t_3^5 - 2231542016640000000t_2^4\\
&- 287728673929542000000t_2^3t_3 - 246965801016869100000t_2^2t_3^2\\
&- 109818018101695500000t_2t_3^3 - 7060738412025000t_3^4\\
&+ 135566177510880000000000000t_2^3 + 433843541357670112500000000t_2^2t_3\\
&- 662569101476807962500000000t_2t_3^2 + 571919811374025000000000t_3^3\\
&- 4117822641892980000000000000000t_2^2\\
&- 3270773656259838098437500000000000t_2t_3\\
&- 231627523606480125000000000000000t_3^2\\
&+ 625394313737496337500000000000000000t_2\\
&+ 469045735303122253125000000000000000t_3\\
&- 37992704559552902503125000000000000000000 = 0
\end{aligned} \tag{34}
$$

$$
\begin{aligned}
a_{111} &= \frac{2}{9}\,(A^2C - 6BC + 9D),\\
a_{112} &= \frac{1}{9}\,(2B^3 + 4ABC + 12C^2 + 3AD),\\
a_{113} = a_{122} &= \frac{1}{9}\left(AB^3 + \frac{4}{3}\,A^2BC + 4B^2C + 6AC^2 + 3BD\right),\\
a_{123} &= \frac{1}{18}\left(2B^4 + 4AB^2C + \frac{4}{3}\,A^2C^2 + 4BC^2 + 3ABD + 12CD\right),\\
a_{133} &= \frac{1}{18}\left(AB^4 + \frac{4}{3}\,A^2B^2C + \frac{16}{3}\,B^3C + \frac{26}{3}\,ABC^2 + 8C^3 + 3B^2D + 2ACD\right),\\
a_{222} &= \frac{1}{9}\left(3B^4 + 6AB^2C + \frac{8}{3}\,A^2C^2 + 2BC^2 - 3CD\right),\\
a_{223} &= \frac{1}{18}\left(-\frac{2}{3}\,B^3C - \frac{4}{3}\,ABC^2 - 4C^3 + 9B^2D + 8ACD\right),\\
a_{233} &= \frac{1}{18}\left(B^5 + 2AB^3C + \frac{8}{9}\,A^2BC^2 + \frac{2}{3}\,B^2C^2 - BCD + 9D^2\right),\\
a_{333} &= \frac{1}{36}\left(-2B^4C - 4AB^2C^2 - \frac{16}{9}\,A^2C^3 - \frac{4}{3}\,BC^3 + 9B^3D + 12ABCD + 20C^2D\right)
\end{aligned}
$$

References

[1] Lubjana Beshaj, *Reduction theory of binary forms*, Advances on superelliptic curves and their applications, NATO Sci. Peace Secur. Ser. D Inf. Commun. Secur., vol. 41, IOS, Amsterdam, 2015, pp. 84–116. MR3525574

[2] L. Beshaj, *Minimal Weierstrass equations for genus 2 curves* (201612), available at 1612.08318.

[3] Lubjana Beshaj and Tony Shaska, *Algebraic curves, http://www.algcurves.org*, 2017.

[4] Oskar Bolza, *On binary sextics with linear transformations into themselves*, Amer. J. Math. **10** (1887), no. 1, 47–70, DOI 10.2307/2369402. MR1505464

[5] Andrew R. Booker, Jeroen Sijsling, Andrew V. Sutherland, John Voight, and Dan Yasaki, *A database of genus-2 curves over the rational numbers*, LMS J. Comput. Math. **19** (2016), no. suppl. A, 235–254, DOI 10.1112/S146115701600019X. MR3540958

[6] Florian Bouyer and Marco Streng, *Examples of CM curves of genus two defined over the reflex field*, LMS J. Comput. Math. **18** (2015), no. 1, 507–538, DOI 10.1112/S1461157015000121. MR3376741

[7] A. Clebsch and P. Gordan, *Theorie der Abelschen Functionen* (German), Physica-Verlag, Würzburg, 1967. Thesaurus Mathematicae, (Neudrucke wichtiger mathematischer Werke), Band 7. MR0485106

[8] P. Gaudry and É. Schost, *On the invariants of the quotients of the Jacobian of a curve of genus 2*, Applied algebra, algebraic algorithms and error-correcting codes (Melbourne, 2001), Lecture Notes in Comput. Sci., vol. 2227, Springer, Berlin, 2001, pp. 373–386, DOI 10.1007/3-540-45624-4_39. MR1913484

[9] Jun-ichi Igusa, *Arithmetic variety of moduli for genus two*, Ann. of Math. (2) **72** (1960), 612–649, DOI 10.2307/1970233. MR0114819

[10] Jun-ichi Igusa, *On Siegel modular forms of genus two*, Amer. J. Math. **84** (1962), 175–200, DOI 10.2307/2372812. MR0141643

[11] Jun-ichi Igusa, *Modular forms and projective invariants*, Amer. J. Math. **89** (1967), 817–855, DOI 10.2307/2373243. MR0229643

[12] V. Krishnamoorthy, T. Shaska, and H. Völklein, *Invariants of binary forms*, Progress in Galois theory, 2005, pp. 101–122. MR2148462 (2006b:13015)

[13] Qing Liu, *Modèles entiers des courbes hyperelliptiques sur un corps de valuation discrète* (French, with English summary), Trans. Amer. Math. Soc. **348** (1996), no. 11, 4577–4610, DOI 10.1090/S0002-9947-96-01684-4. MR1363944

[14] P. Lockhart, *On the discriminant of a hyperelliptic curve*, Trans. Amer. Math. Soc. **342** (1994), no. 2, 729–752, DOI 10.2307/2154650. MR1195511

[15] A. Malmendier and T. Shaska, *A universal pair of genus-two curves* (2016), available at 1607.08294.

[16] Jean-François Mestre, *Construction de courbes de genre 2 à partir de leurs modules* (French), Effective methods in algebraic geometry (Castiglioncello, 1990), Progr. Math., vol. 94, Birkhäuser Boston, Boston, MA, 1991, pp. 313–334. MR1106431

[17] Rachel Shaska, *Equations of curves with minimal discriminant* (201407), available at 1407.7064.

[18] T. Shaska, *Curves of genus two covering elliptic curves*, ProQuest LLC, Ann Arbor, MI, 2001. Thesis (Ph.D.)–University of Florida. MR2701993

[19] T. Shaska, *Genus 2 curves with* $(3,3)$*-split Jacobian and large automorphism group*, Algorithmic number theory (Sydney, 2002), 2002, pp. 205–218. MR2041085

[20] T. Shaska, *Determining the automorphism group of a hyperelliptic curve*, Proceedings of the 2003 International Symposium on Symbolic and Algebraic Computation, 2003, pp. 248–254 (electronic). MR2035219 (2005c:14037)

[21] T. Shaska and L. Beshaj, *The arithmetic of genus two curves*, Information security, coding theory and related combinatorics, NATO Sci. Peace Secur. Ser. D Inf. Commun. Secur., vol. 29, IOS, Amsterdam, 2011, pp. 59–98. MR2963126

[22] T. Shaska and L. Beshaj, *Heights on algebraic curves*, Advances on superelliptic curves and their applications, NATO Sci. Peace Secur. Ser. D Inf. Commun. Secur., vol. 41, IOS, Amsterdam, 2015, pp. 137–175. MR3525576

[23] T. Shaska and H. Völklein, *Elliptic subfields and automorphisms of genus 2 function fields*, Algebra, arithmetic and geometry with applications (2000), 703–723.

[24] T. Shaska and G. S. Wijesiri, *Theta functions and algebraic curves with automorphisms*, Algebraic aspects of digital communications, NATO Sci. Peace Secur. Ser. D Inf. Commun. Secur., vol. 24, IOS, Amsterdam, 2009, pp. 193–237. MR2605301

[25] Michael Stoll and John E. Cremona, *On the reduction theory of binary forms*, J. Reine Angew. Math. **565** (2003), 79–99. MR2024647 (2005e:11091)

[26] J. Tate, *Algorithm for determining the type of a singular fiber in an elliptic pencil*, Modular functions of one variable, IV (Proc. Internat. Summer School, Univ. Antwerp, Antwerp, 1972), Springer, Berlin, 1975, pp. 33–52. Lecture Notes in Math., Vol. 476. MR0393039

[27] Paul van Wamelen, *Examples of genus two CM curves defined over the rationals*, Math. Comp. **68** (1999), no. 225, 307–320, DOI 10.1090/S0025-5718-99-01020-0. MR1609658

Department of Mathematics, The University of Texas at Austin, Austin, Texas 78712
Email address: beshaj@math.utexas.edu

Departamento de Matemática y Estadística, Universidad de La Frontera, Temuco, Chile
Email address: ruben.hidalgo@ufrontera.cl

Department of Mathematics and Statistics, Oakland University, Mathematics and Science Center, Rochester, Michigan 48309
Email address: kruk@oakland.edu

Department of Mathematics, Utah State University, Logan, Utah 84322
Email address: andreas.malmendier@usu.edu

Departamento de Matemática y Estadística, Universidad de La Frontera, Temuco, Chile.
Email address: saul.quispe@ufrontera.cl

Department of Mathematics and Statistics, Oakland University, Mathematics and Science Center, Rochester, Michigan 48309
Email address: shaska@oakland.edu

Contemporary Mathematics
Volume **703**, 2018
http://dx.doi.org/10.1090/conm/703/14133

Strong arithmetic mirror symmetry and toric isogenies

Christopher Magyar and Ursula Whitcher

Abstract. We say a mirror pair of Calabi-Yau varieties exhibits strong arithmetic mirror symmetry if the number of points on each variety over a finite field is equivalent, modulo the order of that field. We search for strong mirror symmetry in pencils of toric hypersurfaces generated using polar dual pairs of reflexive polytopes. We characterize the pencils of elliptic curves where strong arithmetic mirror symmetry arises, and provide experimental evidence that the phenomenon generalizes to higher dimensions. We also provide experimental evidence that pencils of K3 surfaces with the same Picard-Fuchs equation have related point counts.

1. Introduction

Mathematical formulations of mirror symmetry have led to deep insights regarding relationships between very different geometric spaces. In some cases, mirror pairs of varieties have related arithmetic structure.

Let $X/\mathbb{F}_q$ be an algebraic variety over the finite field of order q, and let $\#X(\mathbb{F}_{q^s})$ be the number of $\mathbb{F}_{q^s}$-rational points on X. Then the *zeta function* of X is the generating function given by

$$Z(X/\mathbb{F}_q, T) := \exp\left(\sum_{s=1}^{\infty} \#X(\mathbb{F}_{q^s})\frac{T^s}{s}\right) \in \mathbb{Q}[[T]].$$

The celebrated Weil conjectures, proved in [**Dwo60**], state that the zeta function is a rational function of T, with the degrees of the numerator and denominator determined by the Betti numbers of X. Mirror symmetry for Calabi-Yau threefolds entails an interchange of the Hodge numbers $h^{1,1}$ and $h^{2,1}$, and thus a relationship between the Betti numbers of mirror Calabi-Yau threefolds. Thus, we may use information about the structure of the zeta function of a Calabi-Yau threefold to make predictions about the structure of the zeta function of its mirror.

The authors of [**CDRV00, CDRV01**] identified a much stronger relationship between zeta functions for a specific mirror family of Calabi-Yau threefolds. Any smooth homogeneous quintic polynomial in five variables defines a Calabi-Yau threefold as a hypersurface in $\mathbb{P}^4$. We may construct the mirror family to these

2010 *Mathematics Subject Classification.* Primary 11G42, 14H52, 14J28.

The first author was supported in part by Student Blugold Commitment Differential Tuition funds through the University of Wisconsin-Eau Claire Student/Faculty Research Collaboration program.

quintics using the Greene-Plesser construction, as follows. Consider the Fermat quintic pencil X_t given by

$$x_0^5 + x_1^5 + x_2^5 + x_3^5 + x_4^5 - 5tx_0x_1x_2x_3x_4 = 0.$$

Each member of the pencil admits a group action by $(\mathbb{Z}/5\mathbb{Z})^3$, which consists of multiplying each coordinate by suitably chosen fifth roots of unity. Taking the quotient by this group action and choosing a minimal resolution of singularities, we obtain a new pencil of Calabi-Yau threefolds, Y_t. Then Y_t is the mirror family to the family of smooth quintics in $\mathbb{P}^4$. The computations in [**CDRV01**] show that $Z(X_t/\mathbb{F}_q, T)$ and $Z(Y_t/\mathbb{F}_q, T)$ share a common quartic factor (depending on t) in their numerator.

One may try to extend the arithmetic mirror symmetry relationships observed for the Fermat quintic pencil in two ways: by considering the Fermat pencil and its Greene-Plesser mirror in other dimensions, or by considering other mirror constructions.

Let X_t be the Fermat quartic pencil. In this case, the zeta function and mirror zeta function have been described explicitly. In [**Kad04**], building on results of Dwork, the zeta function is factored as:

$$Z(X_t/\mathbb{F}_p, T) = \frac{1}{(1-T)(1-pT)(1-p^2T)R_t(T)Q_t(T)}$$

Here $R_t(T)$ is a degree 3 polynomial of the form $(1 \pm pT)(1 - a_tT + p^2T)$, where a_t is a constant, and Q_t is a polynomial of degree 18.

Let Y_t be the mirror family to quartics in $\mathbb{P}^3$ (constructed using Greene-Plesser and the Fermat pencil). Then [**Kad04**] gives the mirror zeta function as:

$$Z(Y_t/\mathbb{F}_p, T) = \frac{1}{(1-T)(1-pT)^{19}(1-p^2T)R_t(T)}.$$

The factor $R_t(T)$ corresponds to periods of the holomorphic form and its derivatives, and is invariant under mirror symmetry. Furthermore, by applying Tate's conjecture one can show that for fields $\mathbb{F}_q$ containing sufficiently many roots of unity, $Z(X_t/\mathbb{F}_q, T) = Z(Y_t/\mathbb{F}_q, T)$ (see [**DKSSVW17**]).

In [**Wan06**], Wan characterized the arithmetic mirror symmetry phenomenon that the authors of [**CDRV01**] observed for the Fermat quintic pencil as *strong arithmetic mirror symmetry*. In Wan's formulation, strong mirror symmetry arises when one has one-parameter families of Calabi-Yau varieties X_t and Y_t such that X_t and Y_t are mirrors for each t where both are smooth. In particular, if we generalize the Greene-Plesser construction to smooth $n+1$-folds in $\mathbb{P}^n$, we obtain strong mirrors in this sense. Wan showed that these families of $n+1$-folds satisfy an arithmetic relationship; we take this relationship as our definition of strong arithmetic mirror symmetry.

DEFINITION 1.1. Let X and W be a mirror pair of smooth Calabi-Yau varieties defined over $\mathbb{Q}$. Suppose that for any prime p where X and W have good reduction and any finite field $\mathbb{F}_q$ of characteristic p and order q, X and W satisfy

$$\#X(\mathbb{F}_q) = \#W(\mathbb{F}_q) \pmod{q}.$$

Then we say X and W exhibit *strong arithmetic mirror symmetry* and are a *strong mirror pair*.

Work of [**AS16**] and [**Yu09**] on Dwork's unit root shows that the point counts (mod q) for the Fermat pencil are related to solutions of the Picard-Fuchs equations for the family. The Picard-Fuchs equation for the mirror pencil is identical to the Fermat pencil's Picard-Fuchs equation, but whether [**Yu09**] and [**AS16**]'s results on unit roots can be generalized to give a new proof of strong arithmetic mirror symmetry for this family is not known. One of our goals in the present work is to provide further evidence for a relationship between Picard-Fuchs equations and point counts.

What is known about arithmetic mirror symmetry for other mirror constructions? There is a natural extrapolation of the Greene-Plesser mirror construction to deformations of diagonal hypersurfaces in weighted projective spaces. Kadir, inspired by calculations for a specific two-parameter family of Calabi-Yau threefolds in a weighted projective space, argued that the zeta functions of any Greene-Plesser mirror pair of Calabi-Yau threefolds will have a shared common factor in their numerator in [**Kad06**]. Motivated by the Greene-Plesser construction, Fu and Wan showed in [**FW06**] that for appropriately chosen groups G,

$$\#X(\mathbb{F}_q) = \#(X/G)(\mathbb{F}_q) \pmod q.$$

However, these results for orbifolds do not extend to the mirror family, where singularities have been resolved. More recent investigations of arithmetic mirror symmetry have focused on the Berglund-Hübsch-Krawitz mirror construction (cf. [**DKSSVW17, AP15**]). This allows an extension of observations made about the mirrors of diagonal hypersurfaces to the mirrors of certain other hypersurfaces in weighted projective spaces.

In the current work, we generalize Fermat pencils in projective space to certain one-parameter families in toric varieties, using Batyrev's mirror symmetry construction. From the toric point of view, the Greene-Plesser mirror of the Fermat pencil is determined by choosing monomials corresponding to the vertices and origin of the polytope for $\mathbb{P}^n$, and the Fermat pencil itself is determined by choosing the vertex and origin monomials of the polar dual polytope. We use the vertices and the origin in polar dual pairs of reflexive polytopes to construct pairs of pencils of hypersurfaces, and experimentally investigate when these pairs exhibit strong arithmetic mirror symmetry, testing small prime numbers. We find evidence for strong mirror symmetry when the pairs of pencils are related by resolutions of finite quotient maps, in a further generalization of the Greene-Plesser mirror construction.

For elliptic curves, we apply the classical Tate's isogeny theorem to describe the appropriate strong arithmetic mirror pairs completely. We identify new mirror pairs of K3 surface pencils which exhibit strong arithmetic mirror symmetry for small primes, and describe conjectural, combinatorial criteria for finding similar examples in higher dimensions. We show that our K3 surface pencils are of high Picard rank. The non-weighted projective space examples fall naturally into two groups with the same Picard-Fuchs equations, which we use to study the geometric structure of these families. Experimentally, we detect relationships between point counts for different K3 surface pencils in the same group, even when these pencils were not generated as a mirror pair. Because members of the same group are related by resolutions of quotients by finite abelian groups, we may view the point-counting relationship as a higher-dimensional generalization of Tate's isogeny theorem.

2. Hypersurfaces in toric varieties

We begin by reviewing the construction of Calabi-Yau varieties as toric hypersurfaces, in order to fix notation. Let N be a lattice isomorphic to $\mathbb{Z}^n$. A *cone* in N is a subset of the real vector space $N_{\mathbb{R}} = N \otimes \mathbb{R}$ generated by nonnegative $\mathbb{R}$-linear combinations of a set of vectors $\{v_1, \ldots, v_m\} \subset N$. We assume that cones are strongly convex, that is, they contain no line through the origin. We say a cone is *simplicial* if its generators are linearly independent over $\mathbb{R}$. A *fan* Σ consists of a finite collection of cones such that each face of a cone in the fan is also in the fan, and any pair of cones in the fan intersects in a common face. We say a fan Σ in $N_{\mathbb{R}}$ is *complete* if the union of all of the cones in Σ is the entire vector space $N_{\mathbb{R}}$, and we say a fan is simplicial if all of its cones are simplicial.

Given a complete fan Σ, we may construct a toric variety in the following way. Let $\Sigma(1) = \{v_1, \ldots, v_q\}$ be lattice points generating the one-dimensional cones of Σ. To each lattice point v_j, we associate a coordinate z_j on $\mathbb{C}^q$. Let $\mathcal{S}$ denote any subset of $\Sigma(1)$ that does *not* span a cone of Σ, and let $\mathcal{V}(\mathcal{S}) \subseteq \mathbb{C}^q$ be the linear subspace defined by setting $z_j = 0$ if the corresponding cone is in $\mathcal{S}$. Define $Z(\Sigma) = \cup_{\mathcal{S}} \mathcal{V}(\mathcal{S})$. Then $(\mathbb{C}^*)^q$ acts on $\mathbb{C}^q - Z(\Sigma)$ by coordinatewise multiplication. Let us write v_j in coordinates as $(v_{j1}, \ldots, v_{jn})$. Let $\phi : (\mathbb{C}^*)^q \to (\mathbb{C}^*)^n$ be given by

$$\phi(t_1, \ldots, t_q) \mapsto \left(\prod_{j=1}^{q} t_j^{v_{j1}}, \ldots, \prod_{j=1}^{q} t_j^{v_{jn}} \right)$$

The kernel of ϕ is isomorphic to $(\mathbb{C}^*)^{q-n} \times A$, where A is a finite abelian group. The toric variety V_Σ associated with the fan Σ is given by

$$V_\Sigma = (\mathbb{C}^q - Z(\Sigma))/\mathrm{Ker}(\phi).$$

The variables z_j may be viewed as homogeneous coordinates on V_Σ, subject to the identifications required by $\mathrm{Ker}(\phi)$.

The dual lattice M of N is given by $\mathrm{Hom}(N, \mathbb{Z})$; it is also isomorphic to $\mathbb{Z}^n$. We write the pairing of $v \in N$ and $w \in M$ as $\langle v, w \rangle$. We will use pairs of polytopes in $N_{\mathbb{R}}$ and $M_{\mathbb{R}}$ to construct pairs of toric varieties. A *lattice polytope* Δ has vertices in a lattice. Let Δ be a lattice polytope in N which has the origin as an interior point. The *polar dual* of Δ is defined as

$$\Delta^\circ = \{w \in M_{\mathbb{R}} \mid \langle v, w \rangle \geq -1 \, \forall \, v \in \Delta\}.$$

If both Δ and Δ° are lattice polytopes, we call them *reflexive polytopes.* We illustrate a two-dimensional pair of reflexive polytopes in Figure 1.

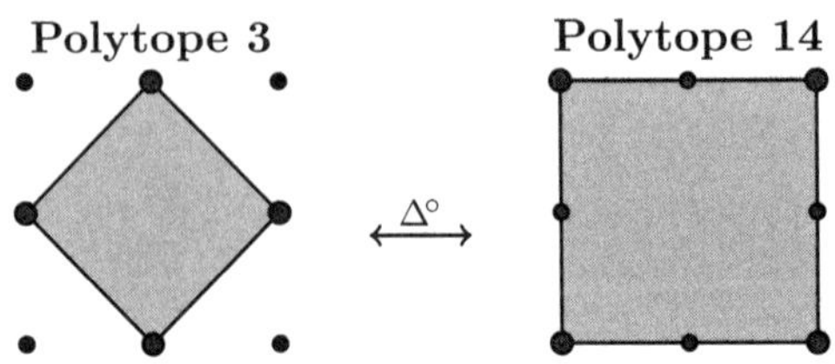

FIGURE 1. Dual reflexive quadrilaterals.

Reflexive polytopes have been classified in two, three, and four dimensions. Under multiplication by elements of $\mathrm{GL}(n, \mathbb{Z})$, there are 16, 4319, and 473 800 776 equivalence classes of reflexive polytopes, respectively (see [**KS00**]).

Given a reflexive polytope Δ, we may obtain a fan R by taking cones over the polytope's faces. The one-dimensional cones of R are then generated by the vertices of Δ, and the n-dimensional cones are given by the facets of Δ. For example, the quadrilateral labeled Polytope 3 in Figure 1 corresponds to the fan that has two-dimensional cones generated by $\{e_1, e_2\}$, $\{e_1, -e_2\}$, $\{-e_1, e_2\}$, and $\{-e_1, -e_2\}$. The resulting toric variety is $\mathbb{P}^1 \times \mathbb{P}^1$.

Computationally, we identify $\mathrm{Ker}(\phi)$ using the matrix Mat_Δ whose columns are given by the vertices of Δ. The action of $(\mathbb{C}^*)^{q-n}$ is given by the kernel of Mat_Δ as an integer matrix, multiplying by Mat_Δ on the left. The free abelian group A is given by the elementary divisors of Mat_Δ. (These are the entries of the Smith normal form on the main diagonal.) For example, if we let Δ be the quadrilateral labeled Polytope 3 in Figure 1 and Δ° its polar dual, both Mat_Δ and $\mathrm{Mat}_{\Delta^\circ}$ have kernels generated by $(1,0,0,1)$ and $(0,1,1,0)$ in $\mathbb{Z}^4$. However, the elementary divisors of Mat_Δ are both 1, while the elementary divisors of $\mathrm{Mat}_{\Delta^\circ}$ are 1 and 2. As we have already observed, Δ determines the toric variety $\mathbb{P}^1 \times \mathbb{P}^1$, while Δ° yields a quotient of $\mathbb{P}^1 \times \mathbb{P}^1$ by a finite abelian group of order 2.

The fan R given by cones over a polytope's faces will be complete, but will not necessarily be simplicial. Thus, the corresponding toric variety may have singularities. However, we can construct a maximal simplicial refinement Σ by using all of the lattice points of Δ to generate one-dimensional cones; such a refinement always exists by results of [**OP91**]. This operation on fans corresponds to resolution of singularities on the level of toric varieties. Anticanonical hypersurfaces in toric varieties obtained from reflexive polytopes are Calabi-Yau varieties. If the dimension of the reflexive polytope is at most four, taking a maximal simplicial refinement Σ guarantees that the general anticanonical hypersurface is smooth. Thus, two-dimensional reflexive polytopes determine families of elliptic curves, three-dimensional reflexive polytopes determine families of K3 surfaces, and four-dimensional reflexive polytopes yield families of Calabi-Yau threefolds. Polar dual pairs of reflexive polytopes correspond to mirror pairs of Calabi-Yau varieties.

EXAMPLE 2.1. Let $e_1, \ldots, e_n$ be generators of N. Let Δ be the simplex with vertices $\{e_1, \ldots, e_n, -e_1 + -e_2 + \cdots + -e_n\}$. The toric variety corresponding to the fan over the faces of Δ is $\mathbb{P}^n$; the anticanonical hypersurfaces are homogeneous polynomials in $n+1$ variables of degree $n+1$. A maximal simplicial refinement of the fan over the faces of Δ° yields Calabi-Yau varieties isomorphic to the mirror varieties Y_t obtained from the Greene-Plesser construction.

Any reflexive simplex defines either a weighted projective space or the resolution of a quotient of a weighted projective space by a finite abelian group. This observation allows us to study weighted projective space generalizations of the Greene-Plesser construction in a toric context.

We may use the lattice points of a reflexive polytope and its polar dual to write an explicit expression for an anticanonical hypersurface in homogeneous coordinates. Let $\{v_k\} \subset \Delta \cap N$ generate the one-dimensional cones of Σ, and choose an arbitrary complex number c_x for each lattice point x of Δ°. Then an anticanonical hypersurface is given by

$$f = \sum_{x \in \Delta^\circ \cap M} c_x \prod_{k=1}^{q} z_k^{\langle v_k, x\rangle + 1}. \tag{2.1}$$

If Δ is the simplex described in Example 2.1, we may use Equation 2 to specify the Fermat pencil in $\mathbb{P}^n$ by choosing special values of the parameters c_x. Specifically, we set the parameters corresponding to the vertices of Δ° equal to one, associate a single parameter t to the origin, and set the parameters corresponding to other lattice points to 0. Combinatorially, this operation makes sense for any reflexive polytope. Thus, we define a one-parameter family of polynomials $f_{\Delta,t}$ by:

$$f_{\Delta,t} = \left(\sum_{x \,\in\, \mathtt{vertices}(\Delta^\circ)} \prod_{k=1}^{q} z_k^{\langle v_k, x\rangle + 1} \right) + t \prod_{k=1}^{q} z_k. \tag{2.2}$$

For a polar dual pair of reflexive polytopes Δ and Δ°, $f_{\Delta,t}$ and $f_{\Delta^\circ,t}$ determine one-parameter subfamilies X_t and Y_t of the mirror families of Calabi-Yau varieties associated to the polytopes. We investigate when X_t and Y_t possess strong arithmetic mirror symmetry in the following sections.

3. Elliptic curves

Polar dual pairs of two-dimensional reflexive polytopes determine mirror families of elliptic curves. Recall that an *isogeny* between two elliptic curves E_1 and E_2 is a (non-constant) rational morphism $\phi : E_1 \to E_2$ that maps the identity element of E_1 to the identity element of E_2. Isogenies of elliptic curves are finite surjective maps. Tate's Isogeny Theorem ([**Tat66**]) relates point counts on isogenous curves: two elliptic curves over a finite field $\mathbb{F}_q$ are isogenous if and only if they have the same number of $\mathbb{F}_q$-rational points.

THEOREM 3.1. *Let Δ and Δ° be a polar dual pair of reflexive polygons, and let Σ and Σ° be maximal simplicial refinements of the fans over the faces of Δ and Δ°, respectively. Let X_t and Y_t be members of the families determined by the polynomials $f_{\Delta,t}$ and $f_{\Delta^\circ,t}$. Then $\#X_t(\mathbb{F}_q) = \#Y_t(\mathbb{F}_q)$ for each rational t where X_t and Y_t are smooth when one of the following holds:*

(1) *The polygons Δ and Δ° are triangles. In this case, either V_Σ or V_{Σ° is a weighted projective space.*
(2) *The polygon Δ is self-dual.*
(3) *The polygons Δ and Δ° are quadrilaterals, and either V_Σ or V_{Σ° is $\mathbb{P}^1 \times \mathbb{P}^1$.*

PROOF. We examine the sixteen equivalence classes of reflexive polygons directly. When Δ and Δ° are triangles, self-dual, or equivalent to the pair illustrated in Figure 1, then either V_{Σ° is a resolution of a finite quotient of V_Σ, or V_Σ is a finite quotient of V_{Σ°. In this case, the quotient map on toric varieties induces an isogeny between each pair of smooth elliptic curves X_t and Y_t. By Tate's Isogeny Theorem, we have $\#X_t(\mathbb{F}_q) = \#Y_t(\mathbb{F}_q)$. □

The cases described in Theorem 3.1 include 10 of the 16 equivalence classes of reflexive polygons. In the remaining 6 cases, we used SageMath ([**S+16**]) to count points on X_t and Y_t in the case that q is a prime number less than 100, and $t = 0, 1, \ldots, q-1$. We found that $\#X_t(\mathbb{F}_q)$ and $\#Y_t(\mathbb{F}_q)$ were different in these cases. Applying Tate's Isogeny Theorem, we see that there is no isogeny between X_t and Y_t. Thus, Theorem 3.1 characterizes the situations where strong arithmetic symmetry arises for $f_{\Delta,t}$ and $f_{\Delta^\circ,t}$ completely.

4. Experimental evidence for strong mirror symmetry

Let us now investigate families of K3 surfaces determined by three-dimensional reflexive polytopes. We wish to generalize the results of Theorem 3.1. Thus, we seek to identify mirror pairs where the corresponding toric varieties are related by a resolution of a finite quotient map.

When we take the fan R over the faces of a polytope Δ with q vertices, we obtain the toric variety $(\mathbb{C}^q - Z(R))/((\mathbb{C}^*)^{q-n} \times A)$, where A is a finite abelian group. The zero set $Z(R)$ is determined by the combinatorial structure of Δ: combinatorially equivalent polytopes, which have the same face poset (and in particular the same number of vertices), will determine the same zero set. As we observed in Section 2, the action of $(\mathbb{C}^*)^{q-n}$ is given by the kernel of Mat_Δ as an integer matrix. We therefore deduce:

LEMMA 4.1. *Let Δ and Γ be combinatorially equivalent reflexive polytopes, and let R and T be the fans over the faces of Δ and Γ respectively. Fix an ordering of the vertices that respects the combinatorial equivalence, and let M_Δ and M_Γ be the matrices with columns given by the vertices in this ordering. Suppose the kernels of Mat_Δ and Mat_Γ are the same submodule of $\mathbb{Z}^q$. Then there exists a toric variety W and finite abelian subgroups of the torus A and B such that $V(R) = W/A$ and $V(T) = W/B$.*

PROOF. Let $W = (\mathbb{C}^q - Z(R))/(\mathbb{C}^*)^{q-n}$, with the action of $(\mathbb{C}^*)^{q-n}$ given by the kernel of Mat_Δ. Then the result follows immediately. □

We searched the database of 4319 three-dimensional reflexive polytopes for polar dual pairs of polytopes satisfying the hypotheses of Lemma 4.1. Combinatorial equivalence is a fairly weak criterion: 1299 of the 4319 three-dimensional reflexive polytopes are combinatorially equivalent to their polar duals. The kernel condition is satisfied far less often: we found 79 self-dual polytopes (including 6 simplices), and 26 pairs of non-self-dual reflexive polytopes with this property. The simplices correspond to Greene-Plesser mirrors in weighted projective spaces, and certain finite quotients of these mirrors. We are particularly interested in new, non-weighted projective space examples. Of the 26 pairs of non-self-dual polytopes that we identified, 5 were not pairs of simplices. These are polytopes 3, 10, 433, 436, 753 and their polar duals, using the numbering in the SageMath database [**S+16**]. Polytopes 3, 753, and their polar duals yield vertex matrices with the same kernels; we classify these polytopes as Group I. Similarly, polytopes 10, 433, 436, and their polar duals also have vertex matrices with the same kernel; we classify these polytopes as Group II. Geometrically, one may write the toric variety V_Σ for any polytope in Group I as a resolution of a finite quotient of the toric variety determined by polytope 3, and one may write the toric variety V_Σ for any polytope in Group II as a resolution of a finite quotient of the toric variety determined by polytope 10. We illustrate Group I and II in Figures 2 and 3.

We use Sage Mathematics Software ([**S+16**]) to count the points on X_t and Y_t for fields $\mathbb{F}_p$ of order 5,7, or 11, where $t = 0, \dots, p-1$. In each case, we generate an explicit list of the points on V_Σ and substitute each point into $f_{\Delta,t}$. We then

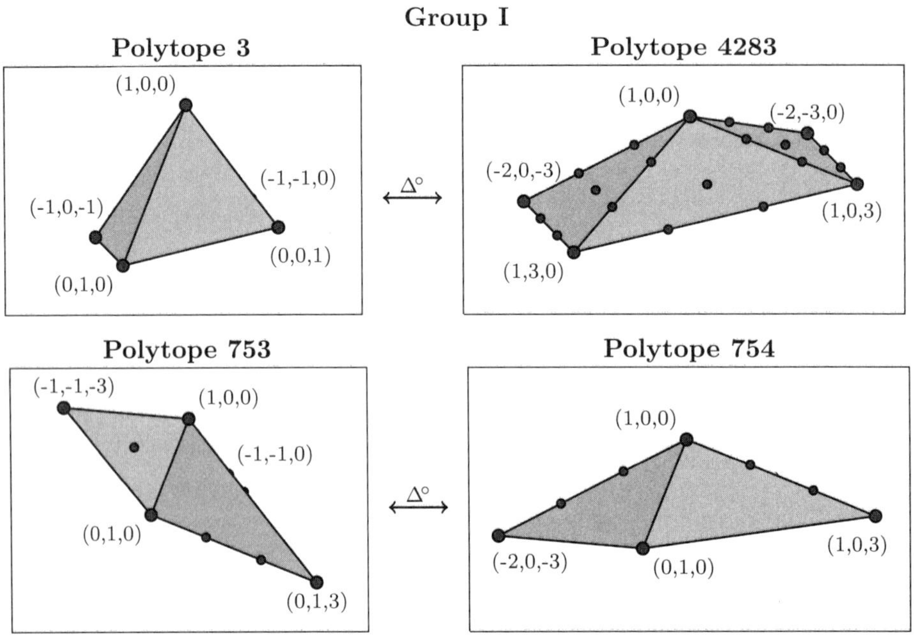

FIGURE 2. Two reflexive polytopes define toric varieties related by quotient maps

compare the point count for $f_{\Delta^\circ,t}$. On a current medium-to-high performance desktop computer, these computations require a minimum of several hours of processor time for each pair Δ and Δ°. We find that for all smooth X_t and Y_t, and $p = 5, 7$, or 11,

$$\#X_t(\mathbb{F}_p) \equiv \#Y_t(\mathbb{F}_p) \pmod p$$

when Δ is in Group I or Group II. Not only do we observe experimental evidence for strong arithmetic mirror symmetry in these cases, we observe further arithmetic relationships between K3 surfaces determined by members of the same group. Indeed, suppose Δ and Γ are both members of either Group I or Group II, let X_t be given by $f_{\Delta,t}$, and let W_t be given by $f_\Gamma(t)$. Suppose X_t and W_t are smooth. Then our experimental evidence shows

$$\#X_t(\mathbb{F}_p) \equiv \#W_t(\mathbb{F}_p) \pmod p$$

for fields $\mathbb{F}_p$ of order 5,7, or 11, where $t = 0, \ldots, p-1$.

We conjecture that strong arithmetic mirror symmetry will hold for similar toric examples in higher dimensions.

CONJECTURE 4.2. *Let Δ and Γ be a pair of reflexive polytopes satisfying the hypotheses of Lemma 4.1, and let X_t and W_t be the hypersurfaces determined by $f_{\Delta,t}$ and $f_\Gamma(t)$, respectively. Then for any rational t where X_t and Y_t are smooth, strong arithmetic mirror symmetry holds:*

$$\#X_t(\mathbb{F}_q) \equiv \#W_t(\mathbb{F}_q) \pmod q.$$

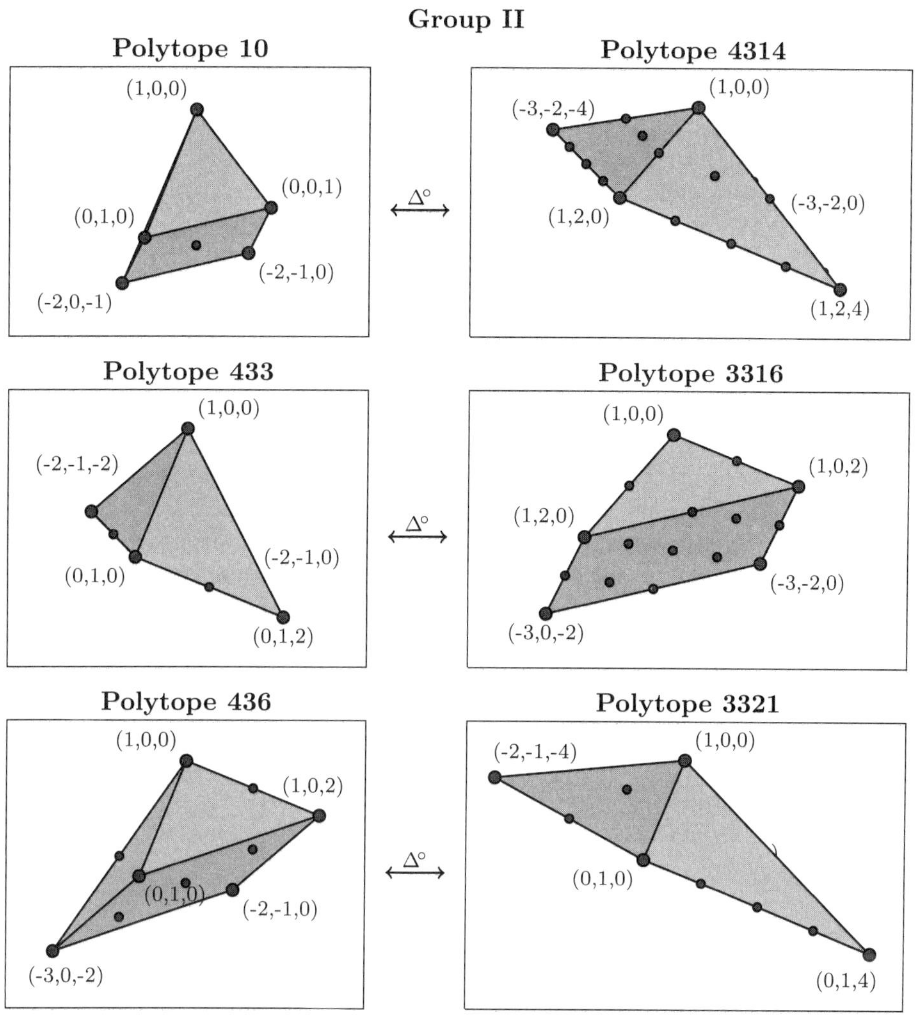

FIGURE 3. Three reflexive polytopes define toric varieties related by quotient maps

5. Picard-Fuchs equations

There is a moduli space of lattice-polarized K3 surfaces. Within this moduli space lie one-parameter families of generically Picard rank 19 K3 surfaces. Because our toric construction yields one-parameter families in a combinatorially natural way, we expect that these are generically Picard rank 19 families. We say a family of K3 surfaces is *lattice-polarized* if there exists a lattice M such that $M \hookrightarrow \mathrm{Pic}(X)$ for every K3 surface X in the family. The dimension of M places a lower bound on the Picard rank of each member of the family. Lattice-polarized Picard rank 19 families have special structure: for example, Picard rank 19 families polarized by the lattice $U \oplus E_8 \oplus E_8 + (-2n)$, where U is a unimodular lattice of signature $(1,1)$ and E_8 is taken to be negative definite, are geometrically related to a product of two elliptic curves with an n-isogeny, via the *Shioda-Inose structure*. More generally, Dolgachev described a mirror map for lattice-polarized families of K3 surfaces in [**D96**].

Computing the general Picard lattice for a specific parametrized family of K3 surfaces is typically a difficult and ad-hoc process. Rather than doing so directly, we will use differential equations to extract information about the geometric structure of our families, following the strategy of [**Dor00**].

A *period* is the integral of a differential form with respect to a specified homology class. Periods of holomorphic forms encode the complex structure of varieties. The *Picard-Fuchs differential equation* of a family of varieties is a differential equation that describes the way the value of a period changes as we move through the family. Solutions to Picard-Fuchs equations for holomorphic forms on Calabi-Yau varieties can be used to define a mirror map, which relates the moduli space of complex deformations of a variety to a moduli space of complexified Kähler deformations of the mirror variety.

We use the Griffiths-Dwork technique to compute the Picard-Fuchs equation for the holomorphic form on the pencil $f_{\Delta,t}$ for each of the reflexive polytopes Δ in Group I and Group II, following the method outlined in [**KLMSW13**]. The quotient maps which relate different toric varieties determined by polytopes in Group I or Group II preserve the holomorphic forms on the K3 surfaces; thus, the Picard-Fuchs equation will be the same for every member of a group. We obtain the pair of third-order differential equations given in Table 1. By [**Dor00**], because the Picard-Fuchs equations are third-order, the Picard rank of a general K3 surface in each family must be 19, as we expected.

TABLE 1. Picard-Fuchs equations

Group	**Picard-Fuchs Equation**
I	$(3t)\,\mathcal{P} + \left(7t^2\right)\mathcal{P}' + \left(6t^3 + 162\right)\mathcal{P}'' + \left(t^4 + 108t\right)\mathcal{P}''' = 0$
II	$(t)\,\mathcal{P} + \left(7t^2\right)\mathcal{P}' + \left(6t^3\right)\mathcal{P}'' + \left(t^4 - 256\right)\mathcal{P}''' = 0$

As we noted earlier, the Picard-Fuchs equation for the Fermat pencil determines its point count $\pmod q$. One might speculate that the Picard-Fuchs equations are enough to predict the relationships between point counts for members of each group.

QUESTION 5.1. *Let V_t and W_t be pencils of Calabi-Yau n-folds, and suppose the Picard-Fuchs equation of V_t is the same as the Picard-Fuchs equation of W_t. Is*

$$\#V_t(\mathbb{F}_q) \equiv \#W_t(\mathbb{F}_q) \pmod q$$

when V_t and W_t are smooth?

We note that the Picard-Fuchs equations computed in Table 1 have a special structure: we can relate these equations to lower-order differential equations which appear to be Picard-Fuchs equations of elliptic curves.

DEFINITION 5.2. Let $L(y)$ be a homogeneous linear differential equation with coefficients in $\mathbb{C}(t)$. Then there exists a homogeneous linear differential equation $M(y) = 0$ with coefficients in $\mathbb{C}(t)$, such that the solution space of $M(y)$ is the $\mathbb{C}$-span of

$$\{\nu_1\nu_2 \mid L(\nu_1) = 0 \text{ and } L(\nu_2) = 0\}\ .$$

We say the differential equation $M(y)$ is the *symmetric square* of L.

In particular, the symmetric square of the second-order linear, homogeneous differential equation

$$a_2 \frac{\partial^2 \omega}{\partial t^2} + a_1 \frac{\partial \omega}{\partial t} + a_0 \omega = 0$$

is

$$\begin{aligned} a_2^2 \frac{\partial^3 \omega}{\partial t^3} + 3a_1 a_2 \frac{\partial^2 \omega}{\partial t^2} + (4a_0 a_2 + 2a_1^2 + a_2 a_1' - a_1 a_2') \frac{\partial \omega}{\partial t} \\ + (4a_0 a_1 + 2a_0' a_2 - 2a_0 a_2')\omega = 0 \end{aligned} \tag{5.1}$$

where primes denote derivatives with respect to t.

THEOREM 5.3. [**Dor00**, Theorem 5] *The Picard-Fuchs equation of a family of Picard rank-19 lattice-polarized K3 surfaces is a third-order ordinary differential equation which can be written as the symmetric square of a second-order homogeneous linear Fuchsian differential equation.*

The Picard-Fuchs equations for the families in Group I and Group II are symmetric squares. We give the corresponding symmetric square roots, which are second-order differential equations, in Table 2.

TABLE 2. Symmetric square roots of Picard-Fuchs equations

Group	Symmetric Square Root
I	$\left(\frac{t^2}{4}\right) \mathcal{P} + \left(2t^3 + 54\right) \mathcal{P}' + \left(t^4 + 108t\right) \mathcal{P}'' = 0$
II	$\left(\frac{t^2}{4}\right) \mathcal{P} + \left(2t^3\right) \mathcal{P}' + \left(t^4 - 256\right) \mathcal{P}'' = 0$

DEFINITION 5.4. The *projective normal form* of an order k Fuchsian differential equation

$$f_0(t)\,\mathcal{P} + f_1(t)\,\mathcal{P}' + \cdots + f_{k-1}(t)\,\mathcal{P}^{(k-1)} + \mathcal{P}^{(k)} = 0$$

is the unique order k Fuchsian differential equation without a $(k-1)$ order derivative

$$g_0(t)\,\mathcal{R} + g_1(t)\,\mathcal{R}' + \cdots + g_{k-2}(t)\,\mathcal{R}^{(k-2)} + \mathcal{R}^{(k)}, \qquad g_i(t) \in \mathbb{C}(t)$$

obtained by scaling the fundamental solutions by the kth root of the Wronskian.

In [**Dor00**], Doran showed that the mirror map of a one-parameter family of Calabi-Yau varieties is determined by the projective normal form of its Picard-Fuchs equation.

In the case of a second-order Fuchsian differential equation

$$f_0(t)\,\mathcal{P} + f_1(t)\,\mathcal{P}' + \mathcal{P}'' = 0$$

the projective normal form is simply

$$\left(f_0(t) - \frac{1}{2} f_1'(t) - \frac{1}{4} f_1(t)^2\right) \mathcal{R} + \mathcal{R}'' = 0.$$

We compute the projective normal forms for our symmetric square roots in Table 3.

We saw in § 3 that mirror relationships for elliptic curves can be described as isogenies, which preserve arithmetic information. One might ask whether the strong arithmetic mirror symmetry relationship we observe for members of Group I and

TABLE 3. Projective normal forms of symmetric square roots

Group	Projective Normal Form
I	$\left(\frac{t^6 - 540t^3 + 8748}{4t^2\left(t^3+108\right)^2}\right)\mathcal{R} + \mathcal{R}'' = 0$
II	$\left(\frac{t^2\left(t^4+2816\right)}{4\left(t^2+16\right)^2\left(t+4\right)^2\left(t-4\right)^2}\right)\mathcal{R} + \mathcal{R}'' = 0$

Group II ultimately derives from a property of elliptic curves: is there a geometric construction relating our families of K3 surfaces to families of elliptic curves with Picard-Fuchs equations described above?

Acknowledgments

The authors thank Volker Braun and the participants of Sage Days 50 for many interesting conversations, the AIM SQuaRE program for an inspirational research environment, and the anonymous referees for suggestions which materially improved our exposition.

References

[AS16] A. Adolphson and S. Sperber, *Distinguished-root formulas for generalized Calabi-Yau hypersurfaces*, preprint, 2016. arXiv:1602.03578 [math.AG]

[AP15] Marco Aldi and Andrija Peruničić, *p-adic Berglund-Hübsch duality*, Adv. Theor. Math. Phys. **19** (2015), no. 5, 1115–1139, DOI 10.4310/ATMP.2015.v19.n5.a5. MR3487652

[Bat94] Victor V. Batyrev, *Dual polyhedra and mirror symmetry for Calabi-Yau hypersurfaces in toric varieties*, J. Algebraic Geom. **3** (1994), no. 3, 493–535. MR1269718

[CDRV00] Philip Candelas, Xenia de la Ossa, and Fernando Rodriguez-Villegas, *Calabi-Yau manifolds over finite fields. II*, Calabi-Yau varieties and mirror symmetry (Toronto, ON, 2001), Fields Inst. Commun., vol. 38, Amer. Math. Soc., Providence, RI, 2003, pp. 121–157. MR2019149

[CDRV01] Philip Candelas, Xenia de la Ossa, and Fernando Rodriguez-Villegas, *Calabi-Yau manifolds over finite fields. II*, Calabi-Yau varieties and mirror symmetry (Toronto, ON, 2001), Fields Inst. Commun., vol. 38, Amer. Math. Soc., Providence, RI, 2003, pp. 121–157. MR2019149

[D96] I. V. Dolgachev, *Mirror symmetry for lattice polarized K3 surfaces*, J. Math. Sci. **81** (1996), no. 3, 2599–2630, DOI 10.1007/BF02362332. Algebraic geometry, 4. MR1420220

[Dor00] Charles F. Doran, *Picard-Fuchs uniformization and modularity of the mirror map*, Comm. Math. Phys. **212** (2000), no. 3, 625–647, DOI 10.1007/s002200000228. MR1779161

[DKSSVW17] C.F. Doran, T. Kelly, A. Salerno, S. Sperber, J. Voight, and U. Whitcher, *Arithmetic mirror symmetry for symmetric K3 quartic pencils*, preprint 2017. arXiv:1612.09249.

[Dwo60] Bernard Dwork, *On the rationality of the zeta function of an algebraic variety*, Amer. J. Math. **82** (1960), 631–648, DOI 10.2307/2372974. MR0140494

[FW06] Lei Fu and Daqing Wan, *Mirror congruence for rational points on Calabi-Yau varieties*, Asian J. Math. **10** (2006), no. 1, 1–10, DOI 10.4310/AJM.2006.v10.n1.a4. MR2213683

[Kad04] S. Kadir, *The Arithmetic of Calabi–Yau Manifolds and Mirror Symmetry*, Oxford DPhil Thesis, 2004. arXiv: hep-th/0409202

[Kad06] Shabnam Kadir, *Arithmetic mirror symmetry for a two-parameter family of Calabi-Yau manifolds*, Mirror symmetry. V, AMS/IP Stud. Adv. Math., vol. 38, Amer. Math. Soc., Providence, RI, 2006, pp. 35–86. MR2282954

[KLMSW13] Dagan Karp, Jacob Lewis, Daniel Moore, Dmitri Skjorshammer, and Ursula Whitcher, *On a family of K3 surfaces with S_4 symmetry*, Arithmetic and geometry of K3 surfaces and Calabi-Yau threefolds, Fields Inst. Commun., vol. 67, Springer, New York, 2013, pp. 367–386, DOI 10.1007/978-1-4614-6403-7_12. MR3156422

[KS00] Maximilian Kreuzer and Harald Skarke, *Complete classification of reflexive polyhedra in four dimensions*, Adv. Theor. Math. Phys. **4** (2000), no. 6, 1209–1230, DOI 10.4310/ATMP.2000.v4.n6.a2. MR1894855

[S+16] *Sage Mathematics Software (Version 7.0)*, The Sage Developers, 2016, `http://www.sagemath.org`.

[Tat66] John Tate, *Endomorphisms of abelian varieties over finite fields*, Invent. Math. **2** (1966), 134–144, DOI 10.1007/BF01404549. MR0206004

[OP91] Tadao Oda and Hye Sook Park, *Linear Gale transforms and Gel′fand-Kapranov-Zelevinskij decompositions*, Tohoku Math. J. (2) **43** (1991), no. 3, 375–399, DOI 10.2748/tmj/1178227461. MR1117211

[Yu09] Jeng-Daw Yu, *Variation of the unit root along the Dwork family of Calabi-Yau varieties*, Math. Ann. **343** (2009), no. 1, 53–78, DOI 10.1007/s00208-008-0265-9. MR2448441

[Wan06] Daqing Wan, *Mirror symmetry for zeta functions*, Mirror symmetry. V, AMS/IP Stud. Adv. Math., vol. 38, Amer. Math. Soc., Providence, RI, 2006, pp. 159–184. With an appendix by C. Douglas Haessig. MR2282960

Department of Mathematics, University of Wisconsin–Eau Claire, Eau Claire, Wisconsin 54702

Email address: `magyarca@uwec.edu`

Department of Mathematics, University of Wisconsin–Eau Claire, Eau Claire, Wisconsin 54702

Current address: Mathematical Reviews, American Mathematical Society, Ann Arbor, Michigan 48103

Email address: `uaw@umich.edu`

Contemporary Mathematics
Volume **703**, 2018
http://dx.doi.org/10.1090/conm/703/14134

Inose's construction and elliptic $K3$ surfaces with Mordell-Weil rank 15 revisited

Abhinav Kumar and Masato Kuwata

ABSTRACT. We describe two constructions of elliptic $K3$ surfaces starting from the Kummer surface of the Jacobian of a genus 2 curve. These parallel the base-change constructions of Kuwata for the Kummer surface of a product of two elliptic curves. One of these also involves the analogue of an Inose fibration. We use these methods to provide explicit examples of elliptic $K3$ surfaces over the rationals of geometric Mordell-Weil rank 15.

1. Introduction

Elliptic $K3$ surfaces have been the focus of much work in algebraic geometry and number theory over the last few decades. For such a surface over a field of characteristic 0 with at least one singular fiber, the rank of the Mordell-Weil group can be as large as 18. Cox [**C**] proved in 1989, using the surjectivity of the period map for $K3$ surfaces, that any integer between 0 and 18 actually occurs as the rank of the Mordell-Weil group of an elliptic $K3$ surface over $\mathbb{C}$. However, the proof uses transcendental methods, and does not lead to an algebraic construction of examples of elliptic $K3$ surfaces of a given rank.

In 2001 Kuwata constructed explicit examples of elliptic $K3$ surfaces over $\mathbb{Q}$ whose Mordell-Weil group over $\overline{\mathbb{Q}}$ has rank r, where $0 \leq r \leq 18$ and $r \neq 15$ ([**Kw**]). The proof uses Inose's theorem (see Proposition 2.2), which relies on a transcendental argument. For this reason, it was not clear how to find explicit generators of the Mordell-Weil group of such elliptic surfaces until our recent work [**KK**], where we found a systematic method to describe a set of generators (see also [**Kl2**] for earlier progress on this question). Meanwhile, there have been constructions of elliptic $K3$ surfaces with Mordell-Weil rank 15 ([**Kl1**][**TdZ**]) in order to fill the gap of [**Kw**]. However, these constructions are based on somewhat different ideas from that of [**Kw**], to which the method of [**KK**] does not seem to apply to find generators of the Mordell-Weil group.

The goal of this article is to describe some new constructions of elliptic $K3$ surfaces with high Mordell-Weil rank using ideas closely related to the idea of [**Kw**]. The construction of [**Kw**] is based on the Kummer surfaces $Km(E_1 \times E_2)$ associated with the product of two elliptic curves. For such a Kummer surface Shioda and Inose [**SI**] constructed a double cover with certain properties (now called a Shioda-Inose

2010 *Mathematics Subject Classification.* Primary 14J27, 14J28; Secondary 14H40, 11G10.
Key words and phrases. Elliptic $K3$ surface, Jacobian, Kummer surface, Inose fibration.

structure). We say that a $K3$ surface S has a Shioda-Inose structure if S admits an involution fixing the global 2-form on S, such that the quotient is a Kummer surface $Km(A)$, and provided that the rational quotient map $S \dashrightarrow Km(A)$ induces a Hodge isometry $T_S(2) \simeq T_{Km(A)}$ of (scaled) transcendental lattices. When $A = E_1 \times E_2$ is the product of two elliptic curves, S was constructed in [**SI**] as an elliptic surface having two II^* fibers. Shioda and Inose used this construction to prove a beautiful theorem establishing a one-to-one correspondence between singular $K3$ surfaces and even positive definite binary quadratic forms. Later, Inose [**I1**] constructed S explicitly as a quartic surface in $\mathbb{P}^3$ and as the quotient of a Kummer surface by an involution. Thus, we have a chain of rational maps $S \dashrightarrow Km(A) \dashrightarrow S$, which we now call a "Kummer sandwich" ([**Shi4**]). The construction of $Km(E_1 \times E_2) \dashrightarrow S$ in [**I1**] can be viewed as a base change of elliptic surfaces, and it is this fact that [**Kw**] generalized to construct elliptic $K3$ surfaces of high rank. Elliptic $K3$ surfaces of various Mordell-Weil ranks are constructed as a base change from S, which we call the Inose surface.

Our main idea in this article is to replace $A = E_1 \times E_2$ by the Jacobian $J(C)$ of a curve C of genus 2. We propose two different generalizations of Inose's fibration on $Km(E_1 \times E_2)$. A detailed study of Shioda-Inose structures on $K3$ surfaces was carried out by Morrison [**Mo**]. For Kummer surfaces of principally polarized abelian surfaces, double covers with Shioda-Inose structure were studied by Naruki [**N**] and Dolgachev [**GLD**]. In [**Km1**], Kumar explicitly described these surfaces as a family of elliptic $K3$ surfaces with II^* and III^* fibers, whose Weierstrass coefficients are related to invariants of the genus 2 curve C. In this construction, the rational map $S \dashrightarrow Km(A)$ in [**Km1**] is a 2-isogeny between elliptic surfaces. The dual isogeny sets up the sandwich $S \dashrightarrow Km(A) \dashrightarrow S$. Our first construction is based on the observation that $Km(A) \dashrightarrow S$ can also be described as an Inose-style base change. More precisely, there is an elliptic fibration on the Kummer surface (fibration 13 in [**Km2**]) which can be obtained from the Shioda-Inose surface in [**Km1**] as a base change by a simple change of variables. In §3 we give the details of this construction of elliptic $K3$ surfaces of various Mordell-Weil ranks. In particular, we give two numerical examples that have Mordell-Weil rank 15.

Note that fibration 13 in [**Km2**] has one IV^* fiber and one I_0^* fiber, while Inose's fibration on $Km(E_1 \times E_2)$ has two IV^* fibers. There does exist an elliptic fibration on $Km(J(C))$ with two IV^* fibers, and it can be constructed from known fibrations in [**Km2**] using the "2-neighbor method" of Elkies. However, this fibration does not generically have a section. Nevertheless, taking its Jacobian fibration produces an elliptic fibration with section and the same fiber type. Taking a base change, we can construct elliptic $K3$ surfaces with Mordell-Weil rank up to 18. In §4 we give details of this construction, and give a numerical example with Mordell-Weil rank 15.

In future work, we hope to find explicit generators for the Mordell-Weil groups of the elliptic $K3$ surfaces constructed here, along the lines of [**KK**].

1.1. Acknowledgements. We thank Matthias Schütt and the referees for helpful comments on an earlier version of the paper. Kumar was supported in part by NSF CAREER grant DMS-0952486, and by a grant from the MIT Solomon Buchsbaum Research Fund. Kuwata was partially supported by JSPS KAKENHI Grant Number JP26400023, and by the Chuo University Grant for Special Research.

The computer algebra systems `Magma`, `sage`, `gp/PARI`, `Maxima` and `Maple` were used in the calculations for this paper.

2. Preliminaries

By an elliptic surface we mean a smooth projective surface S together with a relatively minimal elliptic fibration $\pi : S \to C$ over a smooth projective curve C, all defined over a field k, which we assume to be a number field throughout. We are interested in the case where S is a $K3$ surface, C is isomorphic to $\mathbb{P}^1$ over k, and π admits a section $\sigma_0 : \mathbb{P}^1 \to S$ also defined over k. Under these assumptions the generic fiber $\mathcal{E}$ of π is an elliptic curve defined over the rational function field $k(t)$. We also assume there is at least one singular fiber (to ensure that the Mordell-Weil group is finitely generated). Denote by $\mathrm{MW}_{\bar{k}}(\mathcal{E})$ the Mordell-Weil lattice $\mathcal{E}(\bar{k}(t))/\mathcal{E}(\bar{k}(t))_{\text{tors}}$. We refer the reader to [**Shi2**] for the theory of Mordell-Weil lattices and [**SS**] for a broader overview of elliptic surfaces.

A theorem of Shioda and Tate connects the Mordell-Weil $\mathcal{E}(\bar{k}(t))$ group with the Picard group or the Néron-Severi group $\mathrm{NS}_{\bar{k}}(S)$ of $\mathcal{E}$ (note that linear equivalence and algebraic equivalence coincide). In particular, we have the Shioda-Tate formula (cf. Shioda[**Shi1**])

$$\operatorname{rank} \mathrm{NS}_{\bar{k}}(S) = 2 + \operatorname{rank} \mathrm{MW}_{\bar{k}}(E) + \sum_{P \in C} (m_P - 1), \tag{2.1}$$

where m_P is the number of (geometrically) irreducible components of the fiber $\pi^{-1}(P)$. We call $\rho(S) = \operatorname{rank} \mathrm{NS}_{\bar{k}}(S)$ the (geometric) Picard number.

REMARK 2.1. In the absence of a section we would say that π is a genus one fibration. Sometimes in the literature these are called elliptic fibrations and those with section are labeled Jacobian elliptic fibrations.

The following theorem of Inose is essential to our construction.

PROPOSITION 2.2 (Inose [**I1**, Cor. 1.2]). *Let S_1 and S_2 be $K3$ surfaces defined over a number field k and $f : S_1 \to S_2$ a rational map of finite degree. Then S_1 and S_2 have the same Picard number.*

PROOF. Inose proves this lemma for $K3$ surfaces over $\mathbb{C}$ using a transcendental argument. Since we know that $\mathrm{NS}_{\mathbb{C}}(S) = \mathrm{NS}_{\bar{k}}(S)$, the conclusion is still valid for our case. □

Due to the transcendental nature of Inose's proof we do not have a very good understanding of how the Néron-Severi lattices of S_1 and S_2 are related. Thus, knowledge of an explicit set of divisors generating the Picard group of S_1, or of the Mordell-Weil group if S_1 is elliptically fibered, does not easily translate to the same for S_2.

Let S be a Kummer surface associated with an abelian surface A. Then,

$$\operatorname{rank} \mathrm{NS}_{\bar{k}}(S) = \operatorname{rank} \mathrm{NS}_{\bar{k}}(A) + 16.$$

PROPOSITION 2.3 (cf. Birkenhake-Lang [**BL**, Prop. 5.5.7]). *Let A be a simple abelian surface. Then, the Picard number $\rho(A) = \operatorname{rank} \mathrm{NS}_{\bar{k}}(S)$ is given as follows.*

$\mathrm{End}_{\mathbb{Q}}(A)$	$\rho(A)$
$\mathbb{Q}$	1
real quadratic field	2
indefinite quaternion algebra	3
CM field of degree 4	2

PROPOSITION 2.4. *Let A be an abelian surface isogenous to the product of two elliptic curves $E_1 \times E_2$. Then, the Picard number $\rho(A)$ is given as follows.*

$A \sim E_1 \times E_2$	$\rho(A)$
$E_1 \not\sim E_2$	2
$E_1 \sim E_2$, *without CM*	3
$E_1 \sim E_2$, *with CM*	4

3. Inose type surface for the Jacobian of a curve of genus 2

Let C be a curve of genus 2 given by the equation

$$y^2 = f(x) = \sum_{i=0}^{6} f_i x^i,$$

where f is a squarefree polynomial. In particular, we allow f_6 to be 0 (but in that case $f_5 \neq 0$). In [**Km1**] it is shown that a generic elliptic $K3$ surface with II^* and III^* fibers has a Shioda-Inose structure such that the quotient is the Kummer surface of the Jacobian of a genus 2 curve. In fact, this correspondence sets up a birational map between the moduli space of such elliptic $K3$ surfaces and $\mathcal{M}_2$, the moduli space of genus 2 curves. It can also be viewed as an isomorphism between the moduli space of $K3$ surfaces lattice polarized by $U \oplus E_8(-1) \oplus E_7(-1)$ and $\mathcal{A}_2$, the moduli space of principally polarized abelian surfaces. For a genus 2 curve C, we can describe the unique associated elliptic $K3$ surface with II^* and III^* fibers as follows.

Let (I_2, I_4, I_6, I_8) be the Igusa-Clebsch invariants (see for example [**Km2**]). Then the corresponding $K3$ surface is given by the Weierstrass equation

$$y^2 = x^3 - t^3\Big(\frac{I_4}{12}t + 1\Big)x + t^5\Big(\frac{I_{10}}{4}t^2 + \frac{I_2I_4 - 3I_6}{108}t + \frac{I_2}{24}\Big) \tag{3.1}$$

It has a type II^* fiber at $t = \infty$ and a type III^* at $t = 0$. We say that C is "general" if the other singular fibers are all I_1.

The surface defined by (3.1) is isomorphic to

$$y^2 = x^3 - \Big(\frac{I_4}{12} + \frac{1}{t}\Big)x + \Big(\frac{I_{10}}{4}t + \frac{I_2I_4 - 3I_6}{108} + \frac{I_2}{24t}\Big),$$

which we call $G^{(1)}$, and define the base change

$$G^{(n)} : y^2 = x^3 - \Big(\frac{I_4}{12} + \frac{1}{t^n}\Big)x + \Big(\frac{I_{10}}{4}t^n + \frac{I_2I_4 - 3I_6}{108} + \frac{I_2}{24t^n}\Big).$$

Compare with the definition of $F^{(n)}$ in [**Shi3**] and [**KK**]. Straightforward calculations show that the Kodaira-Néron model of $G^{(n)}$ is a $K3$ surface for $n = 1, \dots, 4$.

In [**Km1**] the map $G^{(1)} \to Km(J(C))$ is given as a 2-isogeny between two elliptic surfaces. All Jacobian elliptic fibrations (elliptic fibrations with section) on $Km(J(C))$ are classified in [**Km2**]. Fibration 13 in [**Km2**] is

$$y^2 = x^3 - 108t^4\bigl(48t^2 + I_4\bigr)x + 108t^4\bigl(72I_2t^4 + (4I_4I_2 - 12I_6)t^2 + 27I_{10}\bigr),$$

which is isomorphic to $G^{(2)}$ by $(x, y, t) \mapsto \bigl(t^2x/9, t^3y/27, 1/(2t)\bigr)$. Thus, $G^{(2)}$ is isomorphic to $Km(J(C))$, and we have a Kummer sandwich diagram: $G^{(1)} \dashrightarrow G^{(2)} \simeq Km(J(C)) \dashrightarrow G^{(1)}$.

REMARK 3.1. The surface $G^{(1)}$ can be realized as a quartic surface just as Inose's suface in [**I2**]:

$$y^2zw - x^3z + \Bigl(\frac{I_4}{12}w + z\Bigr)xzw - \Bigl(\frac{I_{10}}{4}w^2 + \frac{I_2}{24}z^2\Bigr)w^2 - \Bigl(\frac{I_2I_4 - 3I_6}{108}\Bigr)zw^3 = 0.$$

Therefore we can regard $G^{(1)}$ as a generalization of Inose's surface to the case of Jacobian of the genus 2 curve C. This point of view was also studied in [**CD**].

THEOREM 3.2. *Suppose C is general. Then the rank of the Mordell-Weil group $G^{(n)}(\overline{\mathbb{Q}}(t))$ is given by the table below. Here, $\rho = \rho(J(C))$ is the Picard number of $J(C)$.*

	Singular fibers	*Rank*
$G^{(1)}$	$\mathrm{II}^*, \mathrm{III}^*, 5\mathrm{I}_1$	$\rho - 1$
$G^{(2)}$	$\mathrm{IV}^*, \mathrm{I}_0^*, 10\mathrm{I}_1$	$4 + \rho$
$G^{(3)}$	$\mathrm{I}_0^*, \mathrm{III}, 15\mathrm{I}_1$	$9 + \rho$
$G^{(4)}$	$\mathrm{IV}, 20\mathrm{I}_1$	$12 + \rho$

PROOF. Determination of the types of singular fibers is a straightforward application of Tate's algorithm. By Proposition 2.2, the Picard number of each of the $K3$ surfaces $G^{(1)}, \dots, G^{(4)}$ is equal to that of $Km(J(C))$, namely, $16 + \rho$. Now, calculation of the rank is a straightforward application of the Shioda-Tate formula (2.1). □

In order to obtain elliptic $K3$ surfaces whose Mordell-Weil rank 15, it now suffices to find curves C such that the Picard number of its Jacobian $J(C)$ is 3. We give two different examples below.

REMARK 3.3. As shown in [**Km2**], a set of five sections forms a basis of $G^{(2)}(\overline{\mathbb{Q}}(t))$ if $\rho(J(C)) = 1$. These sections are defined over the field $k(J(C)[2])$.

EXAMPLE 3.4 (QM case). In the case where the Jacobian $J(C)$ is simple, $\rho(J(C)) = 3$ if and only if it has quaternionic multiplication. Writing down an explicit equation of such a curve C is a delicate number theoretical problem. Here, we use an example of [**DR**] which depends on a result of Hashimoto and Tsunogai [**HT**] (see also [**HM**]). Take the curve in [**HT**, Lemma 4.5] and let $\sigma = \tau = \sqrt{-3/2}$. Then we obtain

$$C : y^2 = \Bigl(x^2 + \frac{7}{2}\Bigr)\Bigl(\frac{83}{30}x^4 + 14x^3 - \frac{1519}{30}x^2 + 49x - \frac{1813}{120}\Bigr).$$

It has quaternionic multiplication by a maximal order in the quaternion algebra of discriminant 6, which is the smallest possible discriminant for a non-split quaternion algebra. Calculating Igusa-Clebsh invariants, and simplifying the coefficients by a change of variables, we obtain the elliptic surface

$$G^{(4)} : Y^2 = X^3 + 529200\Big(6 - \frac{5}{t^4}\Big)X - 9261000\Big(4t^4 + 20 - \frac{3431}{t^4}\Big).$$

Simple calculations show that the singular fibers are irreducible except at $t = \infty$ where it has a fiber of type IV. This shows that the rank of $G^{(4)}(\overline{\mathbb{Q}}(t))$ equals 15.

REMARK 3.5. If we take the quadratic twist of $G^{(4)}$ by $\mathbb{Q}(\sqrt{2 \cdot 3 \cdot 5 \cdot 7})$, the elliptic surface becomes simpler:

$$Y^2 = X^3 + 12\Big(6 - \frac{5}{t^4}\Big)X - \Big(4t^4 + 20 - \frac{3431}{t^4}\Big).$$

However, it is possible that the field of definition of the Mordell-Weil group of the original surface may be easier to describe.

EXAMPLE 3.6 (Split case). As a second example, we consider the case where $J(C)$ is isogenous to the product of two elliptic curves. Since we want $\rho(J(C))$ to be equal to 3, we need to consider the case where $J(C) \sim E \times E$ with E without complex multiplication.

Take two elliptic curves

$$\begin{aligned} E_1 &: y^2 + xy + y = x^3 + 4x - 6 && \text{(Cremona label 14a1)},\\ E_2 &: y^2 + xy + y = x^3 - 36x - 70 && \text{(Cremona label 14a2)}, \end{aligned}$$

These both have rational 6-torsion points, and are 2-isogenous to each other. Moreover, we have $E_1[3] \simeq E_2[3] \simeq \mathbb{Z}/3\mathbb{Z} \times \mu_3$ as Galois modules. Take the quadratic twist $E_2^{(-3)}$ of E_2 by -3. Then, the Galois isomorphism $E_1[3] \to E_2[3]$ isometric with respect to the Weil pairing induces an anti-isometry $E_1[3] \to E_2^{(-3)}[3]$. Therefore, by Frey-Kani [**FK**], there exists a curve C of genus 2 admitting morphisms $C \to E_1$ and $C \to E_2^{(-3)}$ of degree 3. To find an explicit equation, we use Shaska's result [**Sha**, §3]. Using the fact that $j(E_1) = (215/28)^3$ and $j(E_2) = (1705/98)^3$, we find that the curve C corresponds to the values $\mathfrak{u} = 107553525/12595352$, $\mathfrak{v} = -11619959625/1032401161$ in equation (8) in [**Sha**]. After making some changes of variables, we obtain the following Weierstrass equation for C:

$$C : y^2 = -96393(13x + 12)(7x - 13)(107x^2 - 273x + 252)(56x^2 + 104x + 31).$$

The maps $\phi_1 : C \to E_1$ and $\phi_2 : C \to E_2^{(-3)}$ are given by

$$\phi_1 : (x, y) \mapsto (x_1, y_1) = \left(-\frac{156260x^3 + 3627x + 86895}{3(7x - 13)(56x^2 + 104x + 31)}, -\frac{4(13x - 6)(214x^2 + 273x + 126)y}{9(7x - 13)^2(56x^2 + 104x + 31)^2}\right),$$

$$\phi_2 : (x, y) \mapsto (x_2, y_2) = \left(\frac{47485x^3 + 4173x^2 - 211380}{(13x + 12)(107x^2 - 273x + 252)}, -\frac{(12(7x + 26))(14x^2 - 52x + 31)y}{(13x + 12)^2(107x^2 - 273x + 252)^2}\right),$$

where we have used the following Weierstrass forms for E_1 and $E_2^{(-3)}$:

$$\begin{aligned} E_1 \quad &: y_1^2 = x_1^3 + 5805x_1 - 285714,\\ E_2^{(-3)} &: y_2^2 = x_2^3 - 5115x_2 + 115414. \end{aligned}$$

The elliptic $K3$ surface $G^{(n)}$ corresponding to this C is given by

$$G^{(n)} : Y^2 = X^3 + 33\Big(2933005 - \frac{1126255812}{t^n}\Big)X - 2\Big(28449792t^n - 8690133815 - \frac{274280846290470}{t^n}\Big).$$

(Here, we have scaled X, Y, and t differently from the definition of $G^{(n)}$ to clear denominators.) It is readily verified that the singular fibers other than the IV fiber at $t = \infty$ are all irreducible. Since E_1 and $E_2^{(-3)}$ are isogenous over $\mathbb{Q}(\sqrt{-3})$ and do not have complex multiplication, the rank of $G^{(4)}(\overline{\mathbb{Q}}(t))$ is exactly 15.

REMARK 3.7. The reader may wonder why we did not use a curve C such that $J(C)$ is $(2,2)$-isogenous to the product $E_1 \times E_2$. The reason is that the resulting elliptic surface $G^{(1)}$ has at least one I_2 fiber other than II^* and III^* fibers. Thus, the rank of $G^{(4)}(\overline{\mathbb{Q}}(t))$ cannot be 15. Similarly, Example 1 in [**Sha**, §5] yields $G^{(1)}$ with two I_2 fibers, and Example 2 yields one with a I_3 fiber.

4. Fibration with two IV^* fibers

A Kummer surface $Km(J(C))$ can be realized as a complete intersection of three quadrics in $\mathbb{P}^5$. It contains two sets of sixteen lines intersecting each other as shown in [**GH**, p.767, Figure 21]. Classically, these sets are labeled *nodes* and *tropes*; in the model of the Kummer surface as a quartic surface in $\mathbb{P}^3$ with sixteen singular points, the (blown-down) nodes are the singular points, while the tropes correspond to the sixteen planes tangent to the Kummer surface along plane conics (which are the transforms of the tropes).

As indicated in Figure 1, we find a configuration of a pair of divisors of type IV^*. We thus have an elliptic fibration on $Km(J(C))$ having these two IV^* fibers. It turns out, however, that such an elliptic fibration does not have a section. Thus, it is not in the list of elliptic fibrations studied in [**Km2**].

The elliptic divisor corresponding to the fibration is $F = 3T_0 + 2N_3 + 2N_4 + 2N_5 + T_3 + T_{14} + T_{25}$.

To obtain an explicit equation of this fibration, we start with Fibration 1 in [**Km2**], two I_0^* fibers and six I_2 fibers, with Mordell-Weil group $\mathbb{Z} \oplus (\mathbb{Z}/2\mathbb{Z})^2$. The zero section in [**Km2**] is taken to be T_2. However, we will take T_3 to be the zero section; this does not affect the Weierstrass equation, however changes the identity components of the reducible fibers, for instance. We can now take a 2-neighbor step to the elliptic fibration corresponding to the fiber $F' = T_3 + N_3 + T_0 + N_4 + T_{14} + N_{23}$ (this fiber class is in the orbit of fibration 10 of [**Km2**]). We omit the details of the neighbor-step, apart from giving the elliptic parameter

$$t_1 = \frac{1}{t-a}\left(\frac{y+y_s}{x-x_s} - 2a(b-1)(c-a)\right)$$

where

$$(x_s, y_s) = \big(-4(a-1)(b-1)t(t-c)(ct-ab), -8(a-1)(b-1)(c-a)(c-b)t^2(t-ab)(t-c)\big)$$

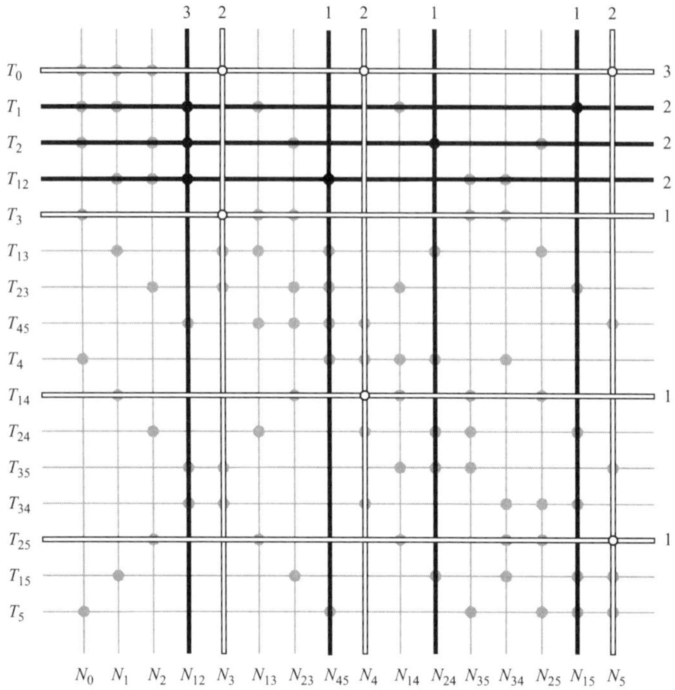

FIGURE 1. Two IV* fibers in the 16–6 configuration of $\mathrm{NS}(Km(J(C)))$

is the equation of the section T_{14}. The new elliptic fibration has Weierstrass equation.

$$
\begin{aligned}
y_1^2 = x_1\big(&x_1^2 + x_1(t_1^4/16 + t_1^3(b-1)(c-a)/2 \\
&+ t_1^2(3b^2c^2 - abc^2 - 4bc^2 + 2ac^2 - 4ab^2c - b^2c + 2a^2bc \\
&+ 3abc + 2bc - 4a^2c + 2ac + 2ab^2 + 2a^2b - 4ab)/2 \\
&+ 2t_1(b-1)b(b-a)(c-1)c(c-a) + b^2(b-a)^2c^2(c-1)^2) \\
&- (a-1)a(b-1)(c-a)(c-b)t_1^3(t_1 + 2bc - 2b)(t_1 + 2bc - 2ac)/2\big)
\end{aligned}
$$

The fiber F' is at ∞, while there is another fiber of type I_6 at $t_1 = 0$, given by $T_1 + N_{12} + T_{12} + N_{45} + T_5 + N_{15}$. There are also I_2 fibers at $t_1 = -2(b-a)c$ and $t_1 = -2b(c-1)$. The first is $T_{25} + N'_{35}$ (in the notation of Fibration 1 of [**Km2**]) and the second has N_{24} as a component. We can now move by another 2-neighbor step to the elliptic fibration with F as a fiber. The elliptic parameter is

$$t_2 = \frac{x_1}{t_1 + 2(b-a)c};$$

we omit the remaining details. The Jacobian of this new fibration is given by

$$Y^2 = X^3 + AX^2 + \Big(B_1 T + B_2 + \frac{B_3}{T}\Big)X + \Big(C_1 T + \frac{C_2}{T}\Big)^2,$$

where $T = t_2$ and

$$\begin{aligned}
A &= 4\big(abc^2 + bc^2 - 2ac^2 + ab^2c - 2b^2c - 2a^2bc + bc + 4a^2c - 2ac + ab^2 - 2a^2b \\
&\quad +ab\big),\\
B_1 &= (a-1)c - a(b-1),\\
B_2 &= -16\big(ab^2c^4 - a^2bc^4 - abc^4 + a^2c^4 + a^2b^3c^3 - ab^3c^3 - b^3c^3 - a^3b^2c^3 \\
&\qquad +2a^2b^2c^3 - 2ab^2c^3 + b^2c^3 + a^2bc^3 + 2abc^3 - 2a^2c^3 - ab^4c^2 + b^4c^2 \\
&\qquad -a^3b^3c^2 - 2a^2b^3c^2 + 6ab^3c^2 - b^3c^2 + a^4b^2c^2 + a^3b^2c^2 - 3a^2b^2c^2 \\
&\qquad -2ab^2c^2 + a^2bc^2 - abc^2 + a^2c^2 + a^2b^4c - ab^4c + 2a^3b^3c - 2a^2b^3c - ab^3c \\
&\qquad -2a^4b^2c + a^3b^2c + 2a^2b^2c + ab^2c - a^2bc - a^3b^3 + a^2b^3 + a^4b^2 - a^3b^2\big),\\
B_3 &= -ab^2c(a-1)(b-1)(c-1)^2(a-b)(b-c)(c-a)\big((a-1)c - a(b-1)\big),\\
C_1 &= 1,\\
C_2 &= ab^2c(a-1)(b-1)(c-1)^2(a-b)(b-c)(c-a).
\end{aligned}$$

It has IV^* fibers at $T = 0$ and ∞.

Define

$$H^{(n)} : Y^2 = X^3 + AX^2 + \Big(B_1 t^n + B_2 + \frac{B_3}{t^n}\Big)X + \Big(C_1 t^n + \frac{C_2}{t^n}\Big)^2.$$

The Kodaira-Néron model of $H^{(n)}$ is a $K3$ surface for $n = 1, 2, 3$.

THEOREM 4.1. *Suppose C is general. Then the rank of the Mordell-Weil group $H^{(n)}(\overline{\mathbb{Q}}(t))$ is given by the table below. Here, $\rho = \rho(J(C))$ is the Picard number of $J(C)$.*

	Singular fibers	*Rank*
$H^{(1)}$	$2\mathrm{IV}^*, 8\mathrm{I}_1$	$2+\rho$
$H^{(2)}$	$2\mathrm{IV}, 12\mathrm{I}_1$	$10+\rho$
$H^{(3)}$	$24\mathrm{I}_1$	$14+\rho$

PROOF. Since $H^{(1)}$ is the Jacobian of an elliptic fibration on $Km(J(C))$, there is a rational map $H^{(1)} \dashrightarrow Km(J(C))$ of finite degree. Thus, by Proposition 2.2, the Picard number of $H^{(1)}$ is equal to $16+\rho$. The rest of the proof is similar to that of Theorem 3.2. □

Thus, $H^{(3)}$ for any simple abelian surface without extra endomorphisms gives an elliptic $K3$ surface of rank 15. However, although a very general abelian surface (in the moduli space of principally polarized abelian surfaces) has this property, and we expect a "random" numerical example to have $\rho = 1$, we must verify it for any putative example explicitly. To do so, we first use Leprévost's criterion [**L**, Lemme 3.1.2]: a genus 2 curve C over $\mathbb{Q}$ has simple Jacobian if there is a prime ℓ of good reduction such that the Galois group of the characteristic polynomial of Frobenius of the reduced curve is the 8-element dihedral group D_4. We can compute the characteristic polynomial of Frobenius in `magma` or `sage`; the latter has an optimized implementation of the p-adic Harvey-Kedlaya algorithm for computing the matrix of Frobenius for a curve over a finite field.

EXAMPLE 4.2. Let $a = -1$, $b = 1/7$, and $c = -6/7$. Then, we can verify that the Jacobian $J(C)$ of the curve

$$C : y^2 = x(x-1)(x+1)\Big(x - \frac{1}{7}\Big)\Big(x + \frac{6}{7}\Big)$$

is simple and $\rho(J(C)) = 1$ by the lemma below. After a certain change of variables, we have

$$H^{(n)} : y^2 = x^3 - \frac{1354}{7}x^2 + \frac{936}{7}\Big(t^n + \frac{42989}{819} + \frac{4}{t^n}\Big)x + \frac{6084}{49}\Big(t^n - \frac{4}{t^n}\Big)^2.$$

The Mordell-Weil rank of $H^{(3)}$ is 15.

LEMMA 4.3. *Let C be the curve of Example 4.2. Then $\rho(J(C)) = 1$.*

PROOF. The characteristic polynomial of Frob_{37} and Frob_{41} on $A = J(C)$ are given by

$$p_{37} = x^4 - 4x^3 + 46x^2 - 148x + 1369$$
$$p_{41} = x^4 + 4x^3 + 6x^2 + 164x + 1681.$$

We can directly verify (for instance in `gp/PARI` or `Magma`) that p_{37} has Galois group D_4. Therefore, the Jacobian is simple. Next, since p_{37} and p_{41} are irreducible, it follows by [**Ta**, Theorem 2] that the endomorphism rings of the reductions $A \otimes \mathbb{F}_{37}$ and $A \otimes \mathbb{F}_{41}$ are orders in quartic fields defined by p_{37} and p_{41}. These fields are easily checked to be linearly disjoint. Since $\mathrm{End}(A) \otimes \mathbb{Q}$ is a subfield of each, it must be $\mathbb{Q}$, and $\mathrm{End}(A)$ must be $\mathbb{Z}$. □

References

[BL] Christina Birkenhake and Herbert Lange, *Complex abelian varieties*, 2nd ed., Grundlehren der Mathematischen Wissenschaften [Fundamental Principles of Mathematical Sciences], vol. 302, Springer-Verlag, Berlin, 2004. MR2062673

[CD] Adrian Clingher and Charles F. Doran, *Lattice polarized K3 surfaces and Siegel modular forms*, Adv. Math. **231** (2012), no. 1, 172–212, DOI 10.1016/j.aim.2012.05.001. MR2935386

[C] David A. Cox, *Mordell-Weil groups of elliptic curves over* $\mathbf{C}(t)$ *with* $p_g = 0$ *or* 1, Duke Math. J. **49** (1982), no. 3, 677–689. MR672502

[DR] Luis V. Dieulefait and Victor Rotger, *On abelian surfaces with potential quaternionic multiplication*, Bull. Belg. Math. Soc. Simon Stevin **12** (2005), no. 4, 617–624. MR2206004

[E] Noam D. Elkies, *Elliptic curves of high rank over* $\mathbb{Q}$ *and* $\mathbb{Q}(t)$, in preparation.

[FK] Gerhard Frey and Ernst Kani, *Curves of genus 2 covering elliptic curves and an arithmetical application*, Arithmetic algebraic geometry (Texel, 1989), Progr. Math., vol. 89, Birkhäuser Boston, Boston, MA, 1991, pp. 153–176, DOI 10.1007/978-1-4612-0457-2_7. MR1085258

[GLD] Federica Galluzzi and Giuseppe Lombardo, *Correspondences between K3 surfaces*, Michigan Math. J. **52** (2004), no. 2, 267–277, DOI 10.1307/mmj/1091112075. With an appendix by Igor Dolgachev. MR2069800

[GH] Phillip Griffiths and Joseph Harris, *Principles of algebraic geometry*, Wiley-Interscience [John Wiley & Sons], New York, 1978. Pure and Applied Mathematics. MR507725

[HM] Ki-ichiro Hashimoto and Naoki Murabayashi, *Shimura curves as intersections of Humbert surfaces and defining equations of QM-curves of genus two*, Tohoku Math. J. (2) **47** (1995), no. 2, 271–296, DOI 10.2748/tmj/1178225596. MR1329525

[HT] Ki-ichiro Hashimoto and Hiroshi Tsunogai, *On the Sato-Tate conjecture for QM-curves of genus two*, Math. Comp. **68** (1999), no. 228, 1649–1662, DOI 10.1090/S0025-5718-99-01061-3. MR1627797

[I1] Hiroshi Inose, *On certain Kummer surfaces which can be realized as non-singular quartic surfaces in* P^3, J. Fac. Sci. Univ. Tokyo Sect. IA Math. **23** (1976), no. 3, 545–560. MR0429915

[I2] Hiroshi Inose, *Defining equations of singular $K3$ surfaces and a notion of isogeny*, Proceedings of the International Symposium on Algebraic Geometry (Kyoto Univ., Kyoto, 1977), Kinokuniya Book Store, Tokyo, 1978, pp. 495–502. MR578868

[Kl1] Remke Kloosterman, *Elliptic $K3$ surfaces with geometric Mordell-Weil rank* 15, Canad. Math. Bull. **50** (2007), no. 2, 215–226, DOI 10.4153/CMB-2007-023-2. MR2317444

[Kl2] Remke Kloosterman, *Explicit sections on Kuwata's elliptic surfaces*, Comment. Math. Univ. St. Pauli **54** (2005), no. 1, 69–86. MR2153956

[Km1] Abhinav Kumar, *$K3$ surfaces associated with curves of genus two*, Int. Math. Res. Not. IMRN **6** (2008), Art. ID rnm165, 26, DOI 10.1093/imrn/rnm165. MR2427457

[Km2] Abhinav Kumar, *Elliptic fibrations on a generic Jacobian Kummer surface*, J. Algebraic Geom. **23** (2014), no. 4, 599–667, DOI 10.1090/S1056-3911-2014-00620-2. MR3263663

[KK] Abhinav Kumar and Masato Kuwata, *Elliptic $K3$ surfaces associated with the product of two elliptic curves: Mordell-Weil lattices and their fields of definition*, Nagoya Math. J. **228** (2017), 124–185, DOI 10.1017/nmj.2016.56. MR3721376

[Kw] Masato Kuwata, *Elliptic $K3$ surfaces with given Mordell-Weil rank*, Comment. Math. Univ. St. Paul. **49** (2000), no. 1, 91–100. MR1777156

[L] Franck Leprévost, *Jacobiennes de certaines courbes de genre 2: torsion et simplicité* (French), J. Théor. Nombres Bordeaux **7** (1995), no. 1, 283–306, DOI 10.5802/jtnb.144. Les Dix-huitièmes Journées Arithmétiques (Bordeaux, 1993). MR1413580

[Mo] David R. Morrison, *On $K3$ surfaces with large Picard number*, Invent. Math. **75** (1984), no. 1, 105–121, DOI 10.1007/BF01403093. MR728142

[N] Isao Naruki, *On metamorphosis of Kummer surfaces*, Hokkaido Math. J. **20** (1991), no. 2, 407–415, DOI 10.14492/hokmj/1381413848. MR1114412

[SS] Matthias Schütt and Tetsuji Shioda, *Elliptic surfaces*, Algebraic geometry in East Asia—Seoul 2008, Adv. Stud. Pure Math., vol. 60, Math. Soc. Japan, Tokyo, 2010, pp. 51–160. MR2732092

[Sha] Tony Shaska, *Genus 2 curves with $(3,3)$-split Jacobian and large automorphism group*, Algorithmic number theory (Sydney, 2002), Lecture Notes in Comput. Sci., vol. 2369, Springer, Berlin, 2002, pp. 205–218, DOI 10.1007/3-540-45455-1_17. MR2041085

[Shi1] Tetsuji Shioda, *On elliptic modular surfaces*, J. Math. Soc. Japan **24** (1972), 20–59, DOI 10.2969/jmsj/02410020. MR0429918

[Shi2] Tetsuji Shioda, *On the Mordell-Weil lattices*, Comment. Math. Univ. St. Paul. **39** (1990), no. 2, 211–240. MR1081832

[Shi3] Tetsuji Shioda, *A note on $K3$ surfaces and sphere packings*, Proc. Japan Acad. Ser. A Math. Sci. **76** (2000), no. 5, 68–72, DOI 10.3792/pjaa.76.68. MR1771143

[Shi4] Tetsuji Shioda, *Kummer sandwich theorem of certain elliptic $K3$ surfaces*, Proc. Japan Acad. Ser. A Math. Sci. **82** (2006), no. 8, 137–140, DOI 10.3792/pjaa.82.137. MR2279280

[SI] Tetsuji Shioda and Hiroshi Inose, *On singular $K3$ surfaces*, Complex analysis and algebraic geometry, Iwanami Shoten, Tokyo, 1977, pp. 119–136. MR0441982

[Ta] John Tate, *Endomorphisms of abelian varieties over finite fields*, Invent. Math. **2** (1966), 134–144, DOI 10.1007/BF01404549. MR0206004

[TdZ] Jaap Top and Frank De Zeeuw, *Explicit elliptic $K3$ surfaces with rank* 15, Rocky Mountain J. Math. **39** (2009), no. 5, 1689–1697, DOI 10.1216/RMJ-2009-39-5-1689. MR2546659

Department of Mathematics, Stony Brook University, Stony Brook, New York 11794

Email address: `thenav@gmail.com`

Faculty of Economics, Chuo University, 742-1 Higashinakano, Hachioji-shi, Tokyo 192-0393, Japan

Email address: `kuwata@tamacc.chuo-u.ac.jp`

Contemporary Mathematics
Volume **703**, 2018
http://dx.doi.org/10.1090/conm/703/14135

Higher-order Weierstrass weights of branch points on superelliptic curves

Caleb McKinley Shor

Abstract. In this paper we consider the problem of calculating the higher-order Weierstrass weight of the branch points of a superelliptic curve C. For any $q > 1$, we give an exact formula for the q-weight of an affine branch point. We also find a formula for the q-weight of a point at infinity in the case where n and d are relatively prime. With these formulas, for any fixed n, we obtain an asymptotic formula for the ratio of the q-weight of the branch points, denoted BW_q, to the total q-weight of points on the curve:

$$\liminf_{d\to\infty} \frac{BW_q}{g(g-1)^2(2q-1)^2} \geq \frac{n+1}{3(n-1)^2(2q-1)^2},$$

with equality when the limit is taken such that $\gcd(n,d) = 1$.

1. Introduction

Let $q \in \mathbb{N}$. A q-Weierstrass point (or higher-order Weierstrass point) is a point P on a curve for which there exist holomorphic q-differentials that have higher than expected orders of vanishing at P. Each q-Weierstrass point has an associated q-weight, denoted $w^{(q)}(P)$, which measures how much higher than expected those orders of vanishing are. A curve of genus $g \geq 2$ has finitely many q-Weierstrass points.

Importantly, the q-weight of a point is invariant under automorphism. Thus, higher-order Weierstrass points are important in the study of automorphisms of algebraic curves. For instance, Lewittes showed in [**6**] that if an automorphism has at least five fixed points, then all of its fixed points are 1-Weierstrass points. Further, Mumford, in [**8**], has suggested that q-Weierstrass points on an algebraic curve are analogous to q-torsion points on an elliptic curve. For more on the history of Weierstrass points, we refer the reader to [**2**]. For background material of Weierstrass points specifically on superelliptic curves, see [**12**].

Let C be a curve of genus $g \geq 2$ of the form $y^n = f(x)$ where $f(x)$ is a separable polynomial of degree $d > n \geq 2$. Such a curve is said to be superelliptic. In the cover $C \to \mathbb{P}^1$, the points above the roots of $f(x)$ are branch points. If $n \nmid d$, the point (or points) above ∞ in the nonsingular model of C is also a branch point. One can show that each branch point is a q-Weierstrass point for all q; in the case

2010 *Mathematics Subject Classification.* Primary 14H55, 11G30.

Key words and phrases. Weierstrass points, higher-order Weierstrass points, superelliptic curves, branch points, numerical semigroups.

where $n = 2$, the branch points are exactly the 1-Weierstrass points. Let B be an affine branch point on C and, if $n \nmid d$, P^∞ a nonsingular branch point at infinity. In [**16**, Theorem 8], Towse calculated the 1-weight of the branch points (affine and at infinity) as a function of n and d. In the case of $\gcd(n,d) = 1$, he found that

$$w^{(1)}(B) = \frac{g(n+1)(d-7)}{12} + (d-1)\sum_{j=1}^{n-1}\left\{\frac{-dj}{n}\right\} j,$$

where $\{x\}$ denotes the fractional part of x, and

$$w^{(1)}(P^\infty) = \frac{(n^2-1)(d^2-1)}{24} - g.$$

Given that the total 1-weight of points on a curve of genus g is $g^3 - g$, he was able to calculate the fraction that the branch points' 1-weight (denoted BW) accounted for as

$$\lim_{d\to\infty} \frac{BW}{g^3-g} \geq \frac{n+1}{3(n-1)^2},$$

with equality when the limit is taken over integers d such that $\gcd(n,d) = 1$.

The goal of this paper is to extend Towse's results to higher-order Weierstrass weights of branch points on a superelliptic curve. To achieve this, we first produce a basis for the space of holomorphic q-differentials on a superelliptic curve C. A common approach to calculate the q-weight is to work with the Wronskian of this basis, a method first described by Hurwitz in [**5**]. However, we take a different approach, instead using results from numerical semigroups and non-representable numbers. In this way we obtain a formula for the q-weight of an affine branch point as a function of n, d, and q. The main result is Theorem 2. In particular, when $\gcd(n,d) = 1$ we find for $q \geq 2$

$$w^{(q)}(B) = \frac{g(n+1)(d-7)}{12} + g + (d-1)\sum_{j=1}^{n-1}\left\{-\frac{(d+1)q+dj}{n}\right\} j.$$

As for the points at infinity, with two examples, we show that if $\gcd(n,d) > 1$, one cannot get a formula for $w^{(q)}(P^\infty)$ based only on n, d, and q. However, if $\gcd(n,d) = 1$ and $q \geq 2$, then in Theorem 3 we have

$$w^{(q)}(P^\infty) = \frac{(n^2-1)(d^2-1)}{24}.$$

Since the total q-weight of points on a curve of genus g is $g(g-1)^2(2q-1)^2$ for $q \geq 2$, we get an similar asymptotic result in Proposition 6 for the proportion of branch points' q-weight (denoted BW_q). We find

$$\lim_{d\to\infty} \frac{BW_q}{g(g-1)^2(2q-1)^2} \geq \frac{n+1}{3(n-1)^2(2q-1)^2},$$

with equality when the limit is taken over integers d such that $\gcd(n,d) = 1$.

The paper is organized as follows. In Section 2, we provide some background material on calculating the q-Weierstrass weight of a point. We also include some notation and results for non-representable integers in numerical semigroups with two generators. In Section 3, we find a basis for the space of holomorphic q-differentials on a curve C by presenting a set of linearly independent holomorphic differentials and then counting them to make sure there are as many as the Riemann-Roch theorem predicts. In Section 4, we have our main results. We find a formula for the q-weight of an affine branch point, and we use that to derive a few corollaries for specific cases of n and d. We also note that, for given n and d, the q-weight of a branch point depends only on the value of q modulo n. We also give examples to show that if $\gcd(n,d) > 1$ the q-weight of a point at infinity cannot necessarily be determined just by knowing n and d. If $\gcd(n,d) = 1$, however, we can calculate the weight. In both cases, we obtain some asymptotic results about the proportion of q-weight that the branch points contain.

2. Preliminaries and notation

2.1. q-Weierstrass points. In this paper, we will follow the approach given in [**13**, Section 2]. We describe the notation and major results for calculating weights of q-Weierstrass points here.

Let k be an algebraically closed field, C be a non-singular projective curve over k of genus $g \geq 2$, and $k(C)$ its function field. For any $f \in k(C)$, let $\operatorname{div}(f)$ denote the divisor associated to f. For any divisor $D = \sum_P n_P P$ and any point P, let $\nu_P(D) = n_P$, and let $\operatorname{ord}_P(f) = \nu_P(\operatorname{div}(f))$.

For any $q \in \mathbb{N}$, let $H^0(C,(\Omega^1)^q)$ be the $\mathbb{C}$-vector space of holomorphic q-differentials on C, a space of dimension

$$d_q = \begin{cases} g & \text{if } q = 1, \\ (g-1)(2q-1) & \text{if } q \geq 2. \end{cases}$$

For P a degree 1 point on C, consider a basis $\{\psi_1, \ldots, \psi_{d_q}\}$ of $H^0(C,(\Omega^1)^q)$ where

$$\operatorname{ord}_P(\psi_1) < \operatorname{ord}_P(\psi_2) < \cdots < \operatorname{ord}_P(\psi_{d_q}).$$

The q-weight of P is

$$w^{(q)}(P) = \sum_{i=1}^{d_q} \operatorname{ord}_P(\psi_i) - \sum_{j=0}^{d_q-1} j.$$

We call the point P a *q-Weierstrass point* if $w^{(q)}(P) > 0$.

PROPOSITION 1. [**4**, III.5.10] *Let C be a curve of genus $g \geq 2$ and let $q \geq 1$. Then the total q-weight of points on C is*

$$\sum_{P \in C} w^{(q)}(P) = (g-1)d_q(2q-1+d_q) = \begin{cases} g^3 - g & \textit{if } q = 1, \\ g(g-1)^2(2q-1)^2 & \textit{if } q \geq 2. \end{cases}$$

2.2. Non-representable numbers. For notation, let $\mathbb{N}_0$ be the set of non-negative integers. Let $a, b \in \mathbb{N}$ and consider the set

$$R(a,b) = \{ax + by : x, y \in \mathbb{N}_0\}.$$

Elements of $R(a,b)$ are called *(a,b)-representable numbers.* The complement of $R(a,b)$ in $\mathbb{N}_0$, denoted $NR(a,b)$, is the set of *(a,b)-representable numbers.* When there is no confusion, we will omit the (a,b) and simply refer to these numbers as representable or non-representable.

The problem of calculating the cardinality of $NR(a,b)$ dates to the late 19th century in [**15**]. Clearly, if $\gcd(a,b) > 1$ then $NR(a,b)$ is an infinite set. It is straightforward to show that the converse is true too. For example, see [**10**, Theorem 1.0.1] for two proofs of the following result.

LEMMA 1. *For $a, b \in \mathbb{N}$, if* $\gcd(a, b) = 1$ *then* $NR(a, b)$ *is a finite set.*

For the rest of this section, we will assume $\gcd(a, b) = 1$. Thus, $NR(a, b)$ is finite and so we can compute its cardinality.

PROPOSITION 2. [**15**, Page 134] *For $a, b \in \mathbb{N}$ with* $\gcd(a, b) = 1$,

$$|NR(a, b)| = \frac{(a-1)(b-1)}{2}.$$

This result is important in the theory of algebraic curves. Suppose a plane curve C is given by the affine equation

$$\alpha_{a,0}x^a + \alpha_{0,b}y^b + \sum_{i,j} \alpha_{i,j}x^i y^j = 0$$

for constants $\alpha_{i,j}$ with $\alpha_{a,0} \cdot \alpha_{0,b} \neq 0$ and where the summation is over non-negative i, j such that $aj + bi < ab$. Such a curve is called a $C_{a,b}$ curve. These curves can be seen as a generalization of elliptic and hyperelliptic curves in Weierstrass form. With the Riemann-Roch Theorem, if the affine part of the curve is non-singular then one can show that the genus of such a curve is exactly $(a-1)(b-1)/2$, the cardinality of $NR(a, b)$. For details, see [**7**] or [**14**].

For the purposes of this paper, we will also need to know the sum of the elements of $NR(a, b)$. This problem was solved in [**1**] using generating functions.

PROPOSITION 3. *For $a, b \in \mathbb{N}$ with* $\gcd(a, b) = 1$,

$$\sum_{n \in NR(a,b)} n = \frac{(a-1)(b-1)(2ab-a-b-1)}{12}.$$

This result was generalized to a formula for the sum of the mth powers of elements of $NR(a, b)$. See [**11**] or [**17**] for details.

3. A basis of holomorphic q-differentials

In this section, we give a basis for the space of holomorphic q-differentials on a superelliptic curve C. The main result of this section is as follows:

THEOREM. *Let C be a curve of genus $g \geq 2$ given in affine coordinates by $y^n = f(x)$, for $f(x)$ a separable polynomial of degree $d > n \geq 2$. For $q \geq 1$, let $H^0(C, (\Omega^1)^q)$ be the space of holomorphic q-differentials on C. Let*

$$\mathfrak{B}_{n,d,q} = \left\{ x^i y^j \left(\frac{\mathrm{d}x}{y^{n-1}} \right)^q : 0 \leq i,\, 0 \leq j < n,\, ni + dj \leq (2g-2)q \right\}.$$

Then $\mathfrak{B}_{n,d,q}$ is a basis for $H^0(C, (\Omega^1)^q)$.

In order to prove this, we need the following results (based on the work in [**16**]) and a useful lemma.

Let $G = \gcd(n, d)$. For g the genus of C, we have $2g - 2 = nd - n - d - G$. Let $\{\alpha_1, \ldots, \alpha_d\}$ denote the d distinct roots of $f(x)$ and let $B_i = (\alpha_i, 0)$ for $i = 1, \ldots, d$. For each non-root ω of $f(x)$, let $P_1^\omega, \ldots, P_n^\omega$ denote the n points on C over $x = \omega$. And let $P_1^\infty, \ldots, P_G^\infty$ denote the G points over ∞ in the non-singular model of C. One then has the following principal divisors.

- $\mathrm{div}(y) = \sum_{j=1}^{d} B_j - \frac{d}{G} \sum_{m=1}^{G} P_m^\infty$,

- $\operatorname{div}(x-\alpha_i) = nB_i - \frac{n}{G}\sum_{m=1}^{G} P_m^\infty$,
- $\operatorname{div}(x-\omega) = \sum_{j=1}^{n} P_j^\omega - \frac{n}{G}\sum_{m=1}^{G} P_m^\infty$.
- $\operatorname{div}(\mathrm{d}x) = (n-1)\sum_{j=1}^{d} B_j - \left(\frac{n}{G}+1\right)\sum_{m=1}^{G} P_m^\infty$,

From these, we see that $\operatorname{div}((\mathrm{d}x/y^{n-1})^q) = \frac{(2g-2)q}{G}\sum_{m=1}^{G} P_m^\infty$.

For integers i, j, let $f_{i,j} = x^i y^j (\mathrm{d}x/y^{n-1})^q$. We want to find conditions on i and j such that $f_{i,j}$ is a holomorphic q-differential. Note that $f_{i,j}$ can have poles only at the points above ∞ if $i, j \geq 0$. In that situation, we find

$$\operatorname{ord}_{P_m^\infty}(f_{i,j}) = \frac{(2g-2)q - (ni+dj)}{G}$$

for each pair (i, j) and each m. Hence $f_{i,j}$ is a holomorphic q-differential as long as $i \geq 0, j \geq 0$, and $ni + dj \leq (2g-2)q$.

LEMMA 2. *Let $n, d, q \in \mathbb{Z}$ with $2 \leq n < d$ and $q \geq 2$. As above, let $G = \gcd(n, d)$ and $2g - 2 = nd - n - d - G$. For all but finitely many triples (n, d, q), one has*

$$(2g-2)q - d(n-1) \geq 0.$$

The exceptional cases are $(n, d, q) \in \{(2, 5, 2), (2, 6, 2)\}$.

PROOF. First, note that $(2g-2)q - d(n-1) = (nd - n - d - G)q - d(n-1) = d(n-1)(q-1) - q(n+G)$. Thus, to show our desired inequality, it is equivalent to show

$$d \geq \left(\frac{q}{q-1}\right)\left(\frac{n+G}{n-1}\right).$$

For notation, let $h(q, n) = \left(\frac{q}{q-1}\right)\left(\frac{n+G}{n-1}\right)$. We aim to show $d \geq h(q, n)$.

For $q \geq 2$, the maximum value of $q/(q-1)$, which occurs when $q = 2$, is 2. For $n \geq 2$, the maximum value of $(n+G)/(n-1)$ occurs when G is largest, so when $G = n$, and when $n = 2$. The maximum value is 4. Thus $h(q, n) \leq 2 \cdot 4 = 8$ so $d \geq h(q, n)$ for all $d \geq 8$.

Now we consider cases of n. If $n \geq 4$, then $G = n$ or $G < n$. If $G = n$, then $n|d$, so $d \geq 2n \geq 8$, so $d \geq h(q, n)$. If $G < n$, then $G \leq n/2$, so $h(q, n) \leq 2(n + n/2)/(n-1) = 3 + 3/(n-1) \leq 4$. Since $d > n$, we have $d > n \geq 4 \geq h(q, n)$, as desired.

If $n = 3$, then $G = 1$ or $G = 3$. If $G = 1$, then $h(q, 3) \leq 2 \cdot 2 = 4$. In this case, since $d > n = 3$, we have $d \geq 4$, so $d \geq h(q, 3)$. If $G = 3$, then $h(q, 3) \leq 2 \cdot 3 = 6$. In this case, $d \geq 2n = 6 \geq h(q, 3)$, as desired.

If $n = 2$, then $G = 1$ or $G = 2$. Note that to have $g \geq 2$, we only consider $d \geq 5$. If $G = 1$ and $q = 2$, then $h(2, 2) = 2 \cdot 3 = 6$. Since $G = 1$, d is odd, so $d \geq h(2, 2)$ for all d except for $d = 5$. If $G = 1$ and $q \geq 3$, then $h(q, 2) \leq (3/2) \cdot 3 = 9/2$, so $d \geq h(q, 2)$ for all $d \geq 5$. If $G = 2$ and $q = 2$, then $h(2, 2) = 2 \cdot 4 = 8$. Since $G = 2$, d is even, so $d \geq h(2, 2)$ for all d except for $d = 6$. If $G = 2$ and $q \geq 3$, then $h(q, 2) \leq (3/2) \cdot 4$, so $d \geq h(q, 2)$ for all $d \geq 6$.

Thus, the only exceptional cases are $(n, d, q) = (2, 5, 2)$ or $(2, 6, 2)$. For all other triples, we find that $d \geq h(q, n)$, or, equivalently, that

$$(2g-2)q - d(n-1) \geq 0,$$

as desired. $\square$

THEOREM 1. *Let C be a curve of genus $g \geq 2$ given in affine coordinates by $y^n = f(x)$, for $f(x)$ a separable polynomial of degree $d > n$. For $q \geq 1$, let $H^0(C, (\Omega^1)^q)$ be the space of holomorphic q-differentials on C. Let*

$$\mathfrak{B}_{n,d,q} = \left\{ x^i y^j \left(\frac{\mathrm{d}x}{y^{n-1}} \right)^q : 0 \leq i,\ 0 \leq j < n,\ ni + dj \leq (2g-2)q \right\}.$$

Then $\mathfrak{B}_{n,d,q}$ is a basis for $H^0(C, (\Omega^1)^q)$.

PROOF. The $q = 1$ case, in a slightly different form, is proved in [**16**]. For completeness, we will first prove the $q \geq 2$ case here and then adapt our argument to cover the $q = 1$ case.

Suppose $q \geq 2$. With the restriction that $0 \leq j < n$, we see that these holomorphic q-differentials are linearly independent. We therefore need to show that $|\mathfrak{B}_{n,d,q}| = d_q = (2q-1)(g-1)$.

We first consider the case where $(n, d, q) \notin \{(2,5,2), (2,6,2)\}$ and let $\mathfrak{B} = \mathfrak{B}_{n,d,q}$. Note that we require $i \geq 0$ and $ni + dj \leq (2g-2)q$, so

$$0 \leq i \leq \left\lfloor \frac{(2g-2)q - dj}{n} \right\rfloor.$$

For each $j = 0, \ldots, n-1$, we have $(2g-2)q - dj \geq (2g-2)q - d(n-1)$, which is non-negative by Lemma 2. (This is why we handle the $(2,5,2)$ and $(2,6,2)$ cases separately.) Thus, to calculate the number of pairs (i,j) in $\mathfrak{B}$, we will let j go from 0 to $n-1$ and count the number of indices i that correspond to each j value. I.e.,

$$|\mathfrak{B}| = \sum_{j=0}^{n-1} \left(1 + \left\lfloor \frac{(2g-2)q - dj}{n} \right\rfloor \right).$$

Since $\lfloor x \rfloor = x - \{x\}$, we simplify the sum to get

$$|\mathfrak{B}| = n + (2g-2)q - \frac{d(n-1)}{2} - \sum_{j=0}^{n-1} \left\{ \frac{(-d-G)q - dj}{n} \right\}.$$

Now, we consider cases of G. If $G = n$, then $n | d$, so $n | ((-d-G)q - dj)$, so each term in the summation is 0. Note that $n - d(n-1)/2 = -(nd - d - 2n)/2 = -(nd - n - d - G)/2 = -(g-1)$. Then $|\mathfrak{B}| = (2g-2)q - (g-1) = (2q-1)(g-1) = d_q$, as desired.

Next, suppose $G \neq n$. Let $n' = n/G$ and $d' = d/G$. Dividing the numerator and denominator by G, the summation equals $\sum_{j=0}^{n-1} \left\{ \frac{(-d'-1)q - d'j}{n'} \right\}$. Since $\gcd(n', d') = 1$, as j goes from 0 to $n'-1$ modulo n', the numerators are distinct modulo n' and therefore in every congruence class exactly once modulo n'. Since $n/n' = G$, this summation equals $G \sum_{k=0}^{n'-1} \frac{k}{n'} = G(n'-1)/2$. All together,

$$|\mathfrak{B}| = n + (2g-2)q - d(n-1)/2 - (n-G)/2.$$

I.e. $|\mathfrak{B}| = n + 2q(g-1) - (1/2)(nd - d + n - G) = 2q(g-1) - (1/2)(nd - d - n - G) = (2q-1)(g-1) = d_q$, as desired.

To complete the proof for $q \geq 2$, we consider the exceptional cases. Suppose $(n, d, q) = (2, 5, 2)$, so $g = 2$ and $d_2 = 3$. Then

$$\mathfrak{B}_{2,5,2} = \left\{ x^i y^j (\mathrm{d}x/y)^2 : i \geq 0, 0 \leq j < 2, 2i + 5j \leq 4. \right\}$$

Thus, $\mathfrak{B}_{2,5,2} = \{(\mathrm{d}x/y)^2,\ x(\mathrm{d}x/y)^2,\ x^2(\mathrm{d}x/y)^2\}$, so $|\mathfrak{B}_{2,5,2}| = 3 = d_2$, as desired.

Suppose $(n,d,q)=(2,6,2)$, so $g=2$ and $d_2=3$. Then

$$\mathfrak{B}_{2,6,2}=\left\{x^iy^j(\mathrm{d}x/y)^2 : i\geq 0, 0\leq j<2, 2i+6j\leq 4.\right\}$$

Thus, $\mathfrak{B}_{2,6,2}=\{(\mathrm{d}x/y)^2,\ x(\mathrm{d}x/y)^2,\ x^2(\mathrm{d}x/y)^2\}$, so $|\mathfrak{B}_{2,6,2}|=3=d_2$, as desired.

Now, suppose $q=1$. Following the approach above, given j we need integers i such that

$$0\leq i\leq\left\lfloor\frac{(2g-2)-dj}{n}\right\rfloor.$$

If $j\leq n-2$, then $(2g-2)-dj\geq(2g-2)-d(n-2)=d-n-G\geq 0$ since $d\geq n$. If $j=n-1$, then $(2g-2)-d(n-1)=-n-G<0$, so there are no such i. Thus, our summation for $|\mathfrak{B}|$ ends at $j=n-2$ instead of $j=n-1$. Since we have a formula for the summation above, we can subtract the $j=n-1$ term out front to get

$$|\mathfrak{B}|=-\left(1+\left\lfloor\frac{-n-G}{n}\right\rfloor\right)+\sum_{j=0}^{n-1}\left(1+\left\lfloor\frac{(2g-2)-dj}{n}\right\rfloor\right).$$

Since $q=1$ the summation equals $g-1$, so $|\mathfrak{B}|=-(1-2)+(g-1)=g=d_1$, as desired. □

4. Weights of branch points

In this section, we use the bases we found in the previous section to calculate the q-weight of the affine branch points and, in the case that $\gcd(n,d)=1$, the point at infinity.

4.1. Weights of affine branch points. Suppose $q\geq 2$. For C given by $y^n=f(x)$ with $f(x)$ separable of degree d, let α be a root of $f(x)$. Then $B=(\alpha,0)$ is an affine branch point of C. Note that we can replace x by $(x-\alpha)$ in our basis $\mathfrak{B}_{n,d,q}$ to produce a new basis $\mathfrak{B}_{n,d,q,\alpha}$. That is,

$$\mathfrak{B}_{n,d,q,\alpha}=\{(x-\alpha)^iy^j(\mathrm{d}x/y^{n-1})^q : i\geq 0,\ 0\leq j<n,\ ni+dj\leq(2g-2)q\}$$

is a basis for $H^0(C,(\Omega^1)^q)$.

Let $f_{i,j,\alpha}=(x-\alpha)^iy^j(\mathrm{d}x/y^{n-1})^q\in\mathfrak{B}_{n,d,q,\alpha}$. Then

$$\nu_B(f_{i,j,\alpha})=ni+j.$$

Since $0\leq j<n$, these valuations are all different, and thus

$$w^{(q)}(B)=\sum_{(i,j)\in S}(ni+j)-\sum_{k=0}^{d_q-1}k,$$

where $S=\{(i,j)\in\mathbb{Z}^2 : i\geq 0,\ 0\leq j<n,\ ni+dj\leq(2g-2)q\}$. We rewrite this as $w^{(q)}(B)=W_1-W_2-W_3$ where

$$W_1=\sum_{(i,j)\in S}(ni+dj),\quad W_2=(d-1)\sum_{(i,j)\in S}j,\quad W_3=\sum_{k=0}^{d_q-1}k. \tag{1}$$

We have

$$W_3=\frac{(d_q-1)(d_q)}{2}=\frac{1}{2}\left((2g-2)^2q^2+(2g-2)(1-2g)q+g(g-1)\right).$$

We will evaluate W_1 and W_2 with the following propositions.

PROPOSITION 4. *Let $n, d, q \in \mathbb{N}$ such that $n < d$ and $q \geq 2$. Then*

$$W_1 = 2(g-1)^2q^2 + (g-1)Gq + \frac{G^2 - 1 - (n-1)(d-1)(2nd-n-d-1)}{12}.$$

We will first sketch the proof in the situation where $\gcd(n, d) = 1$. Afterward, we will prove the theorem for any gcd.

When $\gcd(n, d) = G = 1$, for $(i, j) \in S$, the terms $ni + dj$ are distinct integers from 0 to $(2g-2)q$. From Proposition 2, since $(2g-2)q \geq nd - n - d$ (by Lemma 3 below), all of the $(n-1)(d-1)/2$ (n, d)-non-representable integers are in that interval. The sum of the non-representable integers, as is given in Proposition 3, is $(n-1)(d-1)(2nd-n-d-1)/12$. Thus, if $\gcd(n, d) = 1$, we add up all of the integers from 0 to $(2g-2)q$ and subtract off the non-representable integers to get $W_1 = (2g-2)q((2g-2)q+1)/2 - (n-1)(d-1)(2nd-n-d-1)/12$.

If $\gcd(n, d) > 1$, then the terms $ni + dj$ are no longer distinct, so we need to evaluate the sum more carefully.

PROOF. Let $G = \gcd(n, d)$. First, we observe that $W_1 = G\sum_{(i,j)\in S}(n'i + d'j)$ for $n' = n/G$ and $d' = d/G$. Note that $\gcd(n', d') = 1$. For k from 0 to $G-1$, let

$$S_k = \{(i, j) : i \geq 0,\ kn' \leq j < (k+1)n',\ ni + dj \leq (2g-2)q\}.$$

In particular, S is the disjoint union of the sets S_k. Let $W_{1,k} = \sum_{(i,j)\in S_k}(n'i + d'j)$. Then

$$W_1 = G\sum_{k=0}^{G-1} W_{1,k}.$$

Letting $j' = j - kn'$, we rewrite S_k as

$$S_k = \{(i, j' + kn') : i \geq 0,\ 0 \leq j' < n',\ ni + dj' \leq (2g-2)q - n'dk\}$$

and dividing the last inequality through by G we obtain

$$S_k = \{(i, j' + kn') : i \geq 0,\ 0 \leq j' < n',\ n'i + d'j' \leq \frac{(2g-2)}{G}q - n'd'k\}.$$

Let $m_k = \frac{(2g-2)}{G}q - n'd'k$, the upper bound in S_k. The following lemma will allow us to conclude that all of the (n', d')-non-representable integers are less than m_k.

LEMMA 3. *Let $m_k = \frac{(2g-2)}{G}q - n'd'k$. Then $m_k \geq n'd' - n' - d'$ for all $n, d, q \in \mathbb{N}$ with $0 \leq k \leq G-1$, $n < d$, $g \geq 2$, and $q \geq 2$.*

PROOF. First, note that for $0 \leq k \leq G-1$, $m_k = \frac{(2g-2)}{G}q - n'd'k \geq \frac{(2g-2)}{G}q - n'd'(G-1)$. So we need to show $\frac{(2g-2)}{G}q - n'd'(G-1) \geq n'd' - n' - d'$, which is equivalent to showing $(2g-2)q \geq nd - n - d$. Since $nd - n > nd - n - d$, by Lemma 2, we have $(2g-2)q \geq nd - d > nd - n - d$ for all (n, d, q) combinations except $(2, 5, 2)$ and $(2, 6, 2)$.

We compute the exceptional cases separately. If $(n, d, q) = (2, 5, 2)$, then $g = 2$ and $(2g-2)q = 4 \geq 3 = nd - n - d$. If $(n, d, q) = (2, 6, 2)$, then $g = 2$ and $(2g-2)q = 4 \geq 4 = nd - n - d$. Thus, the bound holds for the exceptional cases as well. □

For S_k, since we are considering $i \geq 0$ and $0 \leq j' < n'$, and since $m_k \geq n'd' - n' - d'$ for all k, our ordered pairs $(i, j' + kn')$ are in one-to-one correspondence with the (n', d')-representable numbers in the interval $[0, m_k]$. And since $m_k \geq n'd' - n' - d'$, all of the $(n'-1)(d'-1)/2$ (n', d')-non-representable numbers are in this interval as well. Thus S_k contains $|S_k| = m_k + 1 - (n'-1)(d'-1)/2$ ordered pairs.

Then

$$\begin{aligned} W_{1,k} &= \sum_{(i,j)\in S_k} (n'i + d'j) \\ &= \sum_{(i,j'+kn')\in S_k} (n'i + d'j' + n'd'k) \\ &= n'd'k \cdot |S_k| + \sum_{(i,j'+kn')\in S_k} (n'i + d'j'). \end{aligned}$$

The summation is the sum of the (n', d')-representable numbers from 0 to m_k. We calculate this by summing all of the integers from 0 to m_k and subtracting the (n', d')-non-representable integers, which all lie in this interval. Using Proposition 3, the summation is $m_k(m_k+1)/2 - (n'-1)(d'-1)(2n'd' - n' - d' - 1)/12$. Thus,

$$\begin{aligned} W_{1,k} &= n'd'k\,(m_k + 1 - (n'-1)(d'-1)/2) \\ &\quad + m_k(m_k+1)/2 - (n'-1)(d'-1)(2n'd' - n' - d' - 1)/12, \end{aligned}$$

so

$$\begin{aligned} W_1 =& G\sum_{k=0}^{G-1} \left[n'd'k\left(m_k + 1 - \frac{(n'-1)(d'-1)}{2}\right) + \frac{m_k(m_k+1)}{2} \right. \\ & \left. - \frac{(n'-1)(d'-1)(2n'd' - n' - d' - 1)}{12}\right]. \end{aligned}$$

To evaluate this sum, we need the following calculations which are straightforward to compute.

- $\sum\limits_{k=0}^{G-1} m_k = (2g-2)q - \frac{n'd'G(G-1)}{2}$.
- $\sum\limits_{k=0}^{G-1} m_k^2 = \frac{(2g-2)^2}{G}q^2 - (2g-2)(G-1)d'n'q + \frac{d'^2 n'^2 (G-1)G(2G-1)}{6}$
- $\sum\limits_{k=0}^{G-1} km_k = (g-1)(G-1)q - \frac{d'n'(G-1)G(2G-1)}{6}$.

Simplifying the resulting expression, we find

$$W_1 = \frac{(2g-2)^2}{2}q^2 + (g-1)Gq + \frac{G^2 - 1 - (n-1)(d-1)(2nd - n - d - 1)}{12},$$

which completes the proof of Proposition 4. □

PROPOSITION 5. *Let $n, d, q \in \mathbb{N}$ such that $n < d$ and $q \geq 2$. Let*

$$D(a,b,c) = \sum_{j=0}^{c-1} \left\{ \frac{a + bj}{c} \right\} j.$$

Then

$$W_2 = (d-1)\left((n-1)\left((g-1)q + \frac{-2nd + 3n + d}{6}\right) - D(-(d+G)q, -d, n)\right).$$

PROOF. We will use Lemma 2, so we first assume $(n, d, q) \notin \{(2,5,2),(2,6,2)\}$. For $W_2 = (d-1)\sum_{(i,j)\in S} j$, we have

$$W_2 = (d-1)\sum_{j=0}^{n-1}\sum_{i=0}^{I_j} j,$$

for $I_j = \left\lfloor \frac{(2g-2)q-dj}{n} \right\rfloor$. By Lemma 2, $I_j \geq 0$ so $W_2 = (d-1)\sum_{j=0}^{n-1}(I_j+1)j$. Since $\lfloor x \rfloor = x - \{x\}$,

$$W_2 = (d-1)\sum_{j=0}^{n-1}\left(\frac{(2g-2)q-dj}{n} - \left\{\frac{(2g-2)q-dj}{n}\right\} + 1\right)j.$$

Note that $\left\{\frac{(2g-2)q-dj}{n}\right\} = \left\{\frac{(nd-n-d-G)q-dj}{n}\right\} = \left\{\frac{(-d-G)q-dj}{n}\right\}$.

Expanding out, we get

$$\begin{aligned} W_2 =& (g-1)(d-1)(n-1)q + \frac{(d-1)(n-1)}{6}(3n - d(2n-1)) \\ &- (d-1)\sum_{j=0}^{n-1}\left\{-\frac{(d+G)q+dj}{n}\right\} j, \end{aligned}$$

which can be rearranged to give the desired result.

Finally, if $(n, d, q) \in \{(2,5,2),(2,6,2)\}$, then $S = \{(0,0),(1,0),(2,0)\}$, and so $W_2 = \sum_{(i,j)\in S} j = 0$. We get the same value if we plug each these (n, d, q) triples into the above formula for W_2. □

REMARK. The summation $D(a, b, c)$ is related to a Dedekind sum. There is no closed form for such sums, though there is a reciprocity law. For a general reference, see [**9**].

Finally, we can combine and simplify $W_1 - W_2 - W_3$. Note that the q^2 and q terms (other than in the summation) cancel. With further manipulation, we have our main result.

THEOREM 2. *Let C be given in affine coordinates by $y^n = f(x)$ for $f(x)$ a separable polynomial of degree $d > n$. Let $G = \gcd(n, d)$, and let $q \in \mathbb{Z}$ with $q \geq 2$. For any root α of $f(x)$, let $B = (\alpha, 0)$ be a branch point.*

The q-weight of B is $w^{(q)}(B) =$

$$\begin{aligned} w^{(q)}(B) =& \frac{1}{24}\left((n-1)(d-1)(n+1)(d-7) + 12g(G+1) + 5(G^2-1)\right) \\ &+ (d-1)\cdot D(-(d+G)q, -d, n) \end{aligned}$$

Note that, for given values of n and d, the q-weight of B depends only on the value of q modulo n.

We will give results for some combinations of n and d in the corollaries below. First, we consider the case where $\gcd(n, d) = 1$.

COROLLARY 1. *If $\gcd(n, d) = 1$,*

$$w^{(q)}(B) = \frac{g}{12}(n+1)(d-7) + g + (d-1)\cdot D(-(d+1)q, -d, n).$$

Fix n and d (with any gcd). If one varies q, then one sees the value of $w^{(q)}(B)$ depends only on the congruence class of q modulo n/G. Further, if $d \equiv -G \pmod{n}$, then the summation term simplifies to $\sum_{j=0}^{n-1} \left\{ \frac{Gj}{n} \right\} j$, for which there is a closed form.

COROLLARY 2. *If $d \equiv -G \pmod{n}$, then $w^{(q)}(B)$ doesn't depend on q. In particular,*

$$w^{(q)}(B) = \frac{1}{24}\Big((n-1)(d-1)(n+1)(d-7) + 12g(G+1) + 5(G^2-1) + 2(d-1)(n-G)(3n+n'-2) \Big).$$

PROOF. The summation term is $\sum_{j=0}^{n-1} \left\{ \frac{Gj}{n} \right\} j, = \sum_{j=0}^{n-1} \left\{ \frac{j}{n'} \right\} j$. Each j can be written uniquely as $j = j' + kn'$ for $0 \le k < G$ and $0 \le j' < n'$. Thus, the summation is $\sum_{k=0}^{G-1} \sum_{j'=0}^{n'-1} \left\{ \frac{j'}{n'} \right\} j = \sum_{k=0}^{G-1} \sum_{j'=0}^{n'-1} \left(\frac{j'^2}{n'} + j'k \right)$, which simplifies to $(n-G)(3n+n'-2)/12$. □

Combining the two corollaries above, we obtain the following.

COROLLARY 3. *If $d \equiv -1 \pmod{n}$, then*

$$w^{(q)}(B) = \frac{g(n+1)(d+1)}{12} = \frac{(n^2-1)(d^2-1)}{24}$$

for all $q \ge 2$.

COROLLARY 4. *If $n \mid d$, then*

$$w^{(q)}(B) = \frac{(n^2-1)(d^2-2d)}{24}.$$

PROOF. If $n \mid d$, then $G = n$ and $n \mid ((d+G)q + dj)$ for all j, so the summation is zero. Since $2g - 2 = nd - n - d - n$, we have $g = \frac{(d-2)(n-1)}{2}$. Plugging in, the result follows. □

4.2. Weights of points at infinity. If $n \mid d$, then there are n points at infinity in the smooth model of C, so these points are not branch points. However, we can still investigate their q-weights. If $\gcd(n,d) > 1$, then we need to know more about $f(x)$ to determine $w^{(q)}(P_m^\infty)$. We give a few examples to illustrate this.

In [**3**], the authors consider curves of the form $y^2 = f(x) = x^6 + ax^4 + bx^2 + 1$, where a, b are parameters and $f(x)$ is separable. In the non-singular models of these curves, there are $G = \gcd(n,d) = 2$ points at infinity P_1^∞ and P_2^∞. If $4b = a^2$, then $w^{(3)}(P_1^\infty) = w^{(3)}(P_2^\infty) = 2$. If $4b \ne a^2$, then $w^{(3)}(P_1^\infty) = w^{(3)}(P_2^\infty) = 0$.

In [**13**, Lemma 4 and Proposition 3], the authors consider hyperelliptic curves of genus 3 of the form $y^2 = f(x)$ where $\deg(f) = 8$. In the non-singular models of these curves, there are $G = \gcd(n,d) = 2$ points at infinity P_1^∞ and P_2^∞. If C is given by $y^2 = x^8 + x^6 + 16x^4 + x^2 + 1$, then $w^{(2)}(P_1^\infty) = w^{(2)}(P_2^\infty) = 1$. If C is given by $y^2 = x^8 + x^4 + 1$, then $w^{(2)}(P_1^\infty) = w^{(2)}(P_2^\infty) = 3$.

Thus, simply knowing n and d is not enough to calculate the q-weight of the points at infinity. However, there are some cases where we can get a result.

First, if $d = n+1$, then the lone point at infinity is a nonsingular branch point, so it will have the same q-weight as the affine branch points. By Corollary 3, since

$d \equiv -1 \pmod{n}$, $w^{(q)}(B) = \frac{(n^2-1)(d^2-1)}{24}$ for $q \geq 2$, so we will have $w^{(q)}(P^\infty) = \frac{(n^2-1)(d^2-1)}{24}$ for $q \geq 2$ as well. This is a special case of the more general result when $\gcd(n,d) = 1$.

THEOREM 3. *Suppose C is a curve of genus $g \geq 2$ given by the affine equation $y^n = f(x)$ for $f(x)$ a separable polynomial of degree d where $n < d$ and $\gcd(n,d) = 1$. Let P_1^∞ be the lone point at infinity in the non-singular model of C. Then*

$$w^{(q)}(P_1^\infty) = \begin{cases} \dfrac{g(n+1)(d+1)}{12} - g = \dfrac{(n^2-1)(d^2-1)}{24} - g & \text{if } q = 1, \\ \dfrac{g(n+1)(d+1)}{12} = \dfrac{(n^2-1)(d^2-1)}{24} & \text{if } q \geq 2. \end{cases}$$

PROOF. For $q = 1$, the formula is given at the end of the proof of [**16**, Theorem 8].

For $q \geq 2$ and $G = 1$, let $\mathfrak{B}_{n,d,q}$ be as in Section 3, and again let $S = \{(i,j) \in \mathbb{Z}^2 : i \geq 0, 0 \leq j < n, ni + dj \leq (2g-2)q\}$. Then $f_{i,j} \in \mathfrak{B}_{n,d,q}$ if and only if $(i,j) \in S$. Recall that $\operatorname{ord}_{P_m^\infty}(f_{i,j}) = (2g-2)q - (ni + dj)$. These orders of vanishing are unique, so

$$w^{(q)}(P_m^\infty) = \left(\sum_{(i,j)\in S} \operatorname{ord}_{P_m^\infty}(f_{i,j}) \right) - \sum_{k=0}^{d_q - 1} k.$$

Since $|S| = d_q$,

$$w^{(q)}(P_m^\infty) = d_q(2g-2)q - \left(\sum_{(i,j)\in S} (ni + dj) \right) - \frac{(d_q - 1)d_q}{2}.$$

The summation, which we called W_1 in Equation 1, is evaluated in Proposition 4. Plugging this and d_q in, the expression simplifies to $w^{(q)}(P_m^\infty) = \frac{(n^2-1)(d^2-1)}{24}$. □

4.3. Branch weight. In the case where $\gcd(n,d) = 1$, we can calculate the total q-weight of the branch points (both affine and at infinity) for $q \geq 2$, which we denote BW_q.

COROLLARY 5. *Suppose $\gcd(n,d) = 1$, so $g = \frac{(n-1)(d-1)}{2}$. Then the total branch q-weight is given by $BW_q = d \cdot w^{(q)}(B) + w^{(q)}(P_1^\infty) =$*

$$d\left(\frac{g}{12}(n+1)(d-7) + g + (d-1)\cdot D(-(d+1)q, -d, n)\right) + g\frac{(n+1)(d+1)}{12}.$$

Rewritten in terms of g, we get

$$BW_q = \frac{n+1}{3(n-1)^2}\left(g^3 - 2g^2(n-1) - g(n-1)^2\right) + d(d-1)\cdot D(-(d+1)q, -d, n).$$

From Proposition 1, we know the total weight of the q-Weierstrass points, for $q \geq 2$, is $g(g-1)^2(2q-1)^2$. We can now calculate the proportion of q-weight of the branch points.

PROPOSITION 6. *Fix n and let $q \geq 2$. Then*

$$\liminf_{d\to\infty} \frac{BW_q}{g(g-1)^2(2q-1)^2} \geq \frac{n+1}{3(n-1)^2(2q-1)^2}.$$

If we restrict to values of d that are relatively prime to n then

$$\lim_{d\to\infty,(n,d)=1} \frac{BW_q}{g(g-1)^2(2q-1)^2} = \frac{n+1}{3(n-1)^2(2q-1)^2}.$$

PROOF. For general n and d, since we do not have an exact formula for the q-weight of the points at infinity, we can only say $BW_q \geq d \cdot w^{(q)}(B)$. Using the result from Theorem 2, since

$$\sum_{j=0}^{n-1} \left\{ -\frac{(d+G)q+dj}{n} \right\} j \leq \sum_{j=0}^{n-1} \frac{n-1}{n} j = \frac{(n-1)^2}{2},$$

in terms of d, the dominant term of $d \cdot w^{(q)}(B)$ is $d^3 \frac{(n-1)(n+1)}{24}$. Since g is on the order of $d(n-1)/2$, the dominant term of the denominator is $d^3 \frac{(n-1)^3(2q-1)^2}{8}$. The result follows.

For $\gcd(n,d) = 1$, the lone point at infinity has weight $\frac{(d^2-1)(n^2-1)}{24}$. Thus, the dominant term of BW_q is precisely $d^3 \frac{(n-1)(n+1)}{24}$, and we thus have an equality if we take a limit involving integers d such that $\gcd(n,d) = 1$. □

References

[1] Tom C. Brown and Peter Jau-Shyong Shiue, *A remark related to the Frobenius problem*, Fibonacci Quart. **31** (1993), no. 1, 32–36. MR1202340

[2] Andrea Del Centina, *Weierstrass points and their impact in the study of algebraic curves: a historical account from the "Lückensatz" to the 1970s*, Ann. Univ. Ferrara Sez. VII Sci. Mat. **54** (2008), no. 1, 37–59, DOI 10.1007/s11565-008-0037-1. MR2403373

[3] Mohamed Farahat and Fumio Sakai, *The 3-Weierstrass points on genus two curves with extra involutions*, Saitama Math. J. **28** (2011), 1–12 (2012). MR2920720

[4] H. M. Farkas and I. Kra, *Riemann surfaces*, 2nd ed., Graduate Texts in Mathematics, vol. 71, Springer-Verlag, New York, 1992. MR1139765

[5] A. Hurwitz, *Ueber algebraische Gebilde mit eindeutigen Transformationen in sich* (German), Math. Ann. **41** (1892), no. 3, 403–442, DOI 10.1007/BF01443420. MR1510753

[6] Joseph Lewittes, *Automorphisms of compact Riemann surfaces*, Amer. J. Math. **85** (1963), 734–752, DOI 10.2307/2373117. MR0160893

[7] Shinji Miura. Algebraic geometric codes on certain plane curves. *Electronics and Communications in Japan (Part III: Fundamental Electronic Science)*, 76(12):1–13, 1993.

[8] David Mumford, *The red book of varieties and schemes*, Second, expanded edition, Lecture Notes in Mathematics, vol. 1358, Springer-Verlag, Berlin, 1999. Includes the Michigan lectures (1974) on curves and their Jacobians; With contributions by Enrico Arbarello. MR1748380

[9] Hans Rademacher and Emil Grosswald, *Dedekind sums*, The Mathematical Association of America, Washington, D.C., 1972. The Carus Mathematical Monographs, No. 16. MR0357299

[10] J. L. Ramírez Alfonsín, *The Diophantine Frobenius problem*, Oxford Lecture Series in Mathematics and its Applications, vol. 30, Oxford University Press, Oxford, 2005. MR2260521

[11] Øystein J. Rødseth, *A note on T. C. Brown and P. J.-S. Shiue's paper: "A remark related to the Frobenius problem" [Fibonacci Quart.* **31** *(1993), no. 1, 32–36; MR1202340 (93k:11018)]*, Fibonacci Quart. **32** (1994), no. 5, 407–408. MR1300276

[12] C. Shor and T. Shaska, *Weierstrass points of superelliptic curves*, Advances on superelliptic curves and their applications, NATO Sci. Peace Secur. Ser. D Inf. Commun. Secur., vol. 41, IOS, Amsterdam, 2015, pp. 15–46. MR3525571

[13] Tony Shaska and Caleb M. Shor, *2-Weierstrass points of genus 3 hyperelliptic curves with extra involutions*, Comm. Algebra **45** (2017), no. 5, 1879–1892, DOI 10.1080/00927872.2016.1226861. MR3582832

[14] Caleb McKinley Shor, *Genus calculations for towers of functions fields arising from equations of C_{ab} curves*, Albanian J. Math. **5** (2011), no. 1, 31–40. MR2783183

[15] J. J. Sylvester, *On Subvariants, i.e. Semi-Invariants to Binary Quantics of an Unlimited Order*, Amer. J. Math. **5** (1882), no. 1-4, 79–136, DOI 10.2307/2369536. MR1505319

[16] Christopher Towse, *Weierstrass points on cyclic covers of the projective line*, Trans. Amer. Math. Soc. **348** (1996), no. 8, 3355–3378, DOI 10.1090/S0002-9947-96-01649-2. MR1357406
[17] Hans J. H. Tuenter, *The Frobenius problem, sums of powers of integers, and recurrences for the Bernoulli numbers*, J. Number Theory **117** (2006), no. 2, 376–386, DOI 10.1016/j.jnt.2005.06.015. MR2213771

Department of Mathematics, Western New England University, Springfield, Massachusetts 01119

Email address: cshor@wne.edu

Contemporary Mathematics
Volume **703**, 2018
http://dx.doi.org/10.1090/conm/703/14136

Poncelet's porism and projective fibrations

E. Previato

ABSTRACT. Poncelet's porism theorem is used to produce a natural compactification of several moduli spaces. The monodromy of the polygons, viewed as torsion points on a fibration by elliptic curves, can be tested computationally for an action of the full symmetric group. Analogous constructions can be implemented for hyperelliptic fibrations corresponding to higher-dimensional versions of Poncelet's porism.

Introduction

The goal of this note is to present certain configurations of projective geometry (over the complex numbers), compute their invariants in the sense of classical invariant theory, and give them an interpretation in terms of elliptic fibrations, with applications to classical mechanics.

The theory of invariants for (sets of) quadric hypersurfaces has a long and ongoing history. So does Poncelet's Porism, a classical result which affords perhaps the first example of a link between projective geometry and abelian functions. In this paper, I offer a few observations on common aspects of these two theories. In Section 1, I set up and classify an interpretation of the Segre symbol of a pencil of space quadrics in relation to a plane Poncelet configuration. My motivation is the construction of natural projective varieties that represent a natural compactification of certain moduli spaces. For example, while there is no natural elliptic fibration that encodes certain data of the Poncelet configurations [BM, Sec. 1], there is a surface in $\mathbb{P}^3$ which contains all the relevant elliptic curves (and their limits): to construct it, I use a linear deformation of the pencil. The relationship between moduli of curves and these projective configurations allows for the computation of interesting numerical invariants. A second reason for investigating limits is a dynamical application: it was proved by G.D. Birkhoff that Poncelet's porism is equivalent to the integrability of a billiard with elliptical boundary (cf. e.g. [KT]), as are its higher-dimensional generalizations (cf. e.g. [Pr] and, particularly for Cayley's closure condition, [DR1, 2]). Using the idea of this note for these generalizations will give hyperelliptic, as opposed to elliptic, fibrations. In Section 2, I pose the Galois problem for the set of conics in a pencil n-circumscribed to a given conic; the answer relies on a computer program: I identify the objects to be calculated, give the first few examples, and explain the strategy for performing the general computation.

2010 *Mathematics Subject Classification.* Primary 14H52 14J27; Secondary 14L24 14N10.

The bibliography on Poncelet's porism is immense, and still growing: I limit myself strictly to the works used in this note. An excellent set of classical references can be found in the entry "Poncelet's Porism", at *MathWorld*–A Wolfram Web Resource, by Eric W. Weisstein, `http://mathworld.wolfram.com`.

1. Limits and Invariants

A pair of (distinct) quadrics in $\mathbb{P}^n$ can be put in normal form under the action of the group of projective transformations, as follows. If we denote by A, B a pair of symmetric $(n+1)\times(n+1)$ matrices that define the quadrics (by abuse of notation we will call the quadrics "A" and "B" also), then:

THEOREM 1.1. [**HP**, XIII.10]
(i) If the matrix B is nonsingular, there exists a linear transformation of $\mathbb{P}^n$, given by a matrix P, such that ${}^tPAP = A_1$ and ${}^tPBP = B_1$ if and only if $\lambda A + \mu B$ and $\lambda A_1 + \mu B_1$ have the same elementary divisors.
(ii) If the matrix B is singular, if the rank of the $(n+1)\times 2(n+1)$ matrix $[AB]$ is $n+1-r_0$, and if the elementary divisors of $\lambda A + \mu B$ are: $(\alpha_i\lambda+\mu)^{e_i}(i = 1,\dots,r); \lambda^{2f_i+1}(i=1,\dots,s); \lambda^{2g_i}(i=1,\dots,t)$, then by a projective transformation A, B can be brought in a unique way into the form:

$$\sum^{r}\theta_1(\alpha_i,e_i)+\sum^{s}\theta_2(f_i)+\sum^{t}\theta_3(g_i)+\sum^{k}\theta_4(\varrho_i)$$

$$\sum^{r}\phi_1(\alpha_i,e_i)+\sum^{s}\phi_2(f_i)+\sum^{t}\phi_3(g_i)+\sum^{k}\phi_4(\varrho_i)$$

where:

$$\theta_1(\alpha,e)=\alpha\sum_{i=0}^{e-1}X_iX_{e-1-i}+\sum_{i=0}^{e-2}X_{i+1}X_{e-1-i},$$

$$\phi_1(\alpha,e)=\sum_{i=0}^{e-1}X_iX_{e-1-i}$$

$$\theta_2(e)=2X_0X_1+2X_2X_3+\ldots+2X_{2e-2}X_{2e-1}+X_{2e}^2$$
$$\phi_2(e)=2X_1X_2+2X_3X_4+\ldots+2X_{2e-1}X_{2e}$$
$$\theta_3(e)=2X_0X_1+2X_2X_3+\ldots+2X_{2e-2}X_{2e-1}$$
$$\phi_3(e)=2X_1X_2+2X_3X_4+\ldots+2X_{2e-3}X_{2e-2}+X_{2e-1}^2$$
$$\theta_4(e)=2X_0X_1+2X_2X_3+\ldots+2X_{2e-2}X_{2e-1}$$
$$\phi_4(e)=2X_1X_2+2X_3X_4+\ldots+2X_{2e-1}X_{2e}$$

(distinct pairs of forms appearing in the quadrics involve non-overlapping sets of coordinates so that $n+1=\sum^r e_i+\sum^s(2f_i+1)+\sum^t 2g_i+\sum^k(2\varrho_i+1)+r_0$).

However, we will be interested in pencils of quadrics, so we are free to choose the generators A, B; moreover, for reasons explained below, we will want the pencil to contain at least one (hence, all with a finite number of exceptions) smooth quadric. This implies (but is not equivalent to) $r_0 = 0$; notice also that the case $r_0 > 0$ can be analyzed by projecting from the common vertex of the quadrics to a pencil in $\mathbb{P}^{n-r_0}$, for which the corresponding r_0 is zero.

Segre symbol. It can be shown [HP, XIII.11] that a basis of the pencil can be put into the normal form of Theorem 1.2 not involving the (θ_2,ϕ_2) or (θ_3,ϕ_3) pairs (basically by a change of coordinates in $\mathbb{P}^1$ that moves the point $[\lambda, \mu] = [0, 1]$) and in this case, the Segre symbol of the pencil is defined by writing down the numbers $e_1, \dots, e_r$ in any order, except that the values of e_i corresponding to elementary divisors with the same α_i are grouped and enclosed in round brackets; after a semicolon, the numbers $\varrho_1, \dots, \varrho_k$ are listed in any order.

Since we will only be interested in pencils for which there are no θ_4, ϕ_4 parts to the normal form, again for reasons to be explained below, we recall for completeness the invariant theory of pencils of that form.

THEOREM 1.2. [**AM**, 2.2] *Two pencils of quadrics* $\{\lambda A_i + \mu B_i, [\lambda, \mu] \in \mathbb{P}^1\}, i = 1, 2,$ *such that* $\det(\lambda A_i + \mu B_i)$ *is not identically zero, with singular elements corresponding to* $[\lambda_{i,j}, \mu_{i,j}]$ *are projectively equivalent if and only if they have the same Segre symbol and there is an automorphism of* $\mathbb{P}^1$ *taking* $[\lambda_{1,j}, \mu_{1,j}]$ *to* $[\lambda_{2,j}, \mu_{2,j}]$.

THEOREM 1.3. [**AG**][**AM**, 2.2] *The ring of invariants of pairs of* $(n+1)\times(n+1)$ *symmetric matrices under the action of* $SL(n+1, \mathbb{C})$ *is generated by the coefficients of the polynomial* $\det(\lambda A + \mu B)$.

If we let $SL(2, \mathbb{C})$ act on $[\lambda, \mu]$ we can use known facts on binary forms to conclude:

COROLLARY 1.4. [**AM**, 3.4] *The stable (semi-stable) pencils of quadrics in* $\mathbb{P}^n$ *are those for which* $\det(\lambda A + \mu B)$ *is not identically zero and has no root of multiplicity* $\geq \frac{n+1}{2} (> \frac{n+1}{2})$.

A pair of conics in $\mathbb{P}^2$ gives rise to several deep geometric constructions; we briefly recall the relevant ones. The next statement is one version of "Poncelet's Porism".

PONCELET'S CLOSURE THEOREM 1.5. [**BKOR**, 7.11] Given two smooth conics $C, D \subset \mathbb{P}^2$, if for some integer $N \geq 3$ there exists a nontrivial interscribed N-gon between C and D i.e. there exists a $(P_0, L_0) \in C \times D^*, P_0 \in L_0$, such that by taking L_1 to be the second tangent to D from P_0, P_1 the second point on $L_1 \cap C$ and iterating, we obtain $P_N = P_0$, then for any $(P'_0, L'_0) \in C \times D^*, P'_0 \in L'_0$, the construction also closes ater N steps.

For completeness, we recall that the authors of [BKOR] define the trivial case to be that in which $P_{N-j} = P_j$ for all $0 \leq j \leq N$, and it occurs when $N = 2k + 1$ is odd and $L_k \in C^* \cap D^*$, or $N = 2k$ is even and $P_k \in C \cap D$.

ELLIPTIC CURVE ASSOCIATED TO THE CLOSURE THEOREM 1.6. [**BKOR**, 7.12-7.17] The curve $I = \{(P, L) | P \in L\} \subset C \times D^*$ has arithmetic genus 1 and the choice of a point allows us to give it a group structure. Poncelet's iteration is then translation by a group element and the construction closes after N steps if and only if that element has N torsion. Five groups can occur, according to the intersection multiplicities of C and D; a local calculation for the equation of I in two parameters, s for a point P of C, say, and t for a point on D where L is tangent, yields the following table:

Intersection Type	Group	I
1+1+1+1	$\mathbb{C}/\Lambda$	smooth
2+1+1	$\mathbb{C}^*$	node
2+2	$\mathbb{C}^* \times (\mathbb{Z}/2)$	2 components, 2 nodes
3+1	$\mathbb{C}_a$	cusp
4	$\mathbb{C}_a \times (\mathbb{Z}/2)$	tacnode

We can ask whether the different intersection patterns, which correspond to certain multiplicities for the points that give the base locus of the pencil spanned by C and D, can be detected in terms of Segre symbols. We note first that, in order for a Poncelet situation to make sense, the conics of the pencil cannot all be singular, and by inspection [HP, XIII.11] we find that the only such case with $r_0 = 0$ is that of form $\theta_4 = 2X_0X_1, \phi_4 = 2X_1X_2$. Having excluded that case, we can match the curves with the symbols as follows (notice that the curve I may be stable in the sense of moduli, even when the pencil is not).

SEGRE SYMBOLS 1.7. Five cases remain and by inspection we find the following intersection patterns [HP,XII.11]

Intersection Type	Segre Symbol
1+1+1+1	[1,1,1]
2+1+1	[2,1]
2+2	[(1,1),1]
3+1	[3]
4	[(2,1)]

A pencil of quadrics in $\mathbb{P}^{2g+1}$ with Segre symbol $[1,\ldots,1]$ determines a hyperelliptic curve of genus g:

THEOREM 1.8. [**N**, Sec. 1] *If a pencil of quadrics in $\mathbb{P}^{2g+1}$ has generators in normal form: $\sum_0^{2g+1} X_i^2, \sum_0^{2g+1} e_iX_i^2$, then the 2:1 covering of $\mathbb{P}^1 = \{[\lambda,\mu]\}$ that parametrizes the pencil given by the two rulings on the quadric $Q_{[\lambda,\mu]}$ is a hyperellyptic curve X of genus g, which has an affine model $y^2 = \prod_0^{2g+1}(x-e_i)$.*

However, when $g = 1$ the intersection of two generators of the pencil is also a curve of genus 1 in $\mathbb{P}^3$. It is isomorphic to the curve given in 1.8, and rather than check this directly we quote another theorem that implies it, because we will need this result also. The original proof was given by several people independently; references can be found in [K] whose further dynamical applications we use below. With the same notation as in Theorem 1.8,

THEOREM 1.9. *(cf. [K]) The variety of $\mathbb{P}^{g-1}$s contained in the intersection of the quadrics*
$\sum_0^{2g+1} x_i^2, \sum_0^{2g+1} e_ix_i^2$ is isomorphic to JacX.

COROLLARY 1.10. *When $g = 1$, the genus-1 curve given by equation $y^2 = \prod_0^{2g+1}(x-e_i)$ (Th. 1.8) is isomorphic to the intersection of the two quadric surfaces $\sum_0^{2g+1} X_i^2, \sum_0^{2g+1} e_iX_i^2$.*

But we may consider all other possible Segre symbols, and the base loci of these pencils, again in the $g = 1$ case. We will relate them to Poncelet's curves. Indeed, it was observed (several times, independently) that a pencil of quadrics in $\mathbb{P}^3$ gives rise to an equivalent version of Poncelet's theorem by projection. Again, we only cite a recent account, best suited to our purposes, when older references may be found. Another recent reference where such an example is worked out is [CCS].

THEOREM 1.11. [**BB**] *Let $Q_1, Q_2 \subset \mathbb{P}^3$ be quadrics of rank ≥ 3 such that their intersection curve is either a smooth elliptic curve or the union of two conics meeting in two distinct points. Fix rulings $\mathcal{R}_i$ on $Q_i (i = 1, 2)$. Suppose that there exists a sequence of distinct lines $L_1, \dots, L_{2n+1} = L_1$ such that the line L_j belongs to $\mathcal{R}_1$, resp. $\mathcal{R}_2$ if j is odd, resp. even, and such that consecutive lines L_j, L_{j+1} intersect each other. Then there exist such sequences of length $2n$ through any point on $Q_1 \cap Q_2$.*

A proof of this theorem is based again on the group law for the curve $Q_1 \cap Q_2$. If the situation obtains, the quadrics Q_1, Q_2 or the rulings $\mathcal{R}_1, \mathcal{R}_2$ are said to be in Poncelet n-position. Notice that the theorem still holds if both rulings $\mathcal{R}_1, \mathcal{R}_2$ are on one of the quadrics, if it is smooth. But if we now assume the pencil to be of Segre type $[1, 1, 1, 1]$, take Q_2 to be smooth and Q_1 to be a cone, with vertex v, then the generic projection from v gives rise to a pair of conics, C and D, where C is the image of Q_1 and D is the ramification locus of the projection of Q_2, for which Poncelet's theorem applies.

THEOREM 1.12. [**BB**, 1.3] *The quadrics Q_1 and Q_2 are in Poncelet n-position if and only if the conic D is n-inscribed in C.*

A suitable projection in $\mathbb{P}^{2g+1}$, from a $\mathbb{P}^{g-1}$ spanned by vertices of cones in a pencil of type $[1, \dots, 1]$, to $\mathbb{P}^{g+1}$, had also been considered in [K], in order to translate the addition law on Jac$X$ into a dynamical flow of lines in $\mathbb{P}^{g+1}$ whereby Jacobi described the completely integrable system of geodesic motion on the g-dimensional ellipsoid (at least, Jacobi treated the $g = 2$ case). Again, a (generalized) Poncelet theorem for g confocal quadrics in $\mathbb{P}^{g+1}$ to be inscribed in another confocal quadric can therefore be implemented by using a point of finite order in Jac$X$ [Pr]. When applied to the higher-genus Poncelet porism, the constructions of this note will provide a link with numerical questions of algebraic geometry [TTZ], which generalize those treated for elliptic fibrations [B].

Our goal is now to extend the analysis to pencils of quadrics in $\mathbb{P}^3$ with other Segre symbols, motivated by two issues. The first is invariant-theoretic: what happens in moduli (the relevant moduli, for instance those of the elliptic curve or the Jacobian, the pencil etc.) in the limit? In particular, what happens to the (possible) 13 Segre symbols in $\mathbb{P}^3$ when translated into the 5 Segre symbols we encountered in $\mathbb{P}^2$? In $\mathbb{P}^3$, a natural compactification of a generic configuration will give a projective variety that will exhibit interesting singularities. Again the concept applies to curves, Jacobians, pencils, or other members of a suitable moduli space. One specific construction is that of an elliptic fibration, which we now describe. First we refine the Poncelet construction, in a way that will be used in Section 2, and that was introduced in [BM].

By fixing four generic points in $\mathbb{P}^2$, which we may normalize to be: $P_0 = [1, 1, 1], P_1 = [-1, 1, 1], P_2 = [1, -1, 1], P_3 = [1, 1, -1]$, we determine a pencil of conics of Segre type $[1, 1, 1]$. As we saw, two smooth conics C, D in the pencil and the choice of an origin determine an elliptic curve $I \subset C \times D^*$. However, a different point of view is to associate to any smooth conic in the pencil, by fixing the order $(P_0, \ldots, P_3)$ an elliptic curve with a level-2 structure, namely the double cover of C branched over P_0, P_1, P_2, P_3, with P_0 representing the origin. In fact:

THEOREM 1.13. [**BM**, (1.1)] *Given an elliptic curve E with a level-2 structure there is a unique (smooth) conic C in the pencil so that the corresponding elliptic curve is isomorphic to E compatibly with the level-2 structure. This gives an identification of the parameter space $\mathbb{P}^1$ of the pencil, minus 3 points, with the modular curve X_2 that parametizes elliptic curves with level-2 structure.*

There does not exist a universal elliptic curve with level-2 structure, but [BM] constructs a rational surface S, which is not "natural" because it depends on the choice of a smooth conic C in the pencil, and is an elliptic fibration whose general fibre is the elliptic curve associated to any smooth conic in the pencil other than C. First, the authors of [BM] blow up the four base points of the pencil in $\mathbb{P}^2$; they pull back to this blow up the incidence correspondence $\mathbb{P}^1 \times \mathbb{P}^1 \to \mathbb{P}^2$ (which is the double cover of $\mathbb{P}^2$ with branch locus C), obtaining a surface T. The rational elliptic fibration S is then the minimal desingularization of T, and is fibered by the elliptic curves that correspond to the given pencil, plus singular fibres of Kodaira type I_0^* and I_2. The sections of S corresponding to points of finite order in the elliptic curves are used in [BM] to count the conics in Poncelet n-position as we'll recall in Section 2. However, S is constructed out of a pencil of Segre type $[1,1,1]$. The previous point of view allows us to deform other symbols; in [BKOR], the normalization is given by fixing the conic D: $Y = X^2$, and taking C to be of the form: $\alpha Y = X^2 + \beta XY + \gamma Y^2, \alpha \neq 0$. These conics have an intersection of multiplicity at least 2 at the origin, and for suitable values of α, β, γ give pencils of the 4 remaining Segre types. A linear deformation of C then, such as $\lambda(Y - X^2) + \mu[s(\alpha Y - X^2 - \beta XY - \gamma Y^2) + tX]$, will provide a 'pencil of pencils', a line in $Gr(2,6)$, such that generically the corresponding curve $I_{[s,t]} \subset C_{[s,t]} \times D^*_{[s,t]}$ is smooth (where C denotes as before the conic corresponding to $[\lambda, \mu] = [0,1]$ and D the one corresponding to [1,0]) while the $t = 0$ curve is singular. While this family does not 'fit' into an elliptic fibration in any natural way that I can see, the corresponding curves in $\mathbb{P}^3$ that will be considered below can be compared with those that fiber the surface S.

With this motivation, in this paper we just compute the Segre symbols which correspond to the projections that could be considered.

CONSTRUCTION 1.14. We consider a projection of this particular type: we choose two generators of a pencil of quadrics in $\mathbb{P}^3$, a cone Q_0 and a smooth quadric Q_1, and we project from a vertex v of Q_0 that does not belong to Q_1. Then: (i) The projection of Q_0 is a conic C, which we can take to be the intersection of Q_0 with a generic plane in $\mathbb{P}^3$, and the ramification locus of the projection of Q_1 is another conic D; (ii) the conic C and all the conics D obtained in this way by varying Q_1 in the pencil (including the singular members) form a pencil of conics. The reason for this particular construction is that we want for a Poncelet polygon in the plane to make sense. As in 1.11, points of C will correspond to lines on Q_0 through the vertex and tangent lines to D will correspond to a plane tangent to Q_1. In particular, this is reason enough to exclude certain pencils in $\mathbb{P}^3$, as explained in the following Remark (1).

REMARK 1.15. (1) The only pencils in $\mathbb{P}^3$ in which all the members are singular are either the one where they have a common vertex, which by intersection reduces to a plane pencil, or the one corresponding to the normal form $\theta_1(a,1) + \theta_4(1), \phi_1(a,1) + \phi_4(1)$ with generators $2X_1X_2 + aX_3^2, 2X_0X_1 + X_3^2$ having the line $X_1 = X_3 = 0$ in common, tangent to the plane $X_1 = 0$ at every point on that line, and with vertex $[1,0,0,0], [0,0,1,0]$, resp. This is not a situation considered in Section 1, because whatever vertex we choose to project from, it will belong to the quadric Q_1.

(2) A different kind of projection is considered in [G] in order to exhibit a geometric model of the elliptic curve. First, the curve $I \subset C \times D^*$ is embedded in $Q \cong \mathbb{P}^1 \times \mathbb{P}^1 \subset \mathbb{P}^3$, then from a point of Q not on I the curve is projected to a smooth cubic, whereas from a point on I the curve is projected to a quartic with two singular points, for which a Poncelet theorem is proved analytically using abelian integrals. We do not consider this type of projection since we are interested in comparing Segre symbols of pencils of quadrics.

(3) The type of the plane pencil will depend on the vertex we choose, even if belonging to the same singular quadric. We give one example, to serve as an illustration of the calculations which we otherwise do not reproduce here.

EXAMPLE 1.16. The pencil [(11)11] has normal-form generators:

$$A = \begin{bmatrix} a &&& \\ & a && \\ && b & \\ &&& c \end{bmatrix} \qquad B = \begin{bmatrix} 1 &&& \\ & 1 && \\ && 1 & \\ &&& 1 \end{bmatrix} \qquad a, b, c \text{ distinct}$$

The curve $A \cap B$ is the union of two conics meeting transversely in two points. The singular quadrics of the pencil are two cones with a 1-point vertex such as

$$\begin{bmatrix} a-c &&& \\ & a-c && \\ && b-c & \\ &&& 0 \end{bmatrix} \text{ and a cone with a line vertex, } \begin{bmatrix} 0 &&& \\ & 0 && \\ && b-a & \\ &&& c-a \end{bmatrix}.$$

The corresponding pencil in $\mathbb{P}^2$, if we choose a vertex $[0,0,0,1]$, is indeed of type [(11)1], because the cone meets a generic $\mathbb{P}^2$, which can be taken to be $X_3 = 0$, in $(a-c)X_0^2 + (a-c)X_1^2 + (b-c)X_2^2 = 0$, and the branch locus of the projection of $\begin{bmatrix} 1 &&& \\ & 1 && \\ && 1 & \\ &&& 1 \end{bmatrix}$ from $[0,0,0,1]$, which consists of the points with $X_3^2 = 0$, is $X_0^2 + X_1^2 + X_2^2$. So, the elliptic curve is the same. However, if we choose a vertex $v = [1,0,0,0]$, and consider the intersection of $\begin{bmatrix} 0 &&& \\ & 0 && \\ && b-a & \\ &&& c-a \end{bmatrix}$ with $X_0 = 0$ and the branch locus of the projection of B, we obtain the pencil $\left\langle \begin{bmatrix} 0 && \\ & b-a & \\ && c-a \end{bmatrix}, \begin{bmatrix} 1 && \\ & 1 & \\ && 1 \end{bmatrix} \right\rangle$, with symbol [111]. Thus, unless the pencil in $\mathbb{P}^3$ is of type [1111], the curve X of Theorem 1.8 may not be the same as a corresponding plane Poncelet curve. Notice that the 3-dimensional Poncelet construction of

Theorem 1.11 does not quite make sense when $A = (b-a)X_2^2 + (c-a)X_3^2$ is the union of two planes, for a point in $A \cap B$ will not belong to a unique line on A; however, for the plane conics $A' = (b-a)X_2^2 + (c-a)X_3^2$ and $B' = X_1^2 + X_2^2 + X_3^2$, say, a Poncelet configuration will make sense: from a point on A', which is the union of two lines, we can send a tangent to B' and iterate; the polygon will never close or one line would be tangent to B', and it can be lifted to a polygon of lines in $\mathbb{P}^3$ through v and tangent to B, by taking the inverse image of the vertices under projection.

As we tabulate the type of elliptic curve to which the 13 possible Segre symbols in $\mathbb{P}^3$ give rise, together with the plane model, which in some cases will depend on the choice of vertex, we record the Kodaira-type notation, (cf. e.g. [BPvdV, V.7]) in view of possible applications to the theory of elliptic fibrations: however, the terminology "Kodaira-type" is only used as an analogy, as this projective surface is not a minimal elliptic fibration, as demonstrated by the table (cf. the comment that follows it).

Symbol	A B	$X = A \cap B$	v	Symbol	$I \subset C \times D^*$
[1,1,1,1]	$\sum_{i=0}^3 a_i X_i^2$ $\sum_{i=0}^3 X_i^2$	I_0	$[1,0,0,0]$	[1,1,1]	I_0
[(1,1),1,1]	$a(X_0^2+X_1^2)+bX_2^2+cX_3^2$ $\sum_{i=0}^3 X_i^2$	I_2	$[0,0,0,1]$	[(1,1),1]	I_2
			[1,0,0,0]	[1,1,1]	I_0
[(1,1,1),1]	$a(X_0^2+X_1^2+X_2^2)+bX_3^2$ $\sum_{i=0}^3 X_i^2$	double conic	$[0,0,0,1]$	[(1,1),1]	I_2
			[1,0,0,0]	[(1,1),1]	I_2
[(1,1),(1,1)]	$a(X_0^2+X_1^2)+b(X_2^2+X_3^2)$ $\sum_{i=0}^3 X_i^2$	I_4	$[1,0,0,0]$	[(1,1),1]	I_2
[2,1,1]	$2aX_0X_1+X_1^2+bX_2^2+cX_3^2$ $2X_0X_1+X_2^2+X_3^2$	I_1	$[0,0,1,0]$	[2,1]	I_1
[(2,1),1]	$a(2X_0X_1+X_2^2)+X_1^2+bX_3^2$ $2X_0X_1+X_2^2+X_3^2$	III	$[0,0,1,0]$	[2,1]	I_1
			[0,0,0,1]	[(2,1)]	III
[2,(1,1)]	$2aX_0X_1+X_1^2+b(X_2^2+X_3^2)$ $2X_0X_1+X_2^2+X_3^2$	I_3	$[0,0,0,1]$	[2,1]	I_1
[(2,1,1)]	$a(2X_0X_1+X_2^2+X_3^2)+X_1^2$ $2X_0X_1+X_2^2+X_3^2$	double singular conic	$[0,0,0,1]$	[(2,1)]	III
[2,2]	$2aX_0X_1+X_1^2+2bX_2X_3+X_3^2$ $2X_0X_1+2X_2X_3$	I_2	no suitable v	—	—
[(2,2)]	$2a(X_0X_1+X_2X_3)+X_1^2+X_3^2$ $2X_0X_1+2X_2X_3$	I_0^*	no suitable v	—	—
[3,1]	$a(2X_0X_2+X_1^2)+2X_1X_2+bX_3^2$ $2X_0X_2+X_1^2+X_3^2$	II	$[0,0,0,1]$	[3]	II
[(3,1)]	$a(2X_0X_2+X_1^2+X_3^2)+2X_1X_2$ $2X_0X_2+X_1^2+X_3^2$	IV	$[0,0,0,1]$	[3]	II
[4]	$2a(X_0X_3+X_1X_2)+2X_1X_3+X_2^2$ $2(X_0X_3+X_1X_2)$	III	no suitable v	—	—

To summarize the main point of this calculation, the advantage of the model in $\mathbb{P}^3$ is that $\mathbb{P}^3$ contains the curves. There is no way to embed $I \subset C \times D^*$ consistently in projective space over a family of pairs (C, D), at least this seems to be the problem confronted in [BM, Section 1] also. In $\mathbb{P}^3$, we can deform a pencil with 'nongeneric' Segre symbol, for example as follows: given two generators A, B of the pencil, it can easily be seen that the pencil $\langle A_t = (1-t)A + tR, B_t = (1-t)B + tT\rangle$ will have symbol $[1,1,1,1]$ for an appropriate choice of R, T. The family of base loci of the pencils $\langle A_t, B_t\rangle$ will be a surface S in $\mathbb{P}^3$. Notice however that it may not be an elliptic fibration, since the curves $X_t = A_t \cap B_t$ may intersect for different t's. The set $\tilde{S} = \{(p,t) | p \in X_t\} \subset \mathbb{P}^3 \times \mathbb{P}^1$, cut out by the same equations, is an elliptic fibration; however, it may fail to be smooth and/or minimal-elliptic, as can be seen from the table above, where for the two symbols [(1,1,1),1] and [(2,1,1)] the curve X does not correspond to a Kodaira type for a smooth minimal-elliptic fibration, in which a multiple fibre cannot be simply connected, e.g. [BPvdV, Lemma III.8.3 and V.7 c)]. It seems that these surfaces might still be of interest, in case they give concrete models of elliptic pencils with few singular fibres [B].

The topic of elliptic fibrations is indeed of relevance to moduli theory (cf. [HL], [Pe], [V]), and there are subtle issues related to compactifications. In [V], an elliptic fibration is defined as a flat family of reduced genus-one curves over a smooth curve, with only irreducible nodal singular fibers, smooth total space, and a choice of section. We will assume the base curve to be $\mathbb{P}^1$. This places them in the open part of the moduli space of rational elliptic surfaces, three compactifications of which are described in [HL]; in turn, [Pe] lists all possible configurations of degenerate fibers on a minimal rational elliptic surface.

More invariants? By analogy with the projection from 3 dimensions to 2, which links two Poncelet theorems, we could project $\mathbb{P}^2$ to $\mathbb{P}^1$ from the vertex of one of the singular conics of a (generic) pencil. We would obtain a set of n points in $\mathbb{P}^1$ corresponding to the vertices of each Poncelet n-gon; it might be interesting to associate the invariants of these n points (or, of the 4 points given by the branch loci of the projection restricted to C and D) to the corresponding pair (C, D), in view of the fact that in the [BM] model a varying $C_{[\lambda,\mu]}$ can be viewed as a point of the modular curve X_2 and the corresponding D's in n-Poncelet position, roughly speaking as n-torsion points of the corresponding elliptic curve.

A second reason for being interested in 'natural' compactifications of Poncelet configurations is the dynamical interpretation that the theorem was given, continuous (geodesic motion) or discrete, as an integrable billiard. As such, the configuration corresponds to a 1-dimensional family of Liouville tori, the real part of an elliptic curve varying in the family. The real monodromy of the systems is then related to the singular curves in the family. Degenerate billiard motion is analyzed in [KT, IV.1, e.g.]; in this analysis the 'confocal' pencil in $\mathbb{P}^{g+1}$ is fixed, but the pencil in $\mathbb{P}^{2g+1}$ that carries the invariants of the elliptic curve does vary. There are indeed quite different ways to encode the invariants in the configuration of motion, and the link between the plane billiard and the points of the abelian variety where the flows are linear (in $\mathbb{P}^3$ for the $g = 1$ case) is provided in Knörrer's work [K] by projecting and dualizing. Notice that in the billiard interpretation the list of possible degenerations is restricted by the fact that choosing a confocal pencil fixes two (g+1 for the higher-dimensional model) of the divisors of the discriminant curve $\det(\lambda A + \mu B)$.

2. Monodromy of n-gons

In this section the two conics C, D of Poncelet's porism will be assumed to meet transversely. If C is n-circumscribed to D we say that C and D are in Poncelet n-position. In Section 1 we alluded to the fact that a given point of the elliptic curve $I \subset C \times D^*$ has order n. To investigate the arithmetic implications of the theorem, Barth and Michel identify the modular curve X_2 that parametrizes elliptic curves with level-2 structure, with the smooth conics of the pencil spanned by C and D, after fixing an order of the base points of the pencil P_0, P_1, P_2, P_3, as in 1.12 above. Moreover, after fixing a smooth conic in the pencil, D say, they construct an elliptic fibration S whose general fibre corresponds to the elliptic curve as an element of X_2, and a plane curve Π'_n which is a birational image of the modular curve $X_{00}(n,2) = X_{00}(n) \times_{\mathbb{P}^1} X_2$; its closure Π_n corresponds, roughly speaking, to the curve $T_n \subset S$ that cuts out on each smooth fibre the n^2 points of n torsion and does not contain any fibre, with the zero section (resp. the 4 sections corresponding to the 2-torsion points if n is even) removed. What is of interest here is that they are able to compute interesting numerical invariants by means of the following observation:

PROPOSITION 2.1. [**BM**, (5.1)] *C and D are in Poncelet n position if and only if the point $T_{P_0}D \cap C$ belongs to the curve Π_n,*

Barth and Michel deduce that the number of conics in the pencil that are n-circumscribed about C is $c(n) = t(n)/4$, where $t(n)$ is the number of primitive n-torsion points of an elliptic curve. In addition, they observe that Cayley's theorem provides an equation for Π_n defined over the rational integers.

My question is then: any generic pencil gives rise to an enumerative problem that has $c(n)$ solutions: what is the Galois group of this problem? One way to set it up is to fix P_1, P_2, P_3 on C, and let P_0 vary; the solution we are considering are then the $c(n)$ points $T_{P_0}D \cap \Pi_n$. Other settings could be, the hyperplane in Gr(2,6) given by the pencil of conics that contain C (pencils of conics are lines in $\mathbb{P}^5$); or the set of ordered 4-tuples of points on a conic. But in order to use the normalized equation of Π_n, given in [BM] for small n, we will fix the pencil, and change the line through P_0. The conic D will change, so we are moving the elliptic curve and, roughly speaking, 'solving' for its $n-$torsion points. I first learned about the Galois problem in the context of enumerative geometry from the beautiful paper of Joe Harris [H1], who points out that some such questions were considered classically by C. Jordan, e.g., and who devises a method of solution based on the identification of the Galois group with the monodromy group of a finite covering of varieties. Guided by the remark at the end of [H1], on the geometric significance of the Galois group, "in every case in which current theory had failed to discern any intrinsic structure in the set of solutions – it is proved here – there is none", I hope that the Galois group is the full symmetric group $\mathcal{S}_{c(n)}$. The method, as devised by Harris, which proved itself highly effective [H2], [GH], would consist of showing two facts:

Step I. The monodromy group is 2-transitive. This is typically achieved by checking that the parameter space of solutions (respectively, those fixed by the stabilizer of a point of the fibre) is an irreducible variety.

Step II. The monodromy group contains a single transposition. This can be achieved by exhibiting a point of the parameter space whose fibre contains exactly one double point and $(c(n)-2)$ simple points, provided the solution space is locally irreducible at the double point.

Here I set up the corresponding objects in the case at hand and list the relevant properties; I implement the above steps in the simplest cases ($n = 3, 4, 5$), by brute force. The full answer will be achieved by computation for each n, and I give the technique for doing so, although of course there may be other methods that elude me.

The Galois group of our problem is the monodromy group of the finite cover $\pi : Y \to X$, where X is the $\mathbb{P}^1$ that parametrizes the lines through P_0, one of the 4 fixed points P_0, P_1, P_2, P_3, and Y is the variety corresponding to the extension field over $\mathbb{C}(X)$ generated by the solutions of the Poncelet $n-$position problem, namely the points $T_{P_0}D \cap \Pi_n$, as we recalled above. A remark at the end of Section 1 in [H1] allows us to consider the case in which Y is not irreducible, which as we will see may occur: "Note that the group M of a map $\pi : Y \to X$ is well-defined even in case Y is reducible. In this case we simply define the Galois groups of π to be M, so that we can discuss G without first checking irreducibility of Y. G is a subgroup of, but is not always equal to, the product of the Galois group G_α of the irreducible components Y_α of Y dominating X; the action of G on the fibres of Y, however, is the same as that of G_α."

In our case the variety Y is a linear system of dimension 1 on the curve Π_n, namely the supports of the divisors cut out by the lines through P_0. Thus, we consider the map:

$$\Pi_n \to \mathbb{P}^1$$

such that a point P of the curve Π_n is mapped to the parameter $[\lambda, \mu]$ of the conic tangent to the line P_0P. The curve Π_n in general is not irreducible, in particular it contains as a component the curve Π_m if $m|n$, but, as in [BM], we can consider the curve Π'_n which cuts the general fibre of the surface S (cf. 1.12) in its primitive n-torsion points. In this way, since Π'_n is a birational image of the modular curve $X_{00}(n,2)$ [BM, (4.10)], we can get transitivity as in Step I. To get Step II, we must find a line through P_0 which is simply tangent to Π'_n at only one point.

Cayley gave an algebraic condition for two conics to be in Poncelet n-position, and the equation g_n of the curve Π_n can be derived from it. The first few equations are worked out in [BM] for the given normalization of the coordinates of the points P_i, and I am able to check that the Galois groups is the full symmetric group in the cases of lowest degree. In general, the formula for g_n "seems however too complicated to be evaluated by hand" [BM, Section 4]; in particular, in [BM] g_n is computed for $3 \leq n \leq 12$ except for $n = 11$ which the authors "did not manage to compute". However, a simple computer program gives g_n for any fixed n from Cayley's formula; specifically, $g_{11} = -1024s_3^6s_2^4s_1^4 + (-64s_3^3s_2^9 + 1536s_3^5s_2^6 + 4096s_3^7s_2^3)s_1^3 + (48s_3^2s_2^{11} - 384s_3^4s_2^8 - 5376s_3^6s_2^5 - 6144s_3^8s_2^2)s_1^2 + (-12s_3s_2^{13} - 32s_3^3s_2^{10} + 1152s_3^5s_2^7 + 6400s_3^7s_2^4 + 4096s_3^9s_2)s_1 + (s_2^{15} + 12s_3^2s_2^{12} - 16s_3^4s_2^9 - 704s_3^6s_2^6 - 2560s_3^8s_2^3 - 1024s_3^{10})$, where the s_i, $i = 1, ..., 3$ are the symmetric functions in the homogeneous coordinates of $\mathbb{P}^2$.

THEOREM 2.2. *The monodromy group of the n-th Poncelet closure is the full symmetric group for $n = 3, 4, 5$.*

PROOF. Case $n = 3$. The equation of Π_3 is $\frac{1}{2}(X_0X_1 + X_1X_2 + X_0X_2)$; we expect indeed 2 solutions, the homogenous pairs $[s, t]$ for which the line $s[1, 1, 1] + t[X_0, X_1, X_2]$ meets Π_3. The Galois groups is $\mathbb{Z}_2$.

Case $n = 4$. The equation of Π_4 is $X_0X_1X_2$, reducible as was to be expected.

Case $n = 5$. The equation of Π_5 is of degree $6 = c(5)$:

$(X_0 + X_1 + X_2)(X_0X_1 + X_1X_2 + X_0X_2)X_0X_1X_2 - \frac{1}{4}(X_0X_1 + X_1X_2 + X_0X_2)^3 - X_0^2X_1^2X_2^2$. To check that it is irreducible and that from $P_0 = [1, 1, 1]$ there can be drawn lines simply tangent to Π_5 at only one point, we can use the smooth cubic computed in [BM]:

$$X_0^3 + X_1^3 + X_2^3 - (X_0^2X_1 + X_0X_1^2 + X_1X_2^2 + X_2X_1^2 + X_0X_2^2 + X_0^2X_2) - 2X_0X_1X_2,$$

via the Cremona transformation:

$$[X_0, X_1X_2] \mapsto [X_1X_2, X_2X_0, X_0X_1],$$

which more generally improves the singularities of Π_n, which occur only at the coordinate points. Lastly, it can be checked that a transitive subgroups of $\mathcal{S}_6$ that contains a simple transposition, in particular those coming from the tangents from P_0 to the given cubic, must coincide with $\mathcal{S}_6$. □

Acknowledgments. I am sincerely thankful to the referees for their perspective on these questions, and the supportive comments that allowed me to lay out a more detailed methodology. My gratitude to the Editors of this volume, for their vision, consideration, and great patience.

References

[AG] O. M. Adamovič and E. O. Golovina, *Invariants of a pair of bilinear forms* (Russian, with English summary), Vestnik Moskov. Univ. Ser. I Mat. Meh. **2** (1977), 15–18. MR0460367

[AM] D. Avritzer and R. Miranda, *Stability of pencils of quadrics in* $\mathbf{P}^4$, Bol. Soc. Mat. Mexicana (3) **5** (1999), no. 2, 281–300. MR1738422

[BB] W. Barth and Th. Bauer, *Poncelet theorems*, Exposition. Math. **14** (1996), no. 2, 125–144. MR1395253

[BM] W. Barth and J. Michel, *Modular curves and Poncelet polygons*, Math. Ann. **295** (1993), no. 1, 25–49, DOI 10.1007/BF01444875. MR1198840

[BPvdV] W. Barth, C. Peters and A. Van de Ven, *Compact complex surfaces*, Ergebnisse der Mathematik und ihrer Grenzgebiete (3) **4**, Springer-Verlag, Berlin, 1984.

[B] Arnaud Beauville, *Les familles stables de courbes elliptiques sur* $\mathbb{P}^1$ *admettant quatre fibres singulières* (French, with English summary), C. R. Acad. Sci. Paris Sér. I Math. **294** (1982), no. 19, 657–660. MR664643

[BKOR] H. J. M. Bos, C. Kers, F. Oort, and D. W. Raven, *Poncelet's closure theorem*, Exposition. Math. **5** (1987), no. 4, 289–364. MR917349

[CCS] S.-J. Chang, B. Crespi, and K.-J. Shi, *Elliptical billiard systems and the full Poncelet's theorem in n dimensions*, J. Math. Phys. **34** (1993), 2242–2256. MR1218986

[DR1] Vladimir Dragović and Milena Radnović, *On periodical trajectories of the billiard systems within an ellipsoid in* $\mathbf{R}^d$ *and generalized Cayley's condition*, J. Math. Phys. **39** (1998), no. 11, 5866–5869, DOI 10.1063/1.532600. MR1653096

[DR2] Vladimir Dragović and Milena Radnović, *Cayley-type conditions for billiards within k quadrics in* $\mathbb{R}^d$, J. Phys. A **37** (2004), no. 4, 1269–1276, DOI 10.1088/0305-4470/37/4/014. MR2043219

[GH] Benedict H. Gross and Joe Harris, *On some geometric constructions related to theta characteristics*, Contributions to automorphic forms, geometry, and number theory, Johns Hopkins Univ. Press, Baltimore, MD, 2004, pp. 279–311. MR2058611

[G] Phillip A. Griffiths, *Variations on a theorem of Abel*, Invent. Math. **35** (1976), 321–390, DOI 10.1007/BF01390145. MR0435074

[H1] Joe Harris, *Galois groups of enumerative problems*, Duke Math. J. **46** (1979), no. 4, 685–724. MR552521

[H2] Joe Harris, *Theta-characteristics on algebraic curves*, Trans. Amer. Math. Soc. **271** (1982), no. 2, 611–638, DOI 10.2307/1998901. MR654853

[HL] Gert Heckman and Eduard Looijenga, *The moduli space of rational elliptic surfaces*, Algebraic geometry 2000, Azumino (Hotaka), Adv. Stud. Pure Math., vol. 36, Math. Soc. Japan, Tokyo, 2002, pp. 185–248. MR1971517

[HP] W. V. D. Hodge and D. Pedoe, *Methods of algebraic geometry. Vol. II*, Cambridge Mathematical Library, Cambridge University Press, Cambridge, 1994. Book III: General theory of algebraic varieties in projective space; Book IV: Quadrics and Grassmann varieties; Reprint of the 1952 original. MR1288306

[K] Horst Knörrer, *Geodesics on the ellipsoid*, Invent. Math. **59** (1980), no. 2, 119–143, DOI 10.1007/BF01390041. MR577358

[KT] Valeriĭ V. Kozlov and Dmitriĭ V. Treshchëv, *Billiards*, Translations of Mathematical Monographs, vol. 89, American Mathematical Society, Providence, RI, 1991. A genetic introduction to the dynamics of systems with impacts; Translated from the Russian by J. R. Schulenberger. MR1118378

[N] P. E. Newstead, *Stable bundles of rank* 2 *and odd degree over a curve of genus* 2, Topology **7** (1968), 205–215, DOI 10.1016/0040-9383(68)90001-3. MR0237500

[Pe] Ulf Persson, *Configurations of Kodaira fibers on rational elliptic surfaces*, Math. Z. **205** (1990), no. 1, 1–47, DOI 10.1007/BF02571223. MR1069483

[Pr] Emma Previato, *Poncelet's theorem in space*, Proc. Amer. Math. Soc. **127** (1999), no. 9, 2547–2556, DOI 10.1090/S0002-9939-99-05307-1. MR1662198

[TTZ] Sheng-Li Tan, Yuping Tu, and Alexis G. Zamora, *On complex surfaces with 5 or 6 semistable singular fibers over* $\mathbb{P}^1$, Math. Z. **249** (2005), no. 2, 427–438, DOI 10.1007/s00209-004-0706-4. MR2115452

[V] Ravi Vakil, *Twelve points on the projective line, branched covers, and rational elliptic fibrations*, Math. Ann. **320** (2001), no. 1, 33–54, DOI 10.1007/PL00004469. MR1835061

Department of Mathematics and Statistics, Boston University, Boston, Massachusetts 02215

Email address: ep@bu.edu

Contemporary Mathematics
Volume **703**, 2018
http://dx.doi.org/10.1090/conm/703/14137

Extending Runge's method for integral points

Aaron Levin

ABSTRACT. We formulate and prove a general version of Runge's method, suited to combination with other methods for integral points on varieties. We give some applications of the main theorem, including results for integral points on curves which recover as a special case some known results on rational points on elliptic curves with prime power denominators.

1. Introduction

An old method of Runge [**Run87**], dating back to 1887, allows one to effectively compute the set of integer solutions to certain polynomial equations. In a more geometric language, Runge's method yields the effective computation of the set of integral points on certain affine curves. Runge's method has since been developed and generalized in various directions, including a general version for curves, due to Bombieri [**Bom83**], which is valid over varying number fields, and a version for integral points on higher-dimensional varieties [**Lev08**]. We will discuss Runge's method and its generalizations in detail in Section 3. Although Runge's method is essentially elementary in nature, it has been recently applied in a variety of interesting contexts, including work of Bilu and Parent [**BP08, BP11a, BP11b**] on a conjecture of Serre concerning Galois representations attached to elliptic curves, work of Le Fourn [**LeF16, LeF17**] on integral points on Siegel modular varieties, work of Corvaja, Sookdeo, Tucker, and Zannier [**CSTZ15**] on a problem in arithmetic dynamics involving integral points in orbits, and work of Bennett and the author on the classical Nagell-Ljunggren equation [**BL15**].

The primary purpose of this article is to state a general version of Runge's method (Theorem 5.15) that is formulated in a way which allows it to be combined with other Diophantine methods for studying integral points. The statement of the main theorem is somewhat technical, but it contains Runge's classical result and its subsequent generalizations as a degenerate special case ($S_0 = \emptyset$). When the main theorem is combined with other effective techniques, like the theory of linear forms in logarithms, the combined result is effective. It seems useful, however, to also combine Runge's method with powerful ineffective techniques (though, of course, one loses effectivity in this case). As an application, we show that some known effective Diophantine results on rational points on elliptic curves with prime power

2010 *Mathematics Subject Classification.* Primary 11G35; Secondary 11D61, 11D72.
Key words and phrases. Runge's method, integral points on varieties.
The author was supported in part by NSF CAREER Grant DMS-1352407.

denominators (Corollaries 7.24 and 7.25) can be quickly derived in the general framework given here.

A brief outline of the paper is as follows. In the next section we recall some basic definitions, primarily involving integral points and heights. In Section 3, we give an overview of Runge's method and its generalizations. In Section 4, we review some consequences of the (effective) Hilbert Nullstellensatz which are essential in proving Runge-type results. In Section 5, we state and prove our main theorem (Theorem 5.15). In Section 6, we discuss the relation between Runge's method and our main theorem. In Section 7, we give some applications to curves, including the previously mentioned Diophantine results on elliptic curves. Finally, in Section 8, we discuss some higher-dimensional applications.

2. Notation and definitions

Let k be a number field and let $\mathcal{O}_k$ denote the ring of integers of k. Let M_k be the canonical set of places of k consisting of one place for each prime ideal $\mathfrak{p}$ of $\mathcal{O}_k$, one place for each real embedding $\sigma : k \to \mathbb{R}$, and one place for each pair of conjugate embeddings $\sigma, \overline{\sigma} : k \to \mathbb{C}$. If S is a finite set of places of k containing the archimedean places, then we let $\mathcal{O}_{k,S}$ be the ring of S-integers of k and $\mathcal{O}_{k,S}^*$ be the group of S-units.

Let k_v denote the completion of k with respect to v. We normalize our absolute values so that $|p|_v = p^{-[k_v:\mathbb{Q}_p]/[k:\mathbb{Q}]}$ if v corresponds to $\mathfrak{p}$ and $\mathfrak{p}|p$, and $|x|_v = |\sigma(x)|^{[k_v:\mathbb{R}]/[k:\mathbb{Q}]}$ if v corresponds to an embedding σ. We let M_k^∞ denote the set of archimedean places of k. With these normalizations, the (absolute logarithmic) height of a point $P = [x_0 : \cdots : x_n] \in \mathbb{P}^n(k)$ is given by

$$h(P) = \sum_{v \in M_k} \log \max(|x_0|_v, \ldots, |x_n|_v).$$

Recall that h is independent both of the choice of k and the choice of coordinates $x_0, \ldots, x_n \in k$ (by the product formula).

If v is a place of k and w is a place of a field extension L of k, then we say that w lies above v, or $w|v$, if w and v define the same topology on k. We let $\mathbb{C}_v$ denote the completion of the algebraic closure $\overline{k_v}$ at v and recall that the absolute value $|\cdot|_v$ extends uniquely to an absolute value on $\mathbb{C}_v$.

Let D be a Cartier divisor on a projective variety X, with both X and D defined over a number field k. Let $\operatorname{Supp} D$ denote the support of D. Let $v \in M_k$. We define a local Weil function (or local height function) for D relative to v to be a function $\lambda_{D,v} : X(\mathbb{C}_v) \setminus \operatorname{Supp} D \to \mathbb{R}$ such that if D is represented locally by (f) on an open set U then

$$\lambda_{D,v}(P) = -\log |f(P)|_v + \alpha_v(P),$$

where α_v is a continuous function on $U(\mathbb{C}_v)$ (in the v-topology). It will sometimes also be convenient to follow the convention that $\lambda_{D,v}(P) = \infty$ if $P \in \operatorname{Supp} D$ (e.g., in Lemma 4.13).

We can naturally view $k \subset k_v \subset \mathbb{C}_v$, and thus view $\lambda_{D,v}$ as a function on $X(k) \setminus \operatorname{Supp} D$. A global Weil function for D over k consists of a collection of local Weil functions, $\lambda_{D,v}$, for $v \in M_k$, where the α_v above satisfy certain reasonable boundedness conditions as v varies (we refer the reader to [**Lan83**] and [**Voj87**] for further details). In this case, the sum of local Weil functions

$$h_D = \sum_{v \in M_k} \lambda_{D,v}$$

gives a height associated to D for all $P \in X(k) \setminus \operatorname{Supp} D$. If $\lambda_{D,v}$, $v \in M_k$, is a global Weil function, L is a finite extension of k, and $w \in M_L$ lies above a place $v \in M_k$, then we may identify $\mathbb{C}_w$ and $\mathbb{C}_v$ and we set

$$\lambda_{D,w} = \frac{[L_w : k_v]}{[L : k]} \lambda_{D,v}.$$

Then

$$h_D(P) = \sum_{w \in M_{k(P)}} \lambda_{D,w}(P)$$

is a height associated to D for all $P \in X(\overline{k}) \setminus \operatorname{Supp} D$.

DEFINITION 2.1. We define an M_k-constant to be a family of real numbers c_v, $v \in M_K$, such that $c_v = 0$ for all but finitely many v. If L is a finite extension of k and $w \in M_L$ lies above a place $v \in M_k$, then we define

$$c_w = \frac{[L_w : k_v]}{[L : k]} c_v.$$

Thus, if $P \in X(k) \setminus \operatorname{Supp} D$, L is a finite extension of k, and $w \in M_L$ lies above a place $v \in M_k$, the conditions $\lambda_{D,v}(P) \leq c_v$ and $\lambda_{D,w}(P) \leq c_w$ are equivalent.

If L is a finite extension of k and $S \subset M_k$, we let $S_L \subset M_L$ denote the set of places of L lying above places in S. We now define the notion of a set of (D, S)-integral points.

DEFINITION 2.2. Let D be an effective Cartier divisor on a projective variety X, with both X and D defined over a number field k. Let $S \subset M_k$ be a finite set of places. Let $R \subset X(\overline{k}) \setminus \operatorname{Supp} D$ and let $\lambda_{D,v}$, $v \in M_k$, be a global Weil function for D over k. Then R is defined to be a (D, S)-integral set of points if there exists an M_k-constant c_v such that

$$\lambda_{D,w}(P) \leq c_w$$

for all $P \in R$ and all places $w \in M_{k(P)} \setminus S_{k(P)}$.

REMARK 2.3. It is customary to require that S contains the set of archimedean places of k. We will find it useful to formally extend the usual definition to arbitrary finite sets of places of k, including the choice $S = \emptyset$.

We will also define a notion of (D, s)-integral points, where we allow the number field and set of places S to vary under the restriction that $|S| \leq s$ and S contains the set of archimedean places.

DEFINITION 2.4. Let D be an effective Cartier divisor on a projective variety X, with both X and D defined over a number field k. Let $\lambda_{D,v}$, $v \in M_k$, be a global Weil function for D over k. Let s be a positive integer. A set $R \subset X(\overline{k}) \setminus \operatorname{Supp} D$ is defined to be a set of (D, s)-integral points if there exists an M_k-constant $c_v, v \in M_k$, such that R is of the form

$$R = \bigcup_{\substack{L \supset k \\ M_L^\infty \subset S \subset M_L \\ |S| \leq s}} R_S, \tag{2.1}$$

where the union runs over all number fields L containing k and each set R_S is a set of (D,S)-integral points with respect to the global Weil function $\lambda_{D,v}$ and the M_k-constant c_v.

REMARK 2.5. Note that since we require the finite set of places $S \subset M_L$ in (2.1) to contain the set of archimedean places, the degree $[L:k]$ in (2.1) is actually bounded. Indeed, in this case we must have $[L:k] \leq 2s$.

We will also need the following definition from algebraic geometry.

DEFINITION 2.6. Let D be a Cartier divisor on a projective variety X over a field k. If $\dim H^0(X, \mathcal{O}(nD)) = 0$ for all $n > 0$ then we define $\kappa(D) = -\infty$. Otherwise, define $\kappa = \kappa(D)$ to be the integer for which

$$\limsup_{n\to\infty} \frac{\dim H^0(X, \mathcal{O}(nD))}{n^\kappa}$$

exists and is nonzero.

It is well-known that $\kappa(D) \in \{-\infty, 0, 1, \ldots, \dim X\}$. We say that D is big if $\kappa(D) = \dim X$ (so $\kappa(D)$ is as large as possible). In particular, ample divisors are big. Many of our results hold for arbitrary projective varieties (or even complete varieties), but for simplicity we will typically assume that our varieties are projective and nonsingular (so, in particular, Cartier and Weil divisors coincide).

3. Runge's method

We now give a brief overview of Runge's method and its subsequent generalizations. In its most basic form, Runge proved:

THEOREM 3.7 (Runge [**Run87**]). *Let $f \in \mathbb{Q}[x,y]$ be an absolutely irreducible polynomial of total degree n. Let f_0 denote the leading form of f, i.e., the sum of the terms of total degree n in f. Suppose that f_0 factors as $f_0 = g_0h_0$, where $g_0, h_0 \in \mathbb{Q}[x,y]$ are nonconstant relatively prime polynomials. Then the set of solutions to*

$$f(x,y) = 0, \quad x, y \in \mathbb{Z},$$

is finite and effectively computable.

Hilliker and Straus [**HS83**] and Walsh [**Wal92**] have given explicit bounds for the integer solutions in Runge's theorem (as well as some of its generalizations). A geometric formulation of Runge's theorem which is valid for arbitrary rings of S-integers is the following.

THEOREM 3.8. *Let C be a nonsingular projective curve defined over a number field k and let $\phi \in k(C)$ be a rational function on C. Let $(\phi)_\infty$ be the divisor of poles of ϕ and let r be the number of irreducible components over k of the support of $(\phi)_\infty$. Let S be a finite set of places of k containing the archimedean places. If $r > |S|$, then the set $\{P \in C(k) \mid \phi(P) \in \mathcal{O}_{k,S}\}$ is finite and effectively computable.*

To see the relation with Theorem 3.7, let C be a projective closure of the affine plane curve defined by $f = 0$ in Theorem 3.7 and let $\pi : C' \to C$ be a normalization. Set $\phi = x \circ \pi$, $K = \mathbb{Q}$, and $S = \{\infty\}$. The factorization condition in Theorem 3.7 implies that the support of $(\phi)_\infty$ has at least two components over $\mathbb{Q}$. Then applying Theorem 3.8 with these parameters yields Theorem 3.7 as a special case.

Building on work of Sprindžuk [**Spr80**], Bombieri [**Bom83**] proved a uniform version of Runge's theorem, allowing the number field k and set of places S to vary.

THEOREM 3.9 (Bombieri). *Under the same hypotheses as Theorem 3.8, the set of points*

$$\bigcup_{\substack{L\supset k, S_L \\ |S_L|<r}} \{P \in C(L) \mid \phi(P) \in \mathcal{O}_{L,S_L}\}$$

is finite and effectively computable, where L ranges over all number fields and S_L over sets of places of L containing the archimedean places.

In higher dimensions, the following version of Runge's theorem was proven in [**Lev08**] (see also Definition 2.4).

THEOREM 3.10 (Levin). *Let X be a nonsingular projective variety defined over a number field k. Let $D = \sum_{i=1}^r D_i$ be a divisor on X, with $D_1, \dots, D_r$ effective divisors defined over k. Suppose that the intersection of any $m+1$ of the supports of the divisors D_i is empty. Let s be a positive integer satisfying*

$$ms < r.$$

Let R be a set of (D,s)-integral points on X. Then the following statements hold.

(a) *If $\kappa(D_i) > 0$ for all i, then R is contained in an effectively computable proper Zariski closed subset $Z \subset X$.*
(b) *If D_i is big for all i, then there exists an effectively computable proper Zariski closed subset $Z \subset X$, independent of R, such that the set $R \setminus Z$ is finite (and effectively computable).*
(c) *If D_i is ample for all i, then R is finite and effectively computable.*

Fixing a set of places S, a refinement of part (a) of Theorem 3.10 was proven by Corvaja, Sookdeo, Tucker, and Zannier.

THEOREM 3.11 (Corvaja, Sookdeo, Tucker, Zannier [**CSTZ15**]). *Let X be a nonsingular projective variety defined over a number field k. Let S be a finite set of places of k containing the archimedean places and let $s = |S|$. Let D be an effective divisor on X. Suppose that for any s points $P_1, \dots, P_s \in X(\overline{k})$, there exists an effective divisor E defined over k such that*

(a) $\operatorname{Supp} E \subset \operatorname{Supp} D$
(b) $\kappa(E) > 0$
(c) $P_1, \dots, P_s \notin \operatorname{Supp} E$.

Then any set of k-rational (D,S)-integral points on X lies in an effectively computable proper closed subset of X.

Finally, we mention the recent "tubular" version of Runge's method due to Le Fourn [**LeF16**, Th. 1]. This version of Runge's method may potentially also be combined with the techniques given here, but we will not explore this.

REMARK 3.12. For the objects in the results of this section to truly be "effectively computable", the objects have to be presented in a computable fashion (e.g., explicit polynomial equations defining the varieties and divisors, explicitly given M_k-constants in Definition 2.2, etc.). Moreover, we assume throughout that one can compute certain fundamental geometric objects that arise in the proofs (e.g., Riemann-Roch bases associated to divisors). In our effective results, we implicitly make these assumptions throughout the paper, without further comment, both to avoid cumbersome statements and to focus on the arithmetic aspects of the problems, as opposed to computational algebraic geometry. In all cases, the geometric computations required to produce explicit results should be clear from the proofs.

4. Two consequences of the Nullstellensatz

A key idea in Runge's method is the intuitive fact that if $Z_1, \ldots, Z_r$ are closed subsets of X without any common point, then no point P in X can be v-adically close to all of the sets Z_i simultaneously. This idea can be made precise using Hilbert's Nullstellensatz and results in the following well-known lemma.

LEMMA 4.13. *Let $D_1, \ldots, D_r$ be effective divisors on a nonsingular projective variety X, all defined over a number field k, such that $\cap_{i=1}^r \operatorname{Supp} D_i = \emptyset$. Let $\lambda_{D_i,v}, v \in M_k$, be global Weil functions for $i = 1, \ldots, r$. Then there exists an effectively computable M_k-constant c_v such that*

$$\min\{\lambda_{D_1,v}(P), \ldots, \lambda_{D_r,v}(P)\} \le c_v$$

for all $v \in M_k$ and all $P \in X(\mathbb{C}_v)$.

The effectiveness of the M_k-constant c_v in Lemma 4.13 (given explicit equations defining $D_1, \ldots, D_r, X$, and the global Weil functions) is a direct consequence of the fact that there are effective versions of the Hilbert Nullstellensatz (e.g., [**MW83**]).

More generally, if a point P is v-adically very close to $D_1, \ldots, D_r$, then it must be be very close to the intersection of the (supports of the) divisors. This observation can be made precise in the fact that $\min\{\lambda_{D_1,v}(P), \ldots, \lambda_{D_r,v}(P)\}$ is a local height associated to the closed subscheme $D_1 \cap D_2 \cap \cdots \cap D_r$ (see [**Sil87**] for more details). Thus, if $\cap_{i=1}^r \operatorname{Supp} D_i = \emptyset$, then $\min\{\lambda_{D_1,v}(P), \ldots, \lambda_{D_r,v}(P)\}$ is a local height associated to the trivial divisor, which is just a restatement of Lemma 4.13.

We record a second version of Lemma 4.13, stated in terms of rational functions instead of local heights (see [**Lev08**, Lemma 2.1]).

LEMMA 4.14. *Let X be a nonsingular projective variety defined over a number field k. Extend each absolute value v in M_k to $\mathbb{C}_v$. Let $\phi_1, \ldots, \phi_r \in k(X)$ be rational functions on X without a common pole. Then there exists an effectively computable M_k-constant c_v such that*

$$\min_{1 \le i \le r} \log |\phi_i(P)|_v \le c_v$$

for all $v \in M_k$ and all $P \in X(\mathbb{C}_v)$.

5. Main theorem

We now state and prove our main result, which will be shown to be a generalization of Runge's method in the next section.

THEOREM 5.15. *Let X be a nonsingular projective variety defined over a number field k. Let $D = \sum_{i=1}^r D_i$ be a divisor on X with $D_1, \ldots, D_r$ effective divisors defined over k. Let S_0 be a finite set of places of k. For a subset $T \subset X(\overline{k})$, let*

$$D_T = \sum_{\substack{i \\ T \cap (\operatorname{Supp} D_i)(\overline{k}) = \emptyset}} D_i.$$

Let s be a positive integer. Let $\mathcal{L}$ be a set of number fields containing k. For each $L \in \mathcal{L}$, let $S_{L,0}$ denote the set of places of L lying above places in S_0. Let $\lambda_{D,v}$, $v \in M_k$, be a global Weil function and let c_v be an M_k-constant. Let

$$R = \bigcup_{\substack{L \in \mathcal{L} \\ S \subset M_L \\ |S \setminus S_{L,0}| \leq s}} \{P \in X(L) \mid \lambda_{D,w}(P) \leq c_w, \forall w \in M_L \setminus S\}.$$

Let

$$\mathcal{D} = \{D_T \mid T \subset X(\overline{k}), |T| \leq s\},$$

a finite set. There exist sets $R_{D'} \subset \cup_{L \in \mathcal{L}} X(L)$, $D' \in \mathcal{D}$, such that each set $R_{D'}$ is a set of (D', S_0)-integral points and

$$R \subset \cup_{D' \in \mathcal{D}} R_{D'}.$$

In particular, if $\dim R_{D'} \leq m$ *for every $D' \in \mathcal{D}$, then* $\dim R \leq m$.

PROOF. For each $i \in \{1, \dots, r\}$, we fix global Weil functions $\lambda_{D_i,v}$, $v \in M_k$. To simplify calculations, we will also, as is always possible, choose the Weil functions so that $\lambda_{D_i,v}(P) \geq 0$ for all $v \in M_k$ and all $P \in X(\mathbb{C}_v) \setminus \operatorname{Supp} D_i$. For any subset $I \subset \{1, \dots, r\}$, we let $\lambda_{\sum_{i \in I} D_i, v} = \sum_{i \in I} \lambda_{D_i,v}$, a global Weil function for $\sum_{i \in I} D_i$ (with this definition, $\lambda_{D,v}$ may differ from the one given in the theorem, in which case we must replace the M_k-constant c_v in the theorem with another appropriate M_k-constant).

Let $\mathcal{I}$ be the set of all sets of indices $I \subset \{1, \dots, r\}$ such that $\cap_{i \in I} \operatorname{Supp} D_i = \emptyset$. By Lemma 4.13, if $I \in \mathcal{I}$, then there exists an M_k-constant $a_{I,v}$ such that

$$\min_{i \in I} \lambda_{D_i,v}(P) \leq a_{I,v}$$

for all $v \in M_k$ and all $P \in X(\mathbb{C}_v)$. Let a_v be the M_k-constant $a_v = \max_{I \in \mathcal{I}} a_{I,v}$.

Let $P \in R$. Let $L \supset k$ and $S \subset M_L$ be such that $P \in X(L)$, $|S \setminus S_{L,0}| \leq s$, and

$$\lambda_{D,w}(P) \leq c_w$$

for all $w \in M_L \setminus S$. For each $w \in S \setminus S_{L,0}$ let

$$J_w = \{j \in \{1, \dots, r\} \mid \lambda_{D_j,w}(P) > a_w\}.$$

By the definition of a_w, $\cap_{j \in J_w} \operatorname{Supp} D_j(\overline{k}) \neq \emptyset$ (if J_w is empty, then by convention we let the intersection be $X(\overline{k})$). For each $w \in S \setminus S_{L,0}$, let $P_w \in \cap_{j \in J_w} \operatorname{Supp} D_j(\overline{k})$ and let $T = \{P_w \mid w \in S \setminus S_{L,0}\}$. If $i \in \{1, \dots, r\}$ and $T \cap (\operatorname{Supp} D_i)(\overline{k}) = \emptyset$, then it follows that

$$\lambda_{D_i,w}(P) \leq a_w$$

for all $w \in S \setminus S_{L,0}$. Let $b_v = \max\{ra_v, c_v\}$, $v \in M_k$.

It follows that

$$R \subset \cup_{D' \in \mathcal{D}} R_{D'},$$

where for $D' \in \mathcal{D}$, $R_{D'}$ is the set of (D', S_0)-integral points

$$\bigcup_{L \in \mathcal{L}} \{P \in X(L) \mid \lambda_{D',w}(P) \leq b_w, \forall w \in M_L \setminus S_{L,0}\}.$$

□

Since the M_k-constant in Lemma 4.13 is effective, if sets of (D', S_0)-integral points are effectively computable for $D' \in \mathcal{D}$, then the set R of Theorem 5.15 will also be effectively computable. We will take advantage of this observation in later results.

6. Relation to Runge's method

In this section we will analyze Theorem 5.15 with the choice $S_0 = \emptyset$. In this case, we will show that Theorem 5.15 recovers Runge's method and its various generalizations described in Section 3.

We first show that when $S_0 = \emptyset$, a set of (D, S_0)-integral points is a subset of $X(\overline{k})$ of bounded height with respect to D.

LEMMA 6.16. *Let X be a nonsingular projective variety and let D be an effective divisor on X, with both X and D defined over a number field k. Let $R \subset X(\overline{k})$ be a set of $(D, \emptyset)$-integral points. Then R is a set of bounded height with respect to D.*

PROOF. If R is a set of $(D, \emptyset)$-integral points, then by definition there exists an M_k-constant c_v such that

$$\lambda_{D,w}(P) \leq c_w$$

for all $P \in R$ and all $w \in M_{k(P)}$. Then we find that for all $P \in R$,

$$h_D(P) = \sum_{w \in M_{k(P)}} \lambda_{D,w}(P) \leq \sum_{w \in M_{k(P)}} c_w = \sum_{v \in M_k} c_v.$$

□

Next, we note some elementary Northcott-type properties of heights.

LEMMA 6.17. *Let D be a divisor on a nonsingular projective variety X, with both X and D defined over a number field k. Let h_D be a height associated to D. Let c be a real number and let d be a positive integer. Let*

$$R = R(X, k, c, d, h_D) = \{P \in X(\overline{k}) \mid [k(P) : k] \leq d, h_D(P) \leq c\}.$$

(a) *If $\kappa(D) > 0$, then R is contained in a proper Zariski closed subset of X.*
(b) *If D is big, then there exists a proper Zariski closed subset $Z \subset X$, independent of c, d, and the choice of h_D, such that the set $R \setminus Z$ is finite.*
(c) *If D is ample, then R is finite.*

PROOF. If $\kappa(D) > 0$, then there exists a nonconstant rational function $\phi \in k(X)$ with all of its poles in D, i.e., $\operatorname{Supp}(\phi)_\infty \subset \operatorname{Supp} D$. Let N be a positive integer such that $(\phi)_\infty \leq ND$ (i.e., $ND - (\phi)_\infty$ is effective). Let $h = h_\infty$ denote the usual absolute logarithmic height on $\mathbb{P}^1$. By functoriality of heights,

$$h(\phi(P)) = h_{\phi^*\infty}(P) + O(1) \leq N h_D(P) + O(1)$$

for all $P \in X(\overline{k}) \setminus \operatorname{Supp} D$. It follows that $\phi(R \setminus \operatorname{Supp} D)$ is a set of bounded height (and bounded degree) on $\mathbb{P}^1$. By Northcott's theorem, $\phi(R \setminus \operatorname{Supp} D)$ is a finite set and so R is contained in the union of $\operatorname{Supp} D$ and finitely many sets of the form $\phi^{-1}(P)$, $P \in \mathbb{P}^1(\overline{k})$.

When D is ample, this is the well-known Northcott property of ample heights and follows from Northcott's theorem on projective space applied to a projective embedding associated to a very ample multiple of D.

If D is big, then by Kodaira's lemma, for some $N > 0$, $ND \sim A + E$ for some ample divisor A and some effective divisor E. Then

$$h_D(P) \leq \frac{1}{N} h_A(P) + O(1)$$

for all $P \in X(\overline{k}) \setminus \operatorname{Supp} E$. By the Northcott property of ample heights, R is contained in the union of a finite set of points and $\operatorname{Supp} E$. □

REMARK 6.18. Assuming that one can carry out basic geometric calculations with respect to X and D (e.g., compute a Riemann-Roch basis with respect to multiples nD of D), it's clear that the sets in Lemma 6.17 are effectively computable, as one ultimately reduces to computing sets of bounded height and bounded degree on some projective space $\mathbb{P}^N$.

We now deduce Theorems 3.10 and 3.11 from Theorem 5.15. The condition $ms < r$ in Theorem 3.10 and the conditions (a)-(c) of Theorem 3.11 both imply that for any subset $T \subset X(\overline{k})$ with $|T| \leq s$, the divisor D_T satisfies $\kappa(D_T) > 0$ or D_T is big/ample appropriately (decomposing D as an appropriate sum of effective divisors in Theorem 3.11). Note also that in Theorem 3.10, the points of R have bounded degree (Remark 2.5). Theorems 3.10 and 3.11 then follow immediately from Theorem 5.15 applied to X, D, s, and $S = \emptyset$, combined with Lemma 6.16 and Lemma 6.17 (under the implicit assumptions of Remark 3.12 and Remark 6.18). In fact, we see that Theorem 3.11 holds in a more general setting where we allow the set of places S (containing the archimedean places) and number field k to vary (as in Theorem 3.10). Note that both Theorem 3.10 and Theorem 3.11 contain Bombieri's version of Runge's method for curves (in the language of divisors).

7. Algebraic curves

We first prove an ineffective, but general, result for curves of positive genus.

THEOREM 7.19. *Let C be a nonsingular projective curve of genus $g \geq 1$ defined over a number field k, $\phi \in k(C)$ a rational function on C, and r the number of irreducible components over k of the set of poles $\operatorname{Supp}(\phi)_\infty$ of ϕ. Let S_0 be a finite set of places of k containing the archimedean places. Then the set of integral points*

$$R = \bigcup_{\substack{S_0 \subset S \subset M_k \\ |S \setminus S_0| \leq r-1}} \{P \in C(k) \mid \phi(P) \in \mathcal{O}_{k,S}\}$$

is finite.

PROOF. Let $D = \sum_{i=1}^r D_i$ be the divisor of poles $(\phi)_\infty$ of ϕ, where $D_1, \ldots, D_r$ are nontrivial effective divisors defined over k with disjoint support. Since a global Weil function for the point at infinity on $\mathbb{P}^1$ is

$$\lambda_{\infty,v}([x_0 : x_1]) = \log \frac{\max\{|x_0|_v, |x_1|_v\}}{|x_1|_v}, \quad v \in M_k,$$

by functoriality, a global Weil function for D is

$$\lambda_{D,v}(P) = \log \max\{|\phi(P)|_v, 1\}, \quad v \in M_k.$$

If $P \in R$ and S is a set of places of k with $S_0 \subset S$, $|S \setminus S_0| \leq r - 1$, and $\phi(P) \in \mathcal{O}_{k,S}$, then $|\phi(P)|_v \leq 1$ for $v \notin S$ and

$$\lambda_{D,v}(P) = 0, \quad \forall v \in M_k \setminus S.$$

Let $s = r-1$. Since $s < r$, for any subset $T \subset C(\overline{k})$ with $|T| \leq s$, the divisor D_T of Theorem 5.15 is nontrivial. By Siegel's theorem, any set of (k-rational) (D_T, S_0)-integral points on C is finite. Thus, in an appropriate application of Theorem 5.15, the set $\cup_{D' \in \mathcal{D}} R_{D'}$ is finite, and hence R is finite as well. □

We note that in Theorem 7.19, and the subsequent results of this section, there is no essential loss of generality in assuming that C is nonsingular. If C is an arbitrary projective curve and $\phi \in k(C)$, let Z be the set of singular points of C, $\pi : \tilde{C} \to C$ be a normalization, and $\tilde{\phi} = \phi \circ \pi$. Then the map $P \mapsto \pi(P)$ gives a bijection between the sets $\{P \in \tilde{C}(k) \mid \tilde{\phi}(P) \in \mathcal{O}_{k,S}\} \setminus \pi^{-1}(Z)$ and $\{P \in C(k) \mid \phi(P) \in \mathcal{O}_{k,S}\} \setminus Z$. Since Z is a finite set, we see that it suffices to study integral points on the nonsingular projective curve $\tilde{C}$.

On the projective line, Theorem 7.19 needs to be modified slightly. In this case, one can also prove an effective result.

THEOREM 7.20. *Let $\phi \in k(x)$ be a rational function with at least three distinct poles in $\mathbb{P}^1(\overline{k})$ and set*

$$\phi\left(\frac{x}{y}\right) = \frac{p(x,y)}{q(x,y)} = \frac{p(x,y)}{q_1(x,y)^{i_1} \cdots q_r(x,y)^{i_r}},$$

where $p, q, q_1, \ldots, q_r \in k[x,y]$ are binary forms, p and q have no factors in common, and $q_1, \ldots, q_r$ are irreducible, pairwise nonproportional binary forms, ordered so that $\deg q_1 \leq \deg q_2 \leq \cdots \leq \deg q_r$. Let $d_i = \deg q_i$, $i = 1, \ldots, r$, and

$$r_0 = \begin{cases} 1 & \text{if } d_1 \geq 3, \\ 2 & \text{if } d_1 \leq 2 \text{ and } d_2 \geq 2, \\ 3 & \text{otherwise.} \end{cases}$$

Let S_0 be a finite set of places of k containing the archimedean places. Then the set of integral points

$$R = \bigcup_{\substack{S_0 \subset S \subset M_k \\ |S \setminus S_0| \leq r - r_0}} \{P \in \mathbb{P}^1(k) \mid \phi(P) \in \mathcal{O}_{k,S}\}$$

is finite and effectively computable.

We give an elementary reduction to Thue-Mahler equations, avoiding the use of Theorem 5.15.

PROOF. We let $\mathrm{Res}(f,g)$ denote the resultant of two binary forms $f(x,y)$ and $g(x,y)$ in $k[x,y]$. Without loss of generality, we may enlarge S_0 so that $\mathcal{O}_{k,S_0}$ is a principal ideal domain, $p, q_1, \ldots, q_r \in \mathcal{O}_{k,S_0}$, $\mathrm{Res}(p,q) \in \mathcal{O}_{k,S_0}^*$, and $\mathrm{Res}(q_i, q_j) \in \mathcal{O}_{k,S_0}^*$, $i \neq j$.

Let S be a finite set of places of k containing S_0 such that $|S \setminus S_0| \leq r - r_0$. Let $P_0 \in \mathbb{P}^1(k)$ with $\phi(P_0) \in \mathcal{O}_{k,S}$. Since $\mathcal{O}_{k,S_0}$ is a principal ideal domain, we may write $P_0 = [x_0 : y_0]$ where $x_0, y_0 \in \mathcal{O}_{k,S_0}$ and x_0 and y_0 are coprime in $\mathcal{O}_{k,S_0}$, that is, the $\mathcal{O}_{k,S_0}$-ideal (x_0, y_0) is the whole ring $\mathcal{O}_{k,S_0}$. By properties of the resultant, there exist homogeneous polynomials $f_1, f_2, g_1, g_2 \in \mathcal{O}_{k,S_0}[x,y]$ such that

$$p(x,y)f_1(x,y) + q(x,y)g_1(x,y) = \mathrm{Res}(p,q)x^{2d-1},$$
$$p(x,y)f_2(x,y) + q(x,y)g_2(x,y) = \mathrm{Res}(p,q)y^{2d-1},$$

where $d = \deg p = \deg q = \deg \phi$. Substituting $x = x_0$ and $y = y_0$, we find the relation of $\mathcal{O}_{k,S_0}$-ideals

$$(p(x_0, y_0), q(x_0, y_0)) \supset (\operatorname{Res}(p, q)x_0^{2d-1}, \operatorname{Res}(p, q)y_0^{2d-1}) = \mathcal{O}_{k,S_0}.$$

So $p(x_0, y_0)$ and $q(x_0, y_0)$ are coprime in $\mathcal{O}_{k,S_0}$, and hence coprime in $\mathcal{O}_{k,S}$. Since $\frac{p(x_0,y_0)}{q(x_0,y_0)} \in \mathcal{O}_{k,S}$, this implies that $q(x_0, y_0) = q_1(x_0, y_0)^{i_1} \cdots q_r(x_0, y_0)^{i_r} \in \mathcal{O}_{k,S}^*$, and hence $q_i(x_0, y_0) \in \mathcal{O}_{k,S}^*$, $i = 1, \ldots, r$. Since $\operatorname{Res}(q_i, q_j) \in \mathcal{O}_{k,S_0}^*$, $i \neq j$, by the same argument as before, $q_i(x_0, y_0)$ and $q_j(x_0, y_0)$ are coprime in $\mathcal{O}_{k,S_0}$ if $i \neq j$. It follows that for each prime $\mathfrak{p}$ of $\mathcal{O}_{k,S_0}$, $\mathfrak{p}$ divides at most one ideal $(q_i(x_0, y_0))$. Since $|S \setminus S_0| \leq r - r_0$, there must be at least r_0 distinct indices $j_1, \ldots, j_{r_0} \in \{1, \ldots, r\}$ such that $q_{j_l}(x_0, y_0) \in \mathcal{O}_{k,S_0}^*$, $l = 1, \ldots, r_0$. By the definition of r_0, $\deg \prod_{l=1}^{r_0} q_{j_l} \geq 3$. Then (x_0, y_0) is a solution to a Thue-Mahler equation

$$q_{j_1}(x, y) \cdots q_{j_{r_0}}(x, y) \in \mathcal{O}_{k,S_0}^*, \quad x, y \in \mathcal{O}_{k,S_0},$$

which is known to have finitely many effectively computable nonproportional solutions (x, y). To each such solution we naturally associate the point $[x : y]$ in $\mathbb{P}^1(k)$. It follows that R is contained in the finite union of points associated to the solutions to a finite number of Thue-Mahler equations.

□

We will also show that Theorem 7.19 is effective when $g = 1$. In fact, we'll prove this as a corollary to a more general result.

Let C be a nonsingular projective curve defined over a number field k and let $\phi \in k(C)$ be a rational function on C. We identify ϕ with the induced morphism $C \to \mathbb{P}^1$ and identify $\mathbb{P}^1(\overline{k})$ with $\overline{k} \cup \{\infty\}$. For $\alpha \in \mathbb{P}^1(\overline{k})$, let $e_1(\alpha), \ldots, e_{n_\alpha}(\alpha)$ be the ramification indices of ϕ above α and let

$$e_\alpha(\phi) = \gcd(e_1(\alpha), \ldots, e_{n_\alpha}(\alpha)).$$

We will use the following effective result of Bilu on integral points on curves.

THEOREM 7.21 (Bilu [**Bil97**]). *Let C be a nonsingular projective curve defined over a number field k and let $\phi \in k(C)$ be a rational function on C. Let S_0 be a finite set of places of k containing the archimedean places. Suppose that*

$$\sum_{\alpha \in \overline{k}} \left(1 - \frac{1}{e_\alpha(\phi)}\right) > 1.$$

Then the set of integral points

$$R = \{P \in C(k) \mid \phi(P) \in \mathcal{O}_{k,S_0}\}$$

is finite and effectively computable.

Bilu gives an explicit bound for the height of a point $\phi(P)$, $P \in R$, in terms of various invariants involving ϕ, C, k, and S_0. Poulakis [**Pou96**], independently, proved an explicit height bound in the case where S_0 consists of the archimedean places of k. Using Bilu's result, we prove the following effective result for integral points on curves.

THEOREM 7.22. *Let C be a nonsingular projective curve defined over a number field k and let $\phi \in k(C)$ be a rational function on C. Let $\mathcal{P} = \operatorname{Supp}(\phi)_\infty$ be the set of poles of ϕ and let $\mathcal{P}_1, \ldots, \mathcal{P}_r$ be the irreducible components of $\mathcal{P}$ over k. Let*

$s < r$ be a positive integer such that for every subset $I \subset \{1,\dots,r\}$, $|I| \leq s$, there exists a nonconstant rational function $\psi_I \in \overline{k}(C)$ satisfying

(a) $\mathcal{P}_i \cap \mathrm{Supp}(\psi_I)_\infty = \emptyset$ *for* $i \in I$.
(b) $\mathrm{Supp}(\psi_I)_\infty \subset \mathcal{P}$.
(c) $\sum_{\alpha\in\overline{k}}\left(1-\frac{1}{e_\alpha(\psi_I)}\right) > 1$.

Let S_0 be a finite set of places of k containing the archimedean places. Then the set of integral points

$$R = \bigcup_{\substack{S_0\subset S\subset M_k\\ |S\setminus S_0|\leq s}} \{P \in C(k) \mid \phi(P) \in \mathcal{O}_{k,S}\}$$

is finite and effectively computable.

We give a direct proof, modeled on the proof of Theorem 5.15, but in the language of rational functions instead of divisors.

PROOF. To clarify the proof and for ease of notation, we first prove the theorem in the case that $\psi_I \in k(C)$ for all I. Let $\mathcal{I} = \{I \subset \{1,\dots,r\} \mid |I| \leq s\}$ be the set of subsets of $\{1,\dots,r\}$ of cardinality at most s. For each $I \in \mathcal{I}$, let $\psi_I \in k(C)$ be a fixed nonconstant rational function as in the statement of the theorem. Let $T = \{\psi_I \mid I \in \mathcal{I}\}$.

Let $\mathcal{U}$ be the set of all subsets $U \subset T$ such that $\cap_{\psi\in U}(\mathrm{Supp}(\psi)_\infty)(\overline{k}) = \emptyset$. By Lemma 4.14, if $U \in \mathcal{U}$, then there exists a constant C_U such that

$$\min_{\psi\in U} \log|\psi(P)|_v \leq C_U$$

for all $v \in M_k$ and all $P \in C(k)$. Let $C_0 = \max_{U\in\mathcal{U}} C_U$.

Let $P \in R$ and let $S \subset M_k$ be such that $S_0 \subset S$, $|S \setminus S_0| \leq s$, and $\phi(P) \in \mathcal{O}_{k,S}$. For each $v \in S \setminus S_0$ let

$$T_v = \{\psi \in T \mid \log|\psi(P)|_v > C_0\}.$$

By the definition of C_0, for each $v \in S \setminus S_0$, we have

$$Z_v = \cap_{\psi\in T_v}(\mathrm{Supp}(\psi)_\infty)(\overline{k}) \neq \emptyset,$$

where if T_v is empty, then by convention we set $Z_v = C(\overline{k})$. For each $v \in S \setminus S_0$, let $i_v \in \{1,\dots,r\}$ be an index such that $\mathcal{P}_{i_v} \cap Z_v \neq \emptyset$. Let $I = \{i_v \mid v \in S \setminus S_0\}$. Then $|I| \leq s$ and from the definition of ψ_I, the set of poles of ψ_I does not contain any Z_v, $v \in S \setminus S_0$. It follows that

$$\log|\psi_I(P)|_v \leq C_0$$

for all $v \in S \setminus S_0$. Let S' be the finite set of nonarchimedean $v \in M_k$ such that

$$\min_{x\in k, |x|_v>1} \log|x|_v \leq C_0.$$

Then clearly $|\psi_I(P)|_v \leq 1$ for $v \in S \setminus (S_0 \cup S')$, i.e., $\psi_I(P)$ is integral at $v \in S \setminus (S_0 \cup S')$.

Since $\mathrm{Supp}(\psi_I)_\infty \subset \mathrm{Supp}(\phi)_\infty$ and $\phi : C \to \mathbb{P}^1$ is a finite morphism, ψ_I is integral over $k[\phi]$. So ψ_I is a root of a monic polynomial $f(x) \in k[\phi][x]$. Let $S_I'' \subset M_k$ be such that $f(x) \in \mathcal{O}_{k,S_I''}[\phi][x]$. Since $\phi(P) \in \mathcal{O}_{k,S}$ and ψ_I satisfies a monic polynomial over $\mathcal{O}_{k,S_I''}[\phi]$, it follows that $\psi_I(P) \in \mathcal{O}_{k,S\cup S_I''}$. In fact, from the remarks above, we have $\psi_I(P) \in \mathcal{O}_{k,S_0\cup S'\cup S''_I}$, where S_0 and S' are fixed and S''_I depends only on I. It follows that

$$R \subset \cup_{I\in\mathcal{I}}\{P \in C(k) \mid \psi_I(P) \in \mathcal{O}_{k,S_0\cup S'\cup S''_I}\}.$$

By Bilu's Theorem 7.21, each of the finitely many sets $\{P \in C(k) \mid \psi_I(P) \in \mathcal{O}_{k,S_0\cup S'\cup S''_I}\}$, $I \in \mathcal{I}$, is finite and effectively computable and thus R is finite and effectively computable.

We now describe the modifications that need to be made in the general case, where ψ_I is not assumed to be defined over k. Let L_I be the field of definition of ψ_I and let $d_I = [L_I : k]$. Let $\sigma_1, \ldots, \sigma_{d_I}$ be the embeddings of L_I in $\overline{k}$ over k. Let $\psi_{I,j} = \sigma_j\psi_I$, $j = 1, \ldots, d_I$, be the conjugates of ψ_I over $k(C)$. Then instead of considering the set of all ψ_I, we consider the larger set

$$T = \{\psi_{I,j} \mid I \in \mathcal{I}, j \in \{1, \ldots, d_I\}\}.$$

Since $\mathcal{P}_1, \ldots, \mathcal{P}_r$ are components over k, for any $I \in \mathcal{I}$ we have

$$\mathcal{P}_i \cap \operatorname{Supp}(\psi_{I,j})_\infty = \emptyset, \quad j = 1, \ldots, d_I.$$

Extend each place $v \in M_k$ to $\overline{k}$ in some fixed way. Then by the same argument as before, for any $P \in R$, with corresponding set of places S, we can find a set of indices $I \in \mathcal{I}$, depending on P, such that

$$\log|\psi_{I,j}(P)|_v \leq C_0$$

for all $v \in S \setminus S_0$ and all $j \in \{1, \ldots, d_I\}$. Then

$$\log|\psi_{I,j}(P)|_v = \log|(\sigma_j\psi_I)(P)|_v = \log|\sigma_j(\psi_I(P))|_v = \log|\psi_I(P)|_{\sigma_j v} \leq C_0$$

for all $v \in S\setminus S_0$ and all $j \in \{1, \ldots, d_I\}$. Let S_{L_I} and $S_{L_I,0}$ denote the set of places of L_I lying above S and S_0, respectively. Then it follows that

$$\log|\psi_I(P)|_w \leq C_0$$

for all $w \in S_{L_I} \setminus S_{L_I,0}$. Now by the same argument as before, there exist S'_{L_I} and S''_{L_I} such that $\psi_I(P) \in \mathcal{O}_{L_I,S_{L_I,0}\cup S'_{L_I}\cup S''_{L_I}}$ and we find that

$$R \subset \cup_{I\in\mathcal{I}}\{P \in C(L_I) \mid \psi_I(P) \in \mathcal{O}_{L_I,S_{L_I,0}\cup S'_{L_I}\cup S''_{L_I}}\},$$

a finite union of effectively computable finite sets by Theorem 7.21. □

We now show that Theorem 7.19 is effective for elliptic curves.

COROLLARY 7.23. *Let E be an elliptic curve defined over a number field k. Let $\phi \in k(E)$ be a rational function on E. Let r be the number of irreducible components over k of the set of poles* $\operatorname{Supp}(\phi)_\infty$ *of ϕ. Let S_0 be a finite set of places of k containing the archimedean places. Then the set of integral points*

$$R = \bigcup_{\substack{S_0\subset S\subset M_k \\ |S\setminus S_0|\leq r-1}} \{P \in E(k) \mid \phi(P) \in \mathcal{O}_{k,S}\}$$

is finite and effectively computable.

PROOF. Let $\mathcal{P}_1, \ldots, \mathcal{P}_r$ be the irreducible components of $\operatorname{Supp}(\phi)_\infty$ over k. Let $s = r - 1$ and let $I \subset \{1, \ldots, r\}$, $|I| \leq s$. Let $i \in \{1, \ldots, r\} \setminus I$ and let $P \in \mathcal{P}_i(\overline{k})$. By Riemann-Roch, there exists a function $\psi_I \in \overline{k}(E)$ with a double pole at P and no other poles. Then $\deg\psi_I = 2$ and by Riemann-Hurwitz, ψ must be ramified (with ramification index 2) above exactly 4 points of $\mathbb{P}^1$ (including P). It follows that

$$\sum_{\alpha\in\overline{k}}\left(1-\frac{1}{e_\alpha(\psi_I)}\right)=\frac{3}{2}>1.$$

Then parts (a)-(c) of Theorem 7.22 are satisfied with $s=r-1$ and the corollary follows. $\square$

As a further application, we derive a result of Everest, Reynolds, and Stevens [**ERS07**] on rational points on an elliptic curve with prime power denominators.

COROLLARY 7.24 (Everest, Reynolds, Stevens). *Let E be an elliptic curve over $\mathbb{Q}$ given by a Weierstrass equation*

$$y^2+a_1xy+a_3y=x^3+a_2x^2+a_4x+a_6,$$

where $a_1,\ldots,a_6\in\mathbb{Z}$. For a (finite) rational point $P\in E(\mathbb{Q})$, write

$$P=\left(\frac{A_P}{B_P^2},\frac{C_P}{B_P^3}\right),$$

where $A_P,B_P,C_P\in\mathbb{Z}$ and $\gcd(B_P,A_PC_P)=1$. Let $\phi: E'\to E$ be an isogeny of elliptic curves over $\mathbb{Q}$ of degree greater than one. Then there are only finitely many points $P\in\phi(E'(\mathbb{Q}))$ such that B_P is a prime power and these points can be effectively computed.

PROOF. Let x be the usual coordinate function on E (corresponding to the given Weierstrass equation) and let $\psi=x\circ\phi$, a rational function on E'. Then the set of poles of ψ contains at least two components over $\mathbb{Q}$ (the origin of E' and at least one other component). Taking $S_0=\{\infty\}\subset M_{\mathbb{Q}}$, we see that the set of points

$$R=\bigcup_{\substack{p\in\mathbb{Z}\\ p\text{ prime}}}\{P\in E'(\mathbb{Q})\mid \psi(P)\in\mathcal{O}_{\mathbb{Q},\{\infty,p\}}\}$$

is finite and effectively computable, which is clearly equivalent to the corollary. $\square$

We also derive a similar result from [**ERS07**] for elliptic curves in the form $u^3+v^3=D$, $D\in\mathbb{Q}$.

COROLLARY 7.25 (Everest, Reynolds, Stevens). *Let E be an elliptic curve defined by an equation*

$$u^3+v^3=D,$$

where $D\in\mathbb{Q}$ is cubefree. For a (finite) rational point $P\in E(\mathbb{Q})$, write

$$P=\left(\frac{A_P}{B_P},\frac{C_P}{B_P}\right),$$

where $A_P,B_P,C_P\in\mathbb{Z}$ and $\gcd(B_P,A_PC_P)=1$. Then there are only finitely many points $P\in E(\mathbb{Q})$ such that B_P is a prime power and these points can be effectively computed.

PROOF. Let ϕ be the rational function given by the coordinate function u on E. From the factorization $u^3 + v^3 = (u+v)(u^2 - uv + v^2)$, we see that the set of poles of ϕ contains two components over $\mathbb{Q}$. Thus, the set

$$R = \bigcup_{\substack{p \in \mathbb{Z} \\ p \text{ prime}}} \{P \in E(\mathbb{Q}) \mid \phi(P) \in \mathcal{O}_{\mathbb{Q},\{\infty,p\}}\}$$

is finite and effectively computable, which is equivalent to the corollary. □

8. Higher-dimensional varieties

We end by giving a few illustrative applications of Theorem 5.15 to integral points on higher-dimensional varieties. For simplicity, we will restrict to proving results on projective space, but Theorem 5.15 could be combined with a multitude of other results (e.g., results on integral points coming from the Corvaja-Zannier Subspace theorem method [**Aut09**, **Aut11**, **CLZ09**, **CZ04**, **Lev09**]).

We fix some notation that we will use throughout the section. Let k be a number field and let S_0 be a finite set of places of k containing the archimedean places. Let $D = \sum_{i=1}^{r} D_i$ be a divisor on $\mathbb{P}^n$, defined over k, such that the divisors D_i are nontrivial effective divisors without components in common (pairwise). We now recall a few results on (D, S)-integral points on projective space.

The next two results can be reduced to arguments involving S-unit equations. We begin by recalling a special case of a result of Vojta.

THEOREM 8.26 (Vojta [**Voj87**, Cor. 2.4.3]). *Suppose that $r \geq n+2$. Then any set of (D, S_0)-integral points in $\mathbb{P}^n(k)$ is not Zariski dense in $\mathbb{P}^n$.*

When the divisors D_i are hyperplanes, we have a stronger result.

THEOREM 8.27. *Suppose that each divisor D_i is a hyperplane defined by a linear form $L_i \in k[x_0, \ldots, x_n]$, $i = 1, \ldots, r$. If the linear forms $L_1, \ldots, L_r$ are linearly dependent, then any set of (D, S_0)-integral points in $\mathbb{P}^n(k)$ is contained in a finite union of hyperplanes in $\mathbb{P}^n$.*

A converse also holds: if $L_1, \ldots, L_r$ are linearly independent and $|S_0| > 1$, then there exists a Zariski dense set of k-rational (D, S_0)-integral points.

Lastly, we recall an effective result on $\mathbb{P}^2$ when $|S_0|$ is small, built on work of Vojta [**Voj83**] on an effective solution to the three-variable S-unit equation with $|S| \leq 3$.

THEOREM 8.28 (Levin [**Lev14**]). *Suppose that $D_1, \ldots, D_r$ are curves over k in $\mathbb{P}^2$ such that the intersection of any $r-1$ of the curves is empty. Suppose that $|S_0| < r$. Then any set of (D, S_0)-integral points in $\mathbb{P}^2(k)$ is contained in an effectively computable proper Zariski closed subset of $\mathbb{P}^2$.*

We now prove results obtained by combining Theorem 5.15 with Theorems 8.26, 8.27, and 8.28. We first fix some more notation. Let

$$f = f_1^{i_1} \cdots f_r^{i_r} \in \mathcal{O}_{k,S_0}[x_0, \ldots, x_n],$$

where $i_1, \ldots, i_r$ are positive integers and $f_1, \ldots, f_r$ are nonconstant irreducible homogeneous polynomials over k that are pairwise nonproportional. For $\alpha \in \mathcal{O}_{k,S_0}$ we define $\omega_{S_0}(\alpha)$ to be the number of distinct primes of $\mathcal{O}_{k,S_0}$ appearing nontrivially in the factorization of the principal ideal $\alpha\mathcal{O}_{k,S_0}$ (or equivalently, the number of distinct primes of $\mathcal{O}_k$ appearing nontrivially in the factorization of the principal fractional ideal $\alpha\mathcal{O}_k$, ignoring primes in S_0). Thus, $\omega_{S_0}(\alpha) = 0$ if and only if α is an S_0-unit. We formulate our results in a manner which generalizes the familiar S-unit equation (which corresponds to the choice $f = x_0 x_1 \cdots x_n(x_0 + \cdots + x_n)$).

Corresponding to Theorem 8.26, we prove:

THEOREM 8.29. *Let s be a positive integer. Suppose that for every subset $T \subset \mathbb{P}^n(\overline{k})$, $|T| = s$, the set*

$$\{i \in \{1, \dots, r\} \mid f_i(P) \neq 0, \forall P \in T\}$$

has cardinality at least $n + 2$. Let

$$R' = \{(x_0, \dots, x_n) \in \mathcal{O}_{k,S_0}^{n+1} \mid \omega_{S_0}(f(x_0, \dots, x_n)) \leq s\}.$$

Then the set $R = \{[x_0 : \cdots : x_n] \in \mathbb{P}^n(k) \mid (x_0, \dots, x_n) \in R'\}$ is not Zariski dense in $\mathbb{P}^n$.

Note that if the hypersurfaces in $\mathbb{P}^n$ defined by $f_1, \dots, f_r$ are in general position, then the hypotheses of the theorem are satisfied when $r \geq (s + 1)n + 2$.

PROOF. Let D_j be the divisor on $\mathbb{P}^n$ defined by f_j, $j = 1, \dots, r$, and let $D = \sum_{j=1}^{r} i_j D_j$ and $d = \deg D = \deg f$. A global Weil function for D is

$$\lambda_{D,v}(P) = \log \frac{\max_i |x_i|_v^d}{|f(P)|_v}, \quad v \in M_k,$$

where $P = [x_0 : \cdots : x_n] \in \mathbb{P}^n(k) \setminus \operatorname{Supp} D$. Let R be the set in Theorem 8.29. Then clearly R is contained in the set R of Theorem 5.15 when $\mathcal{L} = \{k\}$ and $c_v = 0$ for all $v \in M_k$. By our hypotheses and Theorem 8.26, each set $R_{D'}$, $D' \in \mathcal{D}$, of Theorem 5.15 is not Zariski dense in $\mathbb{P}^n$ and the result follows. □

The proofs of the results corresponding to Theorems 8.27 and 8.28 are similar, so we only give the statements. Note that the result in Theorem 8.31 is effective.

THEOREM 8.30. *Suppose that $\deg f_i = 1$, $i = 1, \dots, r$. Let s be a positive integer. Suppose that for every subset $T \subset \mathbb{P}^n(\overline{k})$, $|T| = s$, the set*

$$\{f_i \in \{f_1, \dots, f_r\} \mid f_i(P) \neq 0, \forall P \in T\}$$

is linearly dependent. Let

$$R' = \{(x_0, \dots, x_n) \in \mathcal{O}_{k,S_0}^{n+1} \mid \omega_{S_0}(f(x_0, \dots, x_n)) \leq s\}.$$

Then the set $R = \{[x_0 : \cdots : x_n] \in \mathbb{P}^n(k) \mid (x_0, \dots, x_n) \in R'\}$ is contained in a finite union of hyperplanes in $\mathbb{P}^n$.

THEOREM 8.31. *Suppose that $n = 2$, so that $f_1, \dots, f_r \in k[x_0, x_1, x_2]$. Let s be a positive integer and let $s_0 = |S_0|$. Suppose that for every subset $T \subset \mathbb{P}^2(\overline{k})$, $|T| = s$, the set*

$$\{i \in \{1, \dots, r\} \mid f_i(P) \neq 0, \forall P \in T\}$$

has at least $s_0 + 1$ elements and that for every subset $T \subset \mathbb{P}^2(\overline{k})$, $|T| = s + 1$, the set

$$\{i \in \{1, \dots, r\} \mid f_i(P) \neq 0, \forall P \in T\}$$

has at least two elements. Let

$$R' = \{(x_0, x_1, x_2) \in \mathcal{O}_{k,S_0}^3 \mid \omega_{S_0}(f(x_0, x_1, x_2)) \leq s\}.$$

Then the set $R = \{[x_0 : x_1 : x_2] \in \mathbb{P}^2(k) \mid (x_0, x_1, x_2) \in R'\}$ *is contained in an effectively computable proper Zariski closed subset of* $\mathbb{P}^2$.

References

[Aut09] P. Autissier, *Géométries, points entiers et courbes entières* (French, with English and French summaries), Ann. Sci. Éc. Norm. Supér. (4) **42** (2009), no. 2, 221–239, DOI 10.24033/asens.2094. MR2518077

[Aut11] P. Autissier, *Sur la non-densité des points entiers* (French, with English and French summaries), Duke Math. J. **158** (2011), no. 1, 13–27, DOI 10.1215/00127094-1276292. MR2794367

[Bil97] Y. Bilu, *Quantitative Siegel's theorem for Galois coverings*, Compositio Math. **106** (1997), no. 2, 125–158, DOI 10.1023/A:1000172615719. MR1457336

[BL15] M. A. Bennett and A. Levin, *The Nagell-Ljunggren equation via Runge's method*, Monatsh. Math. **177** (2015), no. 1, 15–31, DOI 10.1007/s00605-015-0748-1. MR3336330

[Bom83] E. Bombieri, *On Weil's "théorème de décomposition"*, Amer. J. Math. **105** (1983), no. 2, 295–308, DOI 10.2307/2374261. MR701562

[BP08] Y. Bilu and P. Parent, *Integral j-invariants and Cartan structures for elliptic curves* (English, with English and French summaries), C. R. Math. Acad. Sci. Paris **346** (2008), no. 11-12, 599–602, DOI 10.1016/j.crma.2008.04.002. MR2423260

[BP11a] Y. Bilu and P. Parent, *Runge's method and modular curves*, Int. Math. Res. Not. IMRN **2011** (2011), no. 9, 1997–2027.

[BP11b] Y. Bilu and P. Parent, *Serre's uniformity problem in the split Cartan case*, Ann. of Math. (2) **173** (2011), no. 1, 569–584, DOI 10.4007/annals.2011.173.1.13. MR2753610

[CLZ09] P. Corvaja, A. Levin, and U. Zannier, *Integral points on threefolds and other varieties*, Tohoku Math. J. (2) **61** (2009), no. 4, 589–601, DOI 10.2748/tmj/1264084501. MR2598251

[CSTZ15] P. Corvaja, V. Sookdeo, T. J. Tucker, and U. Zannier, *Integral points in two-parameter orbits*, J. Reine Angew. Math. **706** (2015), 19–33, DOI 10.1515/crelle-2013-0060. MR3393361

[CZ04] P. Corvaja and U. Zannier, *On integral points on surfaces*, Ann. of Math. (2) **160** (2004), no. 2, 705–726, DOI 10.4007/annals.2004.160.705. MR2123936

[ERS07] G. Everest, J. Reynolds, and S. Stevens, *On the denominators of rational points on elliptic curves*, Bull. Lond. Math. Soc. **39** (2007), no. 5, 762–770, DOI 10.1112/blms/bdm061. MR2365225

[HS83] D. L. Hilliker and E. G. Straus, *Determination of bounds for the solutions to those binary Diophantine equations that satisfy the hypotheses of Runge's theorem*, Trans. Amer. Math. Soc. **280** (1983), no. 2, 637–657, DOI 10.2307/1999638. MR716842

[Lan83] S. Lang, *Fundamentals of Diophantine geometry*, Springer-Verlag, New York, 1983. MR715605

[LeF16] S. Le Fourn, *A "tubular" variant of Runge's method in all dimensions, with applications to integral points on Siegel modular varieties*, preprint, arXiv:1611.09054, 2016.

[LeF17] S. Le Fourn, *Sur la méthode de Runge et les points entiers de certaines variétés modulaires de Siegel* (French, with English and French summaries), C. R. Math. Acad. Sci. Paris **355** (2017), no. 8, 847–852. MR3693503

[Lev08] A. Levin, *Variations on a theme of Runge: effective determination of integral points on certain varieties* (English, with English and French summaries), J. Théor. Nombres Bordeaux **20** (2008), no. 2, 385–417. MR2477511

[Lev09] A. Levin, *Generalizations of Siegel's and Picard's theorems*, Ann. of Math. (2) **170** (2009), no. 2, 609–655, DOI 10.4007/annals.2009.170.609. MR2552103

[Lev14] A. Levin, *Linear forms in logarithms and integral points on higher-dimensional varieties*, Algebra Number Theory **8** (2014), no. 3, 647–687, DOI 10.2140/ant.2014.8.647. MR3218805

[MW83] D. W. Masser and G. Wüstholz, *Fields of large transcendence degree generated by values of elliptic functions*, Invent. Math. **72** (1983), no. 3, 407–464, DOI 10.1007/BF01398396. MR704399

[Pou96] D. Poulakis, *Estimation effective des points entiers d'une famille de courbes algébriques* (French, with English and French summaries), Ann. Fac. Sci. Toulouse Math. (6) **5** (1996), no. 4, 691–725. MR1473079

[Run87] C. Runge, *Ueber ganzzahlige Lösungen von Gleichungen zwischen zwei Veränderlichen* (German), J. Reine Angew. Math. **100** (1887), 425–435, DOI 10.1515/crll.1887.100.425. MR1580107

[Sil87] J. H. Silverman, *Arithmetic distance functions and height functions in Diophantine geometry*, Math. Ann. **279** (1987), no. 2, 193–216, DOI 10.1007/BF01461718. MR919501

[Spr80] V. G. Sprindžuk, *Reducibility of polynomials and rational points on algebraic curves* (Russian), Dokl. Akad. Nauk SSSR **250** (1980), no. 6, 1327–1330. MR564338

[Voj83] P. Vojta, *Integral points on varieties*, ProQuest LLC, Ann Arbor, MI, 1983. Thesis (Ph.D.)–Harvard University. MR2632888

[Voj87] P. Vojta, *Diophantine approximations and value distribution theory*, Lecture Notes in Mathematics, vol. 1239, Springer-Verlag, Berlin, 1987. MR883451

[Wal92] P. G. Walsh, *A quantitative version of Runge's theorem on Diophantine equations*, Acta Arith. **62** (1992), no. 2, 157–172. MR1183987

Department of Mathematics, Michigan State University, 619 Red Cedar Road, East Lansing, Michigan 48824

Email address: `adlevin@math.msu.edu`

Contemporary Mathematics
Volume **703**, 2018
http://dx.doi.org/10.1090/conm/703/14138

Self-inversive polynomials, curves, and codes

David Joyner and Tony Shaska

Abstract. We study connections between self-inversive and self-reciprocal polynomials, reduction theory of binary forms, minimal models of curves, and formally self-dual codes. We prove that if $\mathcal{X}$ is a superelliptic curve defined over $\mathbb{C}$ and its reduced automorphism group is nontrivial or not isomorphic to a cyclic group, then we can write its equation as $y^n = f(x)$ or $y^n = xf(x)$, where $f(x)$ is a self-inversive or self-reciprocal polynomial. Moreover, we state a conjecture on the coefficients of the zeta polynomial of extremal formally self-dual codes.

1. Introduction

Self-inversive and self-reciprocal polynomials have been studied extensively in the last few decades due to their connections to complex functions and number theory. In this paper we explore the connections between such polynomials to algebraic curves, reduction theory of binary forms, and coding theory. While connections to coding theory have been explored by many authors before, we are not aware of any previous work that explores the connections of self-inversive and self-reciprocal polynomials to superelliptic curves and reduction theory.

In Section 2, we give a geometric introduction to inversive and reciprocal polynomials of a given polynomial. We motivate such definitions via the transformations of the complex plane which is the original motivation to study such polynomials. It is unclear who coined the names inversive, reciprocal, palindromic, and antipalindromic, but it is obvious that inversive come from the inversion $z \mapsto \frac{1}{\bar{z}}$ and reciprocal from the reciprocal map $z \mapsto \frac{1}{z}$ of the complex plane.

We take the point of view of the reduction theory of binary forms. While this is an elegant and beautiful theory for binary quadratics, it is rather technical for higher degree forms. However, the inversion plays an important role on reduction as can be seen from section 2 and from [**5**] and [**2**]. We are not aware of other authors have explored the connection between reduction theory and self-inversive and self-reciprocal polynomials before even though the overlap is quite obvious.

We state some of the main results of self-inversive polynomials including the *middle coefficient conjecture* (2.3) and results on the location of the roots of such polynomials. Self-inversive polynomials over $\mathbb{Q}$, $\mathbb{R}$, and $\mathbb{C}$ are discussed and a few recent results on the height of such polynomials. The normal references here are [**7,13,16–20,22,25**]. Further, we discuss the roots of the self-inversive polynomials. There is a huge amount of literature on this topic including several conjectures. It is the location of such roots that makes self-inversive polynomials interesting in

2010 *Mathematics Subject Classification.* Primary 14Hxx; 11Gxx.

reduction theory, coding theory, and other areas of mathematics. An attempt at a converse to this conjecture is discussed in §2.2.

In Section 3 it is given an account of how self-inversive polynomials can be used to determine minimal polynomials of superelliptic curves with extra automorphisms. This is a new idea spurred by Beshaj's thesis [**2**] and [**1**] and has some interesting relations between two different areas of mathematics, namely the theory of algebraic curves and the theory of self-inversive polynomials. Further details in this direction are planned in [**5**]. In this section we prove that for any superelliptic curve with reduced automorphism group not trivial and not isomorphic to a cyclic group we can write the equation of the curve as $y^n = f(x)$ or $y^n = xf(x)$, where $f(x)$ is a palindromic, antipalindromic, or self-inversive polynomial. Indeed, we can say more since in each case when the automorphism group of the curve we can determine the polynomial $f(x)$ specifically.

In Section 4 we explore connections of self-inversive and self-reciprocal polynomials to reduction theory of binary forms. We show that self-inversive polynomials which have all roots on the unit circle correspond to the totally real forms. The reduction theory for such forms is simpler than for other forms since the Julia quadratic of any degree n form $f(x, y)$ is a factor of a degree $(n-1)(n-2)$ covariant $G_f(x, y)$ given in terms of the partial derivatives of f; see [**1**]. We prove that for f palindromic, G_f is self-inversive and if f is palindromic of odd degree then G_f is palindromic. Moreover, we determine explicitly which self-inversive polynomials f with all roots on the unit circle are reduced.

In Section 5 we discuss the Riemann hypothesis for formal weight enumerators of codes and its relation to the self-inversive polynomials. We state several open problems which relate to Riemann hypotheses for extremal formal weight enumerators of codes.

Most of the results obtained here, with the necessary adjustments, can be extended to curves defined over fields of positive characteristic. In [**21**] equations of superelliptic curves are also determined over such fields. The main question that comes from the connection between self-inversive and self-reciprocal polynomials and reduction theory is whether such polynomials are actually reduced. In other words, if $f(x, y)$ is a primitive form which is self-reciprocal or self-inversive, is it true that $f(x, y)$ is reduced? This question is addressed in [**5**].

2. Self-inversive polynomials

Let $\mathbb{P}^1$ be the Riemann sphere and $\mathrm{GL}_2(\mathbb{C})$ the group of 2×2 matrices with entries in $\mathbb{C}$. Then $\mathrm{GL}_2(\mathbb{C})$ acts on $\mathbb{P}^1$ by linear fractional transformations. This action is a transitive action, i.e. has only one orbit. Consider now the action of $\mathrm{SL}_2(\mathbb{R})$ on the Riemann sphere. This action is not transitive, because for $M = \begin{pmatrix} \alpha & \beta \\ \gamma & \delta \end{pmatrix} \in \mathrm{SL}_2(\mathbb{R})$ we have

$$\mathrm{Img}\ (Mz) = \frac{(\alpha\delta - \beta\gamma)}{|\gamma z + \delta|^2} \mathrm{Img}\ z.$$

Hence, z and Mz have the same sign of imaginary part when $\det(M) = 1$. The action of $\mathrm{SL}_2(\mathbb{R})$ on $\mathbb{P}^1$ has three orbits, namely $\mathbb{R} \cup \infty$, the upper half plane, and the lower-half plane. Let $\mathcal{H}_2$ be the complex upper half plane, i.e.

$$\mathcal{H}_2 = \left\{ z = x + iy \in \mathbb{C} \;\middle|\; \mathrm{Img}\ (z) > 0 \right\} \subset \mathbb{C}.$$

The group $\mathrm{SL}_2(\mathbb{R})$ preserves $\mathcal{H}_2$ and acts transitively on it, since for $g \in \mathrm{SL}_2(\mathbb{R})$ and $z \in \mathcal{H}_2$ we have

$$\mathrm{Img}\ (gz) = \frac{\mathrm{Img}\ z}{|\gamma z + \delta|^2} > 0$$

The modular group $\Gamma = \mathrm{SL}_2(\mathbb{Z})/\{\pm I\}$ also acts on $\mathcal{H}_2$. This action has a fundamental domain $\mathcal{F}$

$$\mathcal{F} = \left\{ z \in \mathcal{H}_2 \,\middle|\, |z|^2 \geq 1 \text{ and } |Re(z)| \leq 1/2 \right\}$$

Consider now all binary quadratic forms with real coefficients. A quadratic form $f \in \mathbb{R}[x, y]$ has two complex roots (conjugate of each other) if f is positive definite. Hence, we have a one to one correspondence between positive definite quadratic forms and points of $\mathcal{H}_2$. For a given $f \in \mathbb{R}[x, y]$, let $\xi(f)$ denote the zero of f in $\mathcal{H}_2$. This is called *zero map*. The positive definite binary form f has minimal coefficients if and only if $\xi(f) \in \mathcal{F}$; see [**1**] for details.

The group $SL_2(\mathbb{R})$ acts on the set of positive definite quadratic forms by linear changes of coordinates. Moreover, the zero map $f \mapsto \xi(f)$ is equivariant under this action. In other words, $\xi(f^M) = \xi(f)^M$, for any $M \in \mathrm{SL}_2(\mathbb{R})$. Hence, to *reduce* a binary quadratic f with integer coefficients we simply compute $\xi(f)$ and then determine $M \in \Gamma$ such that $\xi(f)^M \in \mathcal{F}$. Then, the quadratic f^M has minimal coefficients.

This approach can be generalized to higher degree forms $f \in \mathbb{R}[x, y]$. Then $f(x, y)$ is a product of linear and quadratic factors over $\mathbb{R}$. In studying roots of $f(x, y)$ we are simply concerned with roots in the upper half plane $\mathcal{H}_2$. The zero map can also be defined in this case, but its definition is much more technical. The interested reader can check [**1**] or [**2**] for details.

Hence the problem of finding a form equivalent to f with minimal coefficients becomes equivalent to determine a matrix $M \in \Gamma$ such that $\xi(f)^M \in \mathcal{F}$. The generators of the modular group Γ are the matrices

$$S = \begin{bmatrix} 0 & -1 \\ 1 & 0 \end{bmatrix} \quad \text{and} \quad T = \begin{bmatrix} 1 & 1 \\ 0 & 1 \end{bmatrix}$$

which correspond to transformations $z \to -\frac{1}{z}$ and $z \to z + 1$. Next, we will see the geometry of some of these transformations which play an important role in this process.

Let $\sigma(z) = \frac{1}{z}$ be the **reciprocal map** of the complex plane. Then,

$$\sigma(a + bi) = \frac{1}{|z|^2}\,(a - bi)$$

Hence, on the unit circle $U = \{z \in \mathbb{C},\ |z| = 1\}$ the reciprocal map becomes simply the complex conjugation. From this we see that to the geometric inversion of the unit circle corresponds the **inversion map**

$$\tau : z \to \frac{1}{\bar{z}}$$

which sends points $z \in \mathcal{H}_2$ inside the unit circle U to points in $z' \in \mathcal{H}_2$ with the same argument as z and $|z| \cdot |z'| = 1$. It fixes points on the unit circle U. It is exactly this transformation together with $z \mapsto z + 1$ which we use to "move" points within $\mathcal{H}_2$ and bring them in the fundamental domain. We are interested in forms $f(x, y)$ which are fixed by this transformation. Hence, we are interested in polynomials $f(z, 1)$ whose set of roots is fixed by $\tau(z)$.

For a degree n polynomial $f(z) \in \mathbb{C}[z]$, the **inversive** of f is called the function $f^\star(z) = z^n f\left(\frac{1}{\bar{z}}\right)$. A polynomial f will be called **self-inversive** if $f = f^\star$. We can make this definition more precise.

Let $p(z) \in \mathbb{C}[z]$ such that

$$p(z) = \sum_{i=0}^{n} a_i z^i. \tag{2.1}$$

Then, $p(z)$ is called **self-inversive** if its set of zeroes is fixed by the inversion map $\tau(z) = 1/\bar{z}$. Thus, the set of roots is

$$\left\{ \alpha_1, \dots, \alpha_n, \frac{1}{\bar{\alpha}_1}, \dots, \frac{1}{\bar{\alpha}_n} \right\}$$

and then $p(z)$ is given by

$$p(z) = a_n \prod_{i=1}^{s} \left(z^2 - \left(\alpha_i + \frac{1}{\bar{\alpha}_i} \right) z + \frac{\alpha_i}{\bar{\alpha}_i} \right), \tag{2.2}$$

Let us denote by $\bar{p}(z)$ the *conjugate polynomial* of $p(z)$, namely

$$\bar{p}(z) := \sum_{i=0}^{n} \bar{a}_i z^i.$$

Then, we have the following; see [**20**].

LEMMA 1. *If $p(z)$ be given as in Eq.* (2.1). *The following are equivalent:*

(1) *$p(z)$ is self-inversive*
(2) *For every $z \in \mathbb{C} \setminus \{0\}$,*

$$\bar{a}_n \, p(z) = a_0 \, z^n \, \bar{p}\left(\frac{1}{z}\right)$$

(3) *For every $z \in \mathbb{C} \setminus \{0\}$*

$$p(z) = w \cdot z^n \cdot \bar{p}\left(\frac{1}{z}\right),$$

where $|w| = 1$.
(4) *For $j = 0, 1, \dots, n$,*

$$a_0 \bar{a}_j = \bar{a}_n a_{n-j}$$

Moreover, if $p(z)$ is self inversive then

(1) $|a_i| = |a_{n-i}|$ *for all* $i = 0, \dots, n$.
(2) $\bar{a}_n \left[n\, p(z) - z\, p'(z) \right] = a_0\, z^{n-1}\, \bar{p}'\left(\frac{1}{z}\right)$, *for each* $z \in \mathbb{C}$
(3) $\left| n \cdot \frac{p(z)}{z \cdot p'(z)} - 1 \right| = 1$, *for each* $z \in U$.

Studying roots of the self-inversive polynomials is an old problem which has been studied by many authors. A classical result due to Cohn states that a self-inversive polynomial has all its zeros on the unit circle if and only if all the zeros of its derivative lie in the closed unit disk.

For $p(z) \in \mathbb{C}[z]$ we let $||p||$ denote the *maximum modulus* of $p(z)$ on the unit circle. In [**20**] it is proved the following

THEOREM 1. *If $p(z) = \sum_{i=0}^{n} a_i z^j$, $a_n \neq 0$, is a self-inversive polynomial which has all the zeroes on $|z| = 1$, then*

$$|a_i| \leq \frac{||p||}{2}$$

for each $i \neq \frac{n}{2}$ and $|a_{n/2}| \leq \frac{\sqrt{2}}{2} ||p||$.

From the above theorem we can see that the middle coefficient is special. The *middle coefficient conjecture* says that for $p(z)$ as in the above theorem, it is conjectured that

$$|a_{n/2}| \leq ||p|| \tag{2.3}$$

If n is even then the middle coefficient conjecture is true when $|a_{n/2}| \leq 2|a_n|$; see [**20**, pg. 334] for details.

The following theorem holds; see [**25**], [**16**] for details.

THEOREM 2. *Let $p \in \mathbb{C}[x]$ be a degree n self-inversive polynomial. If*

$$|a_{n-\lambda}| > \frac{1}{2} \frac{n}{n - 2\lambda} \sum_{k=0, k \neq \lambda, k \neq n-\lambda}^{n} |a_k|$$

for some $\lambda < \frac{n}{2}$, then $p(z)$ has exactly $n - 2\lambda$ non-real roots on the unit circle.

If n is even and $\lambda = \frac{n}{2}$, then $p(z)$ has no roots on the unit circle if

$$|a_{n/2}| > 2 \sum_{k=0, k \neq n/2}^{n} |a_k|$$

For a proof see [**25**]. If $\lambda = 0$ this correspond to a result of Lakatos and Losonczi [**16**] which says that a self-inversive polynomial with non-zero discriminant has all roots on the unit circle if

$$|a_n| \geq \frac{1}{2} \sum_{k=1}^{n} |a_k| .$$

There is a huge amount of literature on bounding the roots or the coefficients of polynomials or finding polynomials which have bounded coefficients. Most of that work relates to Mahler measure and related works. There was another approach by Julia [**15**] which did not gain the attention it deserved. Lately there are works of Cremona and Stoll in [**24**], Beshaj [**1, 2**], and others who have extended Julia's method and provide an algorithm of finding the polynomial (up to a coordinate change) with the smallest coefficients. The first paragraph of this section eludes to that approach.

2.1. Reciprocal polynomials. For a degree n polynomial $f(z) \in \mathbb{C}[z]$, its **reciprocal** is called the polynomial $f^{\times}(z) = z^n f\left(\frac{1}{z}\right)$. A polynomial is called **self-reciprocal** or **palindromic** if $f = f^{\times}$ and it is called **anti-palindromic** if $f = -f^{\times}$.

If $p(z) \in \mathbb{C}[z]$ be a polynomial such that its set of roots is fixed by reciprocal map $\sigma(z)$, say

$$S = \left\{ \alpha_1, \dots, \alpha_s, \frac{1}{\alpha_1}, \dots, \frac{1}{\alpha_s} \right\},$$

then $f(z)$ is palindromic or antipalindromic polynomial. Due to the properties of the binomial coefficients the polynomials $P(x) = (x+1)^n$ are palindromic for all positive integers n, while the polynomials $Q(x) = (x-1)^n$ are palindromic when n is even and anti-palindromic when n is odd. Also, cyclotomic polynomials are palindromic.

What if we would like some kind of invariant of the reciprocal map $z \mapsto 1/z$? Consider the transformation

$$\alpha(z) = z + \frac{1}{z}$$

Obviously, $\alpha(1/z) = z$. When considered as a function $\alpha : \mathbb{C} \to \mathbb{C}$ this is a 2 to 1 map since both z and $1/z$ go to the same point. Considered on each one of the three orbits of $SL_2(\mathbb{R})$ in $\mathbb{C}$ we have the following: α sends the upper half-plane $\mathcal{H}_2$ onto the complex plane $\mathbb{C}$ except for $(-\infty, 2]$ and $[2, \infty)$ which are doubly covered by $\mathbb{R} \setminus \{0\}$. We organize such actions in the following Lemma:

LEMMA 2. *For any polynomial $p(z) = \sum_{i=0}^n a_i z^i$ of degree $n = 2s$ the following are equivalent:*

(1) *The coefficients of $p(z)$ satisfy*

$$a_i = a_{n-i}, \qquad \text{for all } i = 0, \dots n.$$

(2) *There exists a polynomial $q(z)$ such that*

$$p(z) = z^s\, q\left(\frac{1}{z}\right)$$

(3) *There exists some polynomial $g(z)$ of degree $m \geq 1$ such that*

$$p(z) = z^m \cdot g\left(z + \frac{1}{z}\right)$$

For a proof see [**13**] among other papers. Hence, any polynomial $f(z)$ satisfying any of the properties of the Lemma is self-reciprocal.

Next we list some properties of palindromic and anti-palindromic polynomials. Their proofs are elementary and we skip the details.

REMARK 1. *Palindromic and anti-palindromic polynomials have the following general properties:*

(1) *For any antipalindromic polynomial $p(z) = \sum_{i=0}^n a_i z^i$*

$$a_i = -a_{n-i}, \qquad \text{for all } i = 0, \dots n.$$

(2) *For any polynomial f, the polynomial $f + f^{\times}$ is palindromic and the polynomial $f - f^{\times}$ is antipalindromic.*
(3) *The product of two palindromic or antipalindromic polynomials is palindromic.*

(4) *The product of a palindromic polynomial and an antipalindromic polynomial is antipalindromic.*
(5) *A palindromic polynomial of odd degree is a multiple of* $x+1$ *(it has -1 as a root) and its quotient by* $x+1$ *is also palindromic.*
(6) *An antipalindromic polynomial is a multiple of* $x-1$ *(it has 1 as a root) and its quotient by* $x-1$ *is palindromic.*
(7) *An antipalindromic polynomial of even degree is a multiple of* x^2-1 *(it has -1 and 1 as a roots) and its quotient by* x^2-1 *is palindromic.*

The following lemma shows an important correspondence among the pairs of roots $\left(\alpha, \frac{1}{\alpha}\right)$ of $f(z)$ and real roots of $g\left(z+\frac{1}{z}\right)$. Polynomials $f(z)$ which have all roots on the units circle correspond to $g\left(z+\frac{1}{z}\right)$ which have all real roots. When homogenized the corresponding forms are called *totally real forms* (cf. Section 4).

LEMMA 3. *Let* $f(z)=\sum_{i=0}^n a_i z^i$ *be a palindromic polynomial and* $g(z) \in \mathbb{C}[z]$ *such that* $f(z)=z^m g(z+1/z)$. *Denote by* S_f *the set of pairs of roots of* $f(z)$ *on* U,

$$S_f = \left\{ \left(\alpha, \frac{1}{\alpha}\right), \text{ such that } |\alpha| = 1 \text{ and } f(\alpha)=0 \right\}$$

and by S_g *the set of roots of* $g(z)$ *in* $[-2,2]$. *There is a one-to-one correspondence between* S_f *and* S_g.

PROOF. The proof is rather elementary. If $|z|=1$ then $z=\cos\theta+i\sin\theta$, for some θ. Then, $\alpha(z)=2\cos\theta$ is in the interval $[-2,2]$. Conversely, if $t\in[-2,2]$ then $t=2\cos\theta$ for some θ. Hence, $t=z+1/z$, where $z=\cos\theta \pm i\sin\theta$. □

Notice that the inversion $z \mapsto 1/z$ induces an involution on the group of symmetries of a palindromic polynomial. Hence, the Galois group of such polynomials is non-trivial. We will see in the next section how such involution among the roots of $f(x)$ induces automorphisms for algebraic curves with affine equation $y^n=f(x)$.

A polynomial $f(z)=\sum_{i=0}^n a_i z^i$ is called **quasi-palindromic** if

$$|a_i| = |a_{n-i}|,$$

for all $i=0,\ldots,n$.

The following Lemma will be used in the next section.

LEMMA 4. *Let* $f, g \in \mathbb{C}[x]$ *with no common factor. If* f *and* g *are self-inversive then* fg *is a self-inversive. If* f *and* g *are quasi-palindromic, then* fg *is quasi-palindromic.*

PROOF. The proof is an immediate consequence of the definitions. Since the set of roots of f and g contain all z and $\frac{1}{\bar{z}}$ (resp. z and $\pm\frac{1}{z}$), then so would contain their union, which is the set of roots of fg. □

REMARK 2. *A polynomial with real coefficients all of whose complex roots lie on the unit circle in the complex plane (all the roots are unimodular) is either palindromic or antipalindromic*

2.2. Self-reciprocal polynomials over the reals. Here is a basic fact about even degree self-reciprocal polynomials; see [**8**], §2.1; see also [**18**]. The degree $d=2n$ polynomial $p(z)$ is self-reciprocal if and only if it can be written

$$p(z) = z^n \cdot (a_n + a_{n+1}\cdot(z+z^{-1}) + \cdots + a_{2n}\cdot(z^n+z^{-n})),$$

if and only if it can be written

$$p(z) = a_{2n} \cdot \prod_{k=1}^{n} (1 - \alpha_k z + z^2), \tag{2.4}$$

for some real $\alpha_k \in \mathbb{R}$.

Note that $g(z) = 1 - \alpha z + z^2$ has roots on the unit circle if and only if the roots are of the form $e^{\pm i\theta}$, for some θ, in which case, $\alpha = 2\cos(\theta)$.

For the rest of this section we denote by $p(z) = \sum_{i=0}^{n} a_i z^i$ a degree n self-reciprocal polynomial, where $n = 2d$ or $n = 2d + 1$. The answer to the following question is unknown at this time: for which increasing sequences $a_0 < a_1 < \dots a_d$ do the roots of the corresponding self-reciprocal polynomial, $p(z) = 0$, lie on the unit circle $|z| = 1$?

If $n = 2d$, which $p(z)$ with $a_0 < a_1 < \dots a_d$, can be written as a product $\prod_{k=1}^{d}(1 - 2\cos(\theta_k)z + z^2)$?

It is clear that, in a product such as (2.4), with all its roots on the unit circle so $-2 \leq \alpha_k \leq 2$, we have

$$0 < a_0 \leq a_1 \leq \cdots \leq a_n, \quad a_{n-i} = a_{n+i}, \tag{2.5}$$

for all $i \in \{0, 1, 2, \dots, n\}$, provided the collection α_js satisfy

$$\alpha_k \leq -1. \tag{2.6}$$

A self-reciprocal polynomial satisfying (2.5) is called *symmetric increasing.* Motivated by Problem 3 below, we look for a bound which is more general than (2.6) and which also implies the polynomial is symmetric increasing. For instance, we observe that the following result can be used inductively to establish a generalization of (2.6).

LEMMA 5. *Let $p(z)$ be as above. To multiply $p(z)$ by $1 - \alpha x + x^2$ $(-2 \leq \alpha \leq 2)$, and still have the new coefficients satisfy a symmetric increasing condition such as in (2.5), we require*

$$(a_i, a_{i+1}, a_{i+2}, a_{i+3}) \cdot (1, -1 - \alpha, 1 + \alpha, -1) \leq 0, \tag{2.7}$$

for all $i \leq d$. In particular, if $a_i = a$, $a_{i+1} = a + \epsilon_1$, $a_{i+2} = a + \epsilon_2$, $a_{i+3} = a + \epsilon_3$ then (2.7) holds if

$$\epsilon_2 \leq \frac{\epsilon_1 + \epsilon_3}{2}.$$

PROOF. This is verified simply by multiplying out $p(z)(1 - \alpha x + x^2)$, so omitted. □

The examples below illustrate how sensitive (2.5) is to the size of the α_js.

EXAMPLE 1. *We have*

$$\begin{aligned}
&(1 + 1.05x + x^2)(1 - 0.28x + x^2)(1 + 1.25x + x^2)\\
&= x^6 + 2.02x^5 + 3.6685x^4 + 3.67250x^3 + 3.6685x^2 + 2.02x + 1,
\end{aligned}$$

which satisfies (2.5), but change the 0.28 *to* 0.3 *and*

$$\begin{aligned}
&(1 + 1.05x + x^2)(1 - 0.30x + x^2)(1 + 1.25x + x^2)\\
&= x^6 + 2x^5 + 3.6225x^4 + 3.60625x^3 + 3.6225x^2 + 2x + 1,
\end{aligned}$$

does not. Similarly, we have

$$(1+1.05x+x^2)(1-0.3x+x^2)(1+1.25x+x^2)(1-0.6x+x^2) =$$

$$x^8+1.4x^7+3.4225x^6+3.43275x^5+5.08125x^4+3.43275x^3+3.4225x^2+1.4x+1,$$

which satisfies (2.5), but change the 0.6 *to* 0.7 *and*

$$(1+1.05x+x^2)(1-0.3x+x^2)(1+1.25x+x^2)(1-0.7x+x^2) =$$

$$x^8+1.3x^7+3.2225x^6+3.0705x^5+4.720625x^4+3.0705x^3+3.2225x^2+1.3x+1,$$

does not.

The polynomial

$$(1+1.5x+x^2)(1+0.2x+x^2)(1+0.1x+x^2)$$
$$=x^6+1.8x^5+3.47x^4+3.63x^3+3.47x^2+1.8x+1$$

satisfies (2.5), as does

$$(1+1.5x+x^2)(1+0.2x+x^2)(1+0.1x+x^2)(1-0.5x+x^2) =$$

$$x^8+1.3x^7+3.57x^6+3.695x^5+5.125x^4+3.695x^3+3.57x^2+1.3x+1$$

but change the 0.5 *to* 0.6 *and the product*

$$(1+1.5x+x^2)(1+0.2x+x^2)(1+0.1x+x^2)(1-0.6x+x^2) =$$

$$x^8+1.2x^7+3.39x^6+3.348x^5+4.762x^4+3.348x^3+3.39x^2+1.2x+1$$

does not.

The polynomial

$$(1+0.1x+x^2)(1+0.2x+x^2)(1+0.3x+x^2)(1+0.92x+x^2) =$$

$$x^8+1.52x^7+4.662x^6+4.6672x^5+7.32952x^4+4.6672x^3+4.662x^2+1.52x+1$$

satisfies (2.5), but change the 0.92 *to* 0.91 *and*

$$(1+0.1x+x^2)(1+0.2x+x^2)(1+0.3x+x^2)(1+0.91x+x^2) =$$

$$x^8+1.51x^7+4.656x^6+4.6361x^5+7.31746x^4+4.6361x^3+4.656x^2+1.51x+1$$

does not.

The above lemma holds, namely the condition (2.7), because

$$(1+0.1x+x^2)(1+0.2x+x^2)(1+0.3x+x^2)(1+0.92x+x^2)(1+0.999x+x^2) =$$

$$x^{10}+2.519x^9+7.18048x^8+10.844538x^7+16.6540528x^6+$$

$$+16.65659048x^5+16.6540528x^4+10.844538x^3+7.18048x^2+2.519x+1$$

satisfies (2.5), but change the 0.999 *to* 0.99 *and*

$$(1+0.1x+x^2)(1+0.2x+x^2)(1+0.3x+x^2)(1+0.92x+x^2)(1+0.99x+x^2) =$$

$$x^{10}+2.51x^9+7.1668x^8+10.80258x^7+16.612048x^6+$$

$$+16.5906248x^5+16.612048x^4+10.80258x^3+7.1668x^2+2.51x+1$$

does not.

3. Superelliptic curves and self-inversive polynomials

The following theorem connects self-reciprocal polynomials with a very special class of algebraic curves, namely superelliptic curves. We follow the definitions and notation as in [**4**].

Fix an integer $g \geq 2$. Let $\mathcal{X}_g$ denote a genus g generic planar curve defined over an algebraically closed field k of characteristic $p \geq 0$. We denote by G the full automorphism group of $\mathcal{X}_g$. Hence, G is a finite group. Denote by K the function field of $\mathcal{X}_g$ and assume that the affine equation of $\mathcal{X}_g$ is given by some polynomial in terms of x and y.

Let $H = \langle \tau \rangle$ be a cyclic subgroup of G such that $|H| = n$ and H is in the center of G, where $n \geq 2$. Moreover, we assume that the quotient curve $\mathcal{X}_g/H$ has genus zero. The **reduced automorphism group of $\mathcal{X}_g$ with respect to H** is called the group $\bar{G} := G/H$, see [**4**].

Assume $k(x)$ is the genus zero subfield of K fixed by H. Hence, $[K : k(x)] = n$. Then, the group $\bar{G}$ is a subgroup of the group of automorphisms of a genus zero field. Hence, $\bar{G} < PGL_2(k)$ and $\bar{G}$ is finite. It is a classical result that every finite subgroup of $PGL_2(k)$ is isomorphic to one of the following: C_m, D_m, A_4, S_4, A_5, *semidirect product of an elementary Abelian group with cyclic group*, $PSL(2, q)$ and $PGL(2, q)$.

The group $\bar{G}$ acts on $k(x)$ via the natural way. The fixed field of this action is a genus 0 field, say $k(z)$. Thus, z is a degree $|\bar{G}| := m$ rational function in x, say $z = \phi(x)$.

LEMMA 6. *Let $\mathcal{X}_g$ be a superelliptic curve of level n with $|Aut\ (\mathcal{X}_g)| > n$. Then, $\mathcal{X}_g$ can be written as*

$$y^n = f(x^s), \quad or \quad y^n = xf(x^s)$$

for some $s > 1$.

The proof goes similar as for the hyperelliptic curves as in [**23**]. Since below we display all equations of such curves in such form then the Lemma is obviously true.

Next we focus on studying the nature of the polynomial $f(x)$ and its connections to self-inversive polynomials. We are assuming that the curves are of characteristic zero, so the reduced automorphism group is cyclic, dihedral, A_4, S_4, or A_5. The list of equations, including the full group of automorphisms, the dimension of the loci, and the ramification of the corresponding covers can be taken from [**21**].

THEOREM 3. *If the reduced automorphism group of a superelliptic curve $\mathcal{X}$ is nontrivial or not isomorphic to a cyclic group, then $\mathcal{X}$ can be written with the affine equation*

$$y^n = f(x) \quad or \quad y^n = x \cdot f(x)$$

where $f(x)$ is a palindromic or antipalindromic polynomial. If the reduced automorphism group is isomorphic to A_5, then $f(x)$ is a quasi-palindromic plynomial.

PROOF. If $\overline{\text{Aut}}\,(\mathcal{X})$ is isomorphic to a dihedral group D_{2m}, then the equation of $\mathcal{X}_g$ can be written as in one of the following cases

$$\begin{aligned} y^n &= F(x) := \prod_{i=1}^{\delta} (x^{2m} + \lambda_i x^m + 1) \\ y^n &= (x^m - 1) \cdot F(x), \\ y^n &= x \cdot F(x), \\ y^n &= (x^{2m} - 1) \cdot F(x), \\ y^n &= x(x^m - 1) \cdot F(x), \\ y^n &= x(x^{2m} - 1) \cdot F(x), \end{aligned}$$

The polynomial $F(x)$ is palindromic from Lemma 2. The polynomials $x^m - 1$ and $x^{2m} - 1$ are antipalindromic. From Lemma 4 the products $(x^m - 1)F(x)$ and $(x^{2m} - 1)F(x)$ are antipalindromic. Hence, if the reduced automorphism group of a superelliptic curve is isomorphic to a dihedral group then the equation of the curve can be written as $y^2 = f(x)$ or $y^2 = xf(x)$, where $f(x)$ can be chosen to be a palindromic or antipalindromic polynomial.

If $\overline{\text{Aut}}\,(\mathcal{X})$ is isomorphic to A_4, then the equation of $\mathcal{X}_g$ can be written as in one of the following cases

$$\begin{aligned} y^n &= G(x) \\ y^n &= (x^4 + 2i\sqrt{3}x^2 + 1) \cdot G(x), \\ y^n &= (x^8 + 14x^4 + 1) \cdot G(x), \\ y^n &= x(x^4 - 1) \cdot G(x), \\ y^n &= x(x^4 - 1)(x^4 + 2i\sqrt{3}x^2 + 1) \cdot G(x), \\ y^n &= x(x^4 - 1)(x^8 + 14x^4 + 1) \cdot G(x), \end{aligned}$$

where

$$G(x) := \prod_{i=1}^{\delta} (x^{12} - \lambda_i x^{10} - 33x^8 + 2\lambda_i x^6 - 33x^4 - \lambda_i x^2 + 1)$$

Notice that every factor of $G(x)$ is palindromic, hence $G(x)$ is also palindromic from Lemma 4. The polynomials $x^4 + 2i\sqrt{3}x^2 + 1$ and $x^8 + 14x^4 + 1$ are palindromic and therefore $(x^4 + 2i\sqrt{3}x^2 + 1)\,G(x)$ and $(x^8 + 14x^4 + 1)\,G(x)$ are palindromic. When multiplied by $x^4 - 1$ such polynomials become antipalindromic since $x^4 - 1$ is antipalindromic. So the equation of the curve can be written as $y^2 = f(x)$ or $y^2 = xf(x)$, where $f(x)$ can be chosen to be a palindromic or antipalindromic polynomial.

If $\overline{\mathrm{Aut}}\ (\mathcal{X})$ is isomorphic to S_4, then the equation of $\mathcal{X}_g$ can be written as in one of the following cases

$$
\begin{aligned}
y^n &= M(x)\\
y^n &= \left(x^8+14x^4+1\right)\cdot M(x)\\
y^n &= x(x^4-1)\cdot M(x)\\
y^n &= \left(x^8+14x^4+1\right)\cdot x(x^4-1)\cdot M(x)\\
y^n &= \left(x^{12}-33x^8-33x^4+1\right)\cdot M(x)\\
y^n &= \left(x^{12}-33x^8-33x^4+1\right)\cdot\left(x^8+14x^4+1\right)\cdot M(x)\\
y^n &= \left(x^{12}-33x^8-33x^4+1\right)\cdot x(x^4-1)\cdot M(x)\\
y^n &= \left(x^{12}-33x^8-33x^4+1\right)\cdot\left(x^8+14x^4+1\right)\cdot x(x^4-1)M(x)
\end{aligned}
$$

where

$$
\begin{aligned}
M(x) = \prod_{i=1}^{\delta}\Big(&x^{24}+\lambda_i x^{20}+(759-4\lambda_i)x^{16}+2(3\lambda_i+1228)x^{12}+(759-4\lambda_i)x^8\\
&+\lambda_i x^4+1\Big)
\end{aligned}
$$

Since every factor of $M(x)$ is palindromic, then $M(x)$ is palindromic. By Lemma 4 we have that the equation of the curve can be written as $y^2=f(x)$ or $y^2=xf(x)$, where $f(x)$ can be chosen to be a palindromic or antipalindromic polynomial. The antipalindromic cases correspond exactly to the cases when x^4-1 appears as a factor.

Let $\overline{\mathrm{Aut}}\ (\mathcal{X})$ is isomorphic to A_5. This case is slightly different from the other cases due to the fact that now the reduced group has an element of order 5 and $f(x)$ will be written as a decomposition of x^5. So the change of coordinates $x\mapsto -x$ will preserve the sign for odd powers and change it for even powers of x.

Let $\Lambda(x)$, $Q(x)$, $\psi(x)$ be as follows

$$
\begin{aligned}
\Lambda(x) = \prod_{i=1}^{\delta}(&x^{60}+a_1x^{55}+a_2x^{50}+a_3x^{45}+a_4x^{40}+a_5x^{35}+a_6x^{30}-a_5x^{25}+a_4x^{20}\\
&-a_3x^{15}+a_2x^{10}-a_1x^5+1)\\
a_1 &= \lambda_i-684\\
a_2 &= 55\lambda_i+157434\\
a_3 &= 1205\lambda_i-12527460\\
a_4 &= 13090\lambda_i+77460495\\
a_5 &= 69585\lambda_i-130689144\\
a_6 &= 134761\lambda_i-33211924\\
Q(x) &= x^{30}+522x^{25}-10005x^{20}-10005x^{10}-522x^5+1,\\
\psi(x) &= x^4+2i\sqrt{3}x^2+1
\end{aligned}
$$

Then, the equation of $\mathcal{X}_g$ can be written as in one of the following cases

$$
\begin{aligned}
y^n &= \Lambda(x)\\
y^n &= x(x^{10}+11x^5-1)\cdot\Lambda(x)\\
y^n &= x(x^{20}-228x^{15}+494x^{10}+228x^5+1)(x^{10}+11x^5-1)\cdot\Lambda(x)\\
y^n &= (x^{20}-228x^{15}+494x^{10}+228x^5+1)\cdot\Lambda(x)\\
y^n &= Q(x)\cdot\Lambda(x)\\
y^n &= x(x^{10}+11x^5-1).\psi(x)\cdot\Lambda(x)\\
y^n &= (x^{20}-228x^{15}+494x^{10}+228x^5+1)\cdot\psi(x)\cdot\Lambda(x)\\
y^n &= (x^{20}-228x^{15}+494x^{10}+228x^5+1)(x(x^{10}+11x^5-1))\cdot\psi(x)\cdot\Lambda(x)
\end{aligned}
$$

Notice that $\Lambda(x)$ is a quasi-palindromic polynomial since all its factors are so. So are $Q(x), \psi(x)$ and the other factors. By Lemma 4 we can say that in this case he equation of the curve can be written as $y^2 = f(x)$ or $y^2 = xf(x)$, where $f(x)$ can be chosen to be a quasi-palindromic polynomial.

This completes the proof of the theorem. □

In [**12**] it is shown that if the group H is unique in G and the reduced group G/H is not cyclic or nontrivial, then the field of moduli is a field of definition for superelliptic curves. In [**3**] and [**5**] it is explored the fact that most palindromic or self-inversive polynomials have minimal coefficients. So it is a natural question to investigate what is the relation between the minimal of definition of such curves, the minimal height as in [**2**], and the palindromic polynomial $f(x)$.

4. Self-reciprocal polynomials and reduction theory

Every stable binary form $f(z, y)$ of degree $n \geq 2$ correspond uniquely to a positive definite quadratic $\mathcal{J}_f$ called Julia quadratic; see [**2**]. Since positive definite quadratics have a unique zero in the upper half plane $\mathcal{H}_2$, then we associate the zero of $\mathcal{J}_f$ to the binary form f. This defines a map ε from the set of degree n binary forms to $\mathcal{H}_2$, which is called the zero map. A binary form $f(z, y)$ is called *reduced* if $\varepsilon(f) \in \mathcal{F}_2$. The size of the coefficients of a reduced binary form is bounded by its Julia invariant $\theta(f)$. If f is a reduced form, we say that f has *minimal coefficients*; see [**2**] for details.

There are no efficient ways to compute the Julia quadratic or the Julia invariant of a binary form of high degree (i.e. degree > 6). Moreover, there is no known method to express the Julia invariant $\theta(f)$ in terms of the generators of the ring of invariants of the degree n binary forms (i.e. transvections of the form). However, as discussed in [**2**] the case when f is totally real is much easier. A form is called *totally real* if it splits over $\mathbb{R}$.

Let $f \in \mathbb{C}[z]$ be a degree $n \geq 2$ polynomial. We denote by f_* the corresponding form (homogenization of f) in $\mathbb{C}[z, y]$. $GL_2(\mathbb{C})$ acts on the space of degree n binary forms. For a matrix $M \in GL_2(\mathbb{C})$ we denote by f_*^M the action of M on f_*. By f^M we denote $f_*^M(z, 1)$.

LEMMA 7. *Let $f \in \mathbb{C}[z]$ and $M = \begin{bmatrix} 1 & -i \\ 1 & i \end{bmatrix}$. Then, f_* is a totally real binary form if and only if f^M has all roots in the unit circle.*

PROOF. The proof is rather elementary. The Möbious transformation $h(z) = Mz$ maps $\mathcal{H}_2$ onto the open unit disk. Moreover, it maps bijectively $U \setminus \{1\}$ to $\mathbb{R}$. □

For reduction of totally real forms see [**2**] and [**5**].

THEOREM 4. *Let $f(z)$ be a self-inversive polynomial. Then the following are equivalent:*

i) all roots of $f(z)$ are on the unit circle

ii) all roots of its derivative $f'(z)$ are on the unit disk

iii) f_*^M is totally real form

PROOF. The equivalence of i) and iii) is the above Lemma. The equivalence of i) and ii) is a result of Cohn. □

It is interesting to see how the reduction is performed in such case. From [**2**] we have a polynomial G_f associated to f. The Julia quadratic J_f is the only quadratic factor of G_f when factored over $\mathbb{R}$. Moreover, Beshaj [**2**] has proved that G_f is very similar to a self-inversive polynomial. We describe briefly below

Let f be a generic totally real form given by

$$f(x, y) = a_n x^n + a_{n-1} x^{n-1} y + \cdots + a_1 x y^{n-1} + a_0 y^n$$

where $a_0, \ldots, a_n$ are transcendentals. Identify the coefficients $a_0, \ldots, a_n$ respectively with $1, \ldots, n+1$. Then the symmetric group S_{n+1} acts on $\mathbb{R}[a_0, \ldots a_n][x, y]$ by permuting $a_0, \ldots, a_n$. For any permutation $\tau \in S_{n+1}$ and $f \in \mathbb{R}[a_0, \ldots a_n][x, y]$ we denote by $\tau(f) = f^\tau$. Then

$$f^\tau(x, y) = \tau(a_n)\, x^n + \tau(a_{n-1})\, x^{n-1} y + \cdots + \tau(a_1)\, x y^{n-1} + \tau(a_0)\, y^n.$$

Define $G(x, y)$ as follows

$$G(x, y) = \frac{x \cdot f_x(-f_y(x, y), f_x(x, y)) + y \cdot f_y(-f_y(x, y), f_x(x, y))}{n\, f(x, y)}. \tag{4.1}$$

Notice that since f is totally real, then $f \in \mathbb{R}[x, y]$. Therefore, $G \in \mathbb{R}[x, y]$. Note also that, for $\sigma \in S_{n+1}$ we have an involution

$$\sigma = \begin{cases} (1, n+1)(2, n) \cdots \left(\frac{n}{2}, \frac{n}{2} + 2\right), & \text{if } n \text{ is even} \\ (1, n+1)(2, n) \cdots \left(\frac{n+1}{2}, \frac{n+3}{2}\right), & \text{if } n \text{ is odd.} \end{cases}$$

Next result describes the properties of $G(x, y)$.

THEOREM 5 (Beshaj). *The polynomial $G(x, y)$ satisfies the following*

i) $G(x, y)$ is a covariant of f of degree $(n-1)$ and order $(n-1)(n-2)$.

ii) $G(x, y)$ has a unique quadratic factor over $\mathbb{R}$, which is the Julia quadratic $\mathcal{J}_f$.

iii) $G^\sigma(x, y) = (-1)^{n-1}\, G(x, y)$. Moreover, if $G_f = \sum_{i=1}^d g_i\, x^i y^{d-i}$, then

$$g_i^\sigma = (-1)^{n-1}\, g_{d-i},$$

for all $i = 0, ..., d$.

Then we have the interesting connection between real forms and self-inversive polynomials.

THEOREM 6. *If f is a palindromic real form then $G_f(x, y)$ is self-inversive. If f is of odd degree then G_f is palindromic.*

PROOF. If f is palindromic, then from Lemma 3, i) we have that $a_i = a_{n-i}$ for all $i = 0, \dots, n$. That means that σ fixes all coefficients of f. Hence, $g_i^\sigma = (-1)^{n-1} g_i$ for all $i = 0, \dots, d$, where $d = \deg G_f$. Thus, G_f is self-inversive. If n is odd, then $g_i^\sigma = g_i$. Hence, G_f is palindromic.

□

We know that G_f has exactly two non-real roots, namely $\varepsilon(f)$ and its conjugate. Consider now G_f^M. Then all real roots of G_f will go to roots on the unit circle of G_f^M and the two non-real roots $\varepsilon(f)$ and its conjugate $\overline{\varepsilon(f)}$ go inside the unit disk as roots of G_f^M.

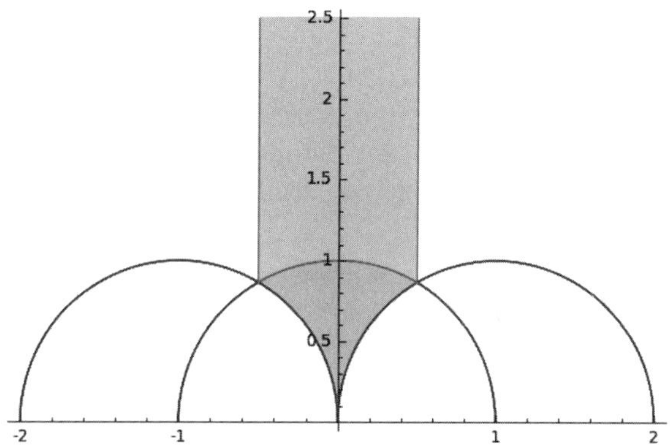

FIGURE 1. The region $\mathcal{T}$

LEMMA 8. *Let f be a self-inversive polynomial with all roots in the unit circle U, f_* its homogenization, $\mathcal{T}$ be the region in the complex plane given by*

$$\mathcal{T} = \{z = a + bi \,|\, a^2 - 2a + b^2 \geq 0,\ a^2 + 2a + b^2 \geq 0\},$$

and $M = \begin{bmatrix} 1 & -i \\ 1 & i \end{bmatrix}$. If $\varepsilon(f_*)^M \in \mathcal{T}$ or $\varepsilon(f_*)^M \in \mathcal{F}_2$, then f^M has minimal coefficients.

PROOF. From Lem. 7 we have that f_*^M is a totally real form. Then $\varepsilon(f_*^M)$ is the image of the zero map in the upper half plane $\mathcal{H}_2$.

If $\varepsilon(f_*)^M \in \mathcal{F}_2$ then f_*^M is reduced and we are done. If $\varepsilon(f_*)^M \in \mathcal{T}$ then let $S = \begin{bmatrix} 0 & 1 \\ 1 & 0 \end{bmatrix}$ and compute $\varepsilon(f_*)^{MS}$. Let $\varepsilon(f_*)^M = a + bi$. Then

$$\varepsilon(f_*)^{MS} = \frac{1}{a+bi} = \frac{a}{a^2+b^2} - \frac{b}{a^2+b^2} i$$

Hence, $|\varepsilon(f_*)^{MS}| \geq 1$ and

$$-\frac{1}{2} \leq \frac{a}{a^2+b^2} \leq \frac{1}{2}$$

Hence, $\varepsilon(f_*)^M \in \mathcal{F}_2$ However, the height of f_*^M does not change under the transformation S. Hence, f_*^M has minimal coefficients. Thus, in both cases f^M has minimal coefficients. □

The region $\mathcal{T}$ is the blue colored region in Fig. 1 and the grey area is the fundamental domain.

5. Self-reciprocal polynomials and codes

The goal of this section is to show how self-reciprocal polynomials are connected to other areas of mathematics, namely whether extremal formal weight enumerators for codes satisfy the Riemann hypothesis. We will follow the setup of [**11**].

For $d \leq n$, denote the weight enumerator of an MDS code C over $\mathbb{F} = GF(q)$ of length n and minimum distance d by $M_{n,d}(x, y)$. The dual $C^\perp$ is also an MDS code of length n and minimum distance $d^\perp = n + 2 - d$. Therefore, for $d \geq 2$, the weight enumerator of $C^\perp$ is $M_{n,n+2-d}(x, y)$. Let $M_{n,n+1} = x^n$. The MDS code with weight enumerator $M_{n,1}$ has dimension $n - d + 1 = n - 1 + 1 = n$, hence $C = \mathbb{F}_q^n$. It is easy to see that $M_{n,n+1}$ is the MacWilliams transform, $(x, y) \mapsto (\frac{x+(q-1)y}{\sqrt{q}}, \frac{x-y}{\sqrt{q}})$, of $M_{n,1}$. We may think of $M_{n,1}$ as the weight enumerator of the zero code.

The set $\{M_{n,1}, M_{n,2}, \dots, M_{n,n-1}, M_{n,n+1}\}$ is a basis for the vector space of homogeneous polynomials of degree n in x, y. Furthermore, this set is closed under the MacWilliams transform; see [**11**] for details.

If C is an $[n, k, d]_q$-code, then one can easily see that

$$A_C(x, y) = \sum_{i=d}^{n+1} a_{i-d} M_{n,i} = a_0 M_{n,d} + \ldots + a_{n+1-d} M_{n,n+1},$$

for some integers a_i as in §4.4.2 in [**14**]. The zeta polynomial of C is defined as

$$P(T) := a_0 + a_1 T + \cdots + a_{n-d+1} T^{n+1-d}.$$

The zeta polynomial $P(T)$ of an $[n, k, d]_q$-code C determines uniquely the weight enumerator of C. The degree of $P(T)$ is at most $n - d + 1$. The quotient

$$Z(t) = \frac{P(T)}{(1 - T)(1 - qT)}$$

is called **the zeta function** of the linear code C. The zeta function of an MDS code

$$\frac{1}{(1 - T)(1 - qT)} = \sum_{j=0}^{\infty} \frac{q^{j+1} - 1}{q - 1} T^j$$

is the rational zeta function over $\mathbb{F}_q$; see [**11**, Cor. 1]. Formally self-dual codes lead to self-reciprocal polynomials. The proof of the following Proposition can be found in [**11**].

PROPOSITION 1. *If $P(T)$ is the zeta polynomial of a formally self-dual code, then $P(T/\sqrt{q})$ is a self-reciprocal polynomial.*

5.1. Riemann zeta function versus zeta function for self-dual codes.

From [**11**] we have that for a self-dual code C,

$$Z(T) = q^{g-1} T^{2g-2} Z(1/qT),$$

which for $z(T) := T^{1-g} Z(T)$, may be written as

$$z(T) = z(1/qT).$$

Now let

$$\zeta_C(s) := Z(q^{-s}) \text{ and } \xi_C(s) := z(q^{-s}).$$

We obtain

$$\xi_C(s) = \xi_C(1 - s),$$

which is the same symmetry equation is analogous to the functional equation for the Riemann zeta function. We note that $\zeta(s)$ and $\xi(s)$ have the same zeros.

The zeroes of the zeta function of a linear code C are useful in understanding possible values of its minimum distance d.

Let C be a linear code with weight distribution vector $(A_0, A_1, \dots, A_n)$. Let $\alpha_1, \dots, \alpha_r$ be the zeros of the zeta polynomial $P(T)$ of C Then

$$d = q - \sum_i \alpha_i^{-1} - \frac{A_{d+1}}{A_d}\frac{d+1}{n-d}.$$

In particular,

$$d \leq q - \sum_i \alpha_i^{-1};$$

see [**11**] for details.

A self-dual code C is said to satisfy *Riemann hypothesis* if the real part of any zero of $\zeta_C(s)$ is $1/2$, or equivalently, the zeros of the zeta polynomial $P_C(T)$ lie on the circle $|T| = 1/\sqrt{q}$, or equivalently, the roots of the self-reciprocal polynomial (see Proposition 1 above) $P_C(T/\sqrt{q})$ lie on the unit circle.

While Riemann hypothesis is satisfied for curves over finite fields, in general it does not hold for linear codes. A result that generates many counterexamples may be found in [**14**]. There is a family of self-dual codes that satisfy the Riemann hypothesis which we are about to discuss. The theory involved in this description holds in more generality than linear codes and their weight enumerators.

5.2. Virtual weight enumerators. A homogeneous polynomial

$$F(x, y) = x^n + \sum_{i=1}^{n} f_i x^{n-i} y^i$$

with complex coefficients is called a *virtual weight enumerator.* The set

$$\{0\} \cup \{i : f_i \neq 0\}$$

is called its *support.* If

$$F(x, y) = x^n + \sum_{i=d}^{n} f_i x^{n-i} y^i, \tag{5.1}$$

with $f_d \neq 0$, then n is called the *length* and d is called the *minimum distance* of $F(x, y)$.

Let C be a self-dual linear $[n, k, d]$-code. Recall that n is even, $k = n/2$ and its weight enumerator satisfies MacWilliams' Identity. A virtual generalization of $A_C(x, y)$ is straightforward. A virtual weight enumerator $F(x, y)$ of even degree that is a solution to MacWilliams' Identity

$$F(x, y) = F\left(\frac{x + (q-1)y}{\sqrt{q}}, \frac{x - y}{\sqrt{q}}\right), \tag{5.2}$$

is called *virtually self dual* over $\mathbb{F}_q$ with *genus* $\gamma(F) = n/2 + 1 - d$. Although a virtual weight enumerator in general does not depend on a prime power q, a virtually self-dual weight enumerator does.

Problem 1. *Find the conditions under which a (self-dual) virtual weight enumerator with positive integer coefficients arises from a (self-dual) linear code.*

The zeta polynomial and the zeta function of a virtual weight enumerator are defined as in the case of codes.

PROPOSITION 2 ([**6**]). *Let $F(x, y)$ be a virtual weight enumerator of length n and minimum distance d. Then, there exists a unique function $P_F(T)$ of degree at most $n - d$ which satisfies the following*

$$\frac{(y(1-T)+xT)^n}{(1-T)(1-qT)} P_F(T) = \dots + \frac{F(x,y)-x^n}{q-1} T^{n-d} + \dots$$

The polynomial $P_F(T)$ and the function

$$Z_F(T) := \frac{P(T)}{(1-T)(1-qT)},$$

are called respectively *the zeta polynomial and the zeta function of the virtual weight enumerator* $F(x, y)$.

A virtual self-dual weight enumerator *satisfies the Riemann hypothesis* if the zeroes of its zeta polynomial $P_F(T)$ lie on the circle $|T| = 1/\sqrt{q}$. There is a family of virtual self-dual weight enumerators that satisfy Riemann hypothesis. It consists of enumerators that have certain divisibility properties.

Let $b > 1$ be an integer. If $\operatorname{supp}(F) \subset b\mathbb{Z}$, then F is called *b-divisible*. Let F given by Eq. (5.1) be a b-divisible, virtually self-dual weight enumerator over $\mathbb{F}_q$. Then $F(x, y)$ is called

Type I: if $q = b = 2,\ 2|n$.
Type II: if $q = 2, b = 4, 8|n$.
Type III: if $q = b = 3, 4|n$.
Type IV: if $q = 4, b = 2, 2|n$.

Then we have the following theorem:

THEOREM 7 (Mallows-Sloane-Duursma). *If $F(x, y)$ is a b-divisible self-dual virtual enumerator with length n and minimum distance d, then*

$$d \le \begin{cases} 2\left[\frac{n}{8}\right] + 2, & \text{if } F \text{ is Type I,} \\ 4\left[\frac{n}{24}\right] + 4, & \text{if } F \text{ is Type II,} \\ 3\left[\frac{n}{12}\right] + 3, & \text{if } F \text{ is Type III,} \\ 2\left[\frac{n}{6}\right] + 2, & \text{if } F \text{ is Type IV.} \end{cases}$$

A virtually self-dual weight enumerator $F(x, y)$ is called *extremal* if the bound in Theorem 7 holds with equality. A linear code C is called b-divisible, extremal, Type I, II, II, IV if and only if its weight enumerator has the corresponding property. The zeta functions of all extremal virtually self-dual weight enumerators are known; see [**10**]. The following result can be found in [**10**].

PROPOSITION 3. *All extremal type IV virtual weight enumerators satisfy the Riemann hypothesis.*

For all other extremal enumerators, Duursma has suggested the following conjecture in [**9**].

PROBLEM 2. *Prove that any extremal virtual self-dual weight enumerators of type I-III satisfies the Riemann hypothesis.*

Let F denote a weight enumerator as in (5.2) and $P_F(T)$ the associated zeta polynomial. Let $p_F(T) = P_F(T/\sqrt{q})$ denote the normalized zeta polynomial. Numerous computations suggest the following result.

PROBLEM 3. *If F is an extremal weight enumerator of Type I, II, II, IV then the normalized zeta polynomial is symmetric increasing. In fact, using the notation of (2.5), if if $a_i = a$, $a_{i+1} = a + \epsilon_1$, $a_{i+2} = a + \epsilon_2$, $a_{i+3} = a + \epsilon_3$ then $\epsilon_2 \leq \frac{\epsilon_1 + \epsilon_3}{2}$.*

Acknowledgment

The authors would like to thank Lubjana Beshaj for helpful conversations and explaining to us the reduction theory of self-inversive and self-reciprocal forms.

References

[1] Lubjana Beshaj, *Reduction theory of binary forms*, Advances on superelliptic curves and their applications, NATO Sci. Peace Secur. Ser. D Inf. Commun. Secur., vol. 41, IOS, Amsterdam, 2015, pp. 84–116. MR3525574

[2] Lubjana Beshaj, *Integral binary forms with minimal height*, ProQuest LLC, Ann Arbor, MI, 2016. Thesis (Ph.D.)–Oakland University. MR3579531

[3] L. Beshaj, *Julia quadratic of superelliptic curves with extra automorphisms*, Algebraic curves and their fibrations in mathematical physics and arithmetic geometry, 2016.

[4] Lubjana Beshaj, Valmira Hoxha, and Tony Shaska, *On superelliptic curves of level n and their quotients, I*, Albanian J. Math. **5** (2011), no. 3, 115–137. MR2846162

[5] L. Beshaj and T. Shaska, *Julia quadratic of self-inversive binary forms*, 2016. in preparation.

[6] Koji Chinen, *Zeta functions for formal weight enumerators and the extremal property*, Proc. Japan Acad. Ser. A Math. Sci. **81** (2005), no. 10, 168–173 (2006). MR2196722

[7] Keith Conrad, *Root on a circle*, 2015.

[8] Stephen A. DiPippo and Everett W. Howe, *Real polynomials with all roots on the unit circle and abelian varieties over finite fields*, J. Number Theory **73** (1998), no. 2, 426–450, DOI 10.1006/jnth.1998.2302. MR1657992

[9] Iwan Duursma, *A Riemann hypothesis analogue for self-dual codes*, Codes and association schemes (Piscataway, NJ, 1999), DIMACS Ser. Discrete Math. Theoret. Comput. Sci., vol. 56, Amer. Math. Soc., Providence, RI, 2001, pp. 115–124. MR1816392

[10] Iwan Duursma, *Extremal weight enumerators and ultraspherical polynomials*, Discrete Math. **268** (2003), no. 1-3, 103–127, DOI 10.1016/S0012-365X(02)00683-0. MR1983272

[11] A. Elezi and T. Shaska, *Weight distributions, zeta functions and riemann hypothesis for linear and algebraic geometry codes*, Arithmetic of superelliptic curves, 2015.

[12] R. Hidalgo and T. Shaska, *On the field of moduli of superelliptic curves*, Algebraic curves and their fibrations in mathematical physics and arithmetic geometry, 2016.

[13] David Joyner, *Zeros of some self-reciprocal polynomials*, Excursions in harmonic analysis. Volume 1, Appl. Numer. Harmon. Anal., Birkhäuser/Springer, New York, 2013, pp. 329–348, DOI 10.1007/978-0-8176-8376-4_17. MR3050347

[14] David Joyner and Jon-Lark Kim, *Selected unsolved problems in coding theory*, Applied and Numerical Harmonic Analysis, Birkhäuser/Springer, New York, 2011. MR2838861

[15] Gaston Julia, *Étude sur les formes binaires non quadratiques à indéterminées réelles, ou complexes, ou à indéterminées conjuguées* (French), NUMDAM, [place of publication not identified], 1917. MR3532882

[16] Piroska Lakatos and László Losonczi, *Self-inversive polynomials whose zeros are on the unit circle*, Publ. Math. Debrecen **65** (2004), no. 3-4, 409–420. MR2107957

[17] Piroska Lakatos, *On polynomials having zeros on the unit circle* (English, with French summary), C. R. Math. Acad. Sci. Soc. R. Can. **24** (2002), no. 2, 91–96. MR1902028

[18] Piroska Lakatos, *On zeros of reciprocal polynomials*, Publ. Math. Debrecen **61** (2002), no. 3-4, 645–661. MR1943722

[19] László Losonczi and Andrzej Schinzel, *Self-inversive polynomials of odd degree*, Ramanujan J. **14** (2007), no. 2, 305–320, DOI 10.1007/s11139-007-9029-5. MR2341855

[20] P. J. O'Hara and R. S. Rodriguez, *Some properties of self-inversive polynomials*, Proc. Amer. Math. Soc. **44** (1974), 331–335, DOI 10.2307/2040432. MR0349967

[21] R. Sanjeewa and T. Shaska, *Determining equations of families of cyclic curves*, Albanian J. Math. **2** (2008), no. 3, 199–213. MR2492096

[22] A. Schinzel, *Self-inversive polynomials with all zeros on the unit circle*, Ramanujan J. **9** (2005), no. 1-2, 19–23, DOI 10.1007/s11139-005-0821-9. MR2166374

[23] T. Shaska, *Determining the automorphism group of a hyperelliptic curve*, Proceedings of the 2003 International Symposium on Symbolic and Algebraic Computation, 2003, pp. 248–254 (electronic). MR2035219

[24] Michael Stoll and John E. Cremona, *On the reduction theory of binary forms*, J. Reine Angew. Math. **565** (2003), 79–99. MR2024647 (2005e:11091)

[25] R. S. Vieira, *On the number of roots of self-inversive polynomials on the complex unit circle* (2015).

Department of Mathematics, US Naval Academy, Annapolis, Maryland 21402
Email address: wdj@usna.edu

Department of Mathematics and Statistics, Oakland University, Rochester, Michigan 48309
Email address: shaska@oakland.edu

Contemporary Mathematics
Volume **703**, 2018
http://dx.doi.org/10.1090/conm/703/14139

Syzygy divisors on Hurwitz spaces

Anand Deopurkar and Anand Patel

Abstract. We describe a sequence of effective divisors on the Hurwitz space $H_{d,g}$ for d dividing $g-1$ and compute their cycle classes on a partial compactification. These divisors arise from vector bundles of syzygies canonically associated to a branched cover. We find that the cycle classes are all proportional to each other. These computations are motivated by the question of determining the effective cone and ultimately the birational type of $H_{d,g}$.

1. Introduction

The Hurwitz space $H_{d,g}$ is the moduli space of simply branched maps $\alpha\colon C \to \mathbf{P}^1$, where C is a smooth algebraic curve of genus g and α is finite of degree d. It is one of the oldest moduli spaces studied in algebraic geometry. Indeed, its idea goes back to the time of Riemann—a time when algebraic curves were thought of primarily as branched covers of the Riemann sphere. It was put on a rigorous modern algebraic footing by Fulton [**12**]. It was compactified by Harris and Mumford [**14**], whose construction was then refined by Mochizuki [**17**] and Abramovich, Corti, and Vistoli [**1**]. We refer the reader to [**21**] for an introduction to Hurwitz spaces.

The Hurwitz spaces have attracted mathematical attention not only because of their intrinsic appeal, but also because of their role in illuminating the geometry of the moduli space M_g. Indeed, it was through the Hurwitz spaces that Riemann [**20**] computed the dimension of M_g and Severi [**22**], building on work of Clebsch [**5**] and Hurwitz [**15**], showed that M_g is irreducible. In more recent times, Harris and Mumford [**14**] used the compactified Hurwitz spaces to carry out a divisor class computation to show that $\overline{M}_g$ is of general type for large g. Hurwitz spaces and their variants have also been of interest outside of algebraic geometry. For example, spaces of branched covers of $\mathbf{P}^1$ with a given Galois group feature prominently in inverse Galois theory [**10, 11**], and spaces of covers of $\mathbf{P}^1$ by $\mathbf{P}^1$ play a key role in dynamics [**23**].

A basic open question about the Hurwitz spaces is the question of placing them in the birational classification of varieties: For which d and g is $H_{d,g}$ rational, unirational, uniruled, rationally connected, or of general type? As with many other open questions about the Hurwitz spaces, the answer is known only for very small or very large d. For $d \leq 5$, the space $H_{d,g}$ is known to be unirational, thanks to a determinantal description of covers of degree up to 5. For $d > \lfloor g/2 \rfloor$, the space $H_{d,g}$ dominates M_g. Since M_g has non-negative Kodaira dimension for $g \geq 22$, in

2010 *Mathematics Subject Classification.* Primary 14H10, 14H51, 14C17.

this case $H_{d,g}$ cannot be uniruled. The intermediate cases are rather mysterious, but remain an active area of research. See, for example, the recent work [**13**] on unirationality results for $d = 6$ and all $g \leq 28$ (and several more, up to $g = 45$).

At the heart of determining the birational type of $\overline{H}_{d,g}$ is the question of understanding its cone of effective divisors. Indeed, modulo an obstruction coming from singularities, saying that $\overline{H}_{d,g}$ is of general type is equivalent to saying that its canonical class lies in the interior of its effective cone. To show this, we need not know the full effective cone; it would suffice to know enough effective divisors, whose classes contain the canonical class in their convex span.

One way to get effective divisors is by using topology. A general point of $H_{d,g}$ parametrizes simply branched covers. In codimension 1, this simple topological picture can specialize in two ways: the cover can develop a ramification point of index 2 or can have two ramification points of index 1 over the same branch point. The two possibilities give two effective divisors on $H_{d,g}$.

The goal of this paper is to describe a number of other effective divisors on $H_{d,g}$ (for d dividing $g-1$) and to compute their classes on a partial compactification $\widetilde{H}_{d,g}$. Their origin is distinctly algebraic, orthogonal to any topological considerations. They are in the spirit of the classical Maroni divisor, and generalize the Casnati–Ekedahl divisors studied by the second author [**18**].

Before describing the divisors, we recall the Maroni divisor. A finite map $\alpha\colon C \to \mathbf{P}^1$ canonically factors as an embedding $\iota\colon C \to \mathbf{P}E$ followed by a projection $\pi\colon \mathbf{P}E \to \mathbf{P}^1$, where E is a vector bundle of rank $d-1$ and degree $g+d-1$ on $\mathbf{P}^1$. In the cases where the rank divides the degree, the bundle E is balanced for generic α—it is a twist of the trivial bundle. The *Maroni divisor* is the locus of α for which it is unbalanced.

Our divisors $\mu_1, \dots, \mu_{d-3}$, which we call *syzygy divisors*, are defined analogously for a sequence of other vector bundles $N_1, \dots, N_{d-3}$ associated with α. Roughly, N_i is the bundle whose fiber at $t \in \mathbf{P}^1$ is the vector space of $(i-1)$th syzygies among the generators of the homogeneous ideal of $C_t \subset \mathbf{P}E_t$. If d divides $g-1$, then N_i has rank dividing the degree, and for generic α, it is balanced. The divisor μ_i is the locus where it is unbalanced.

THEOREM 1.1 (Main). *Suppose d divides $g-1$. Let i be an integer with $1 \leq i \leq d-3$. The locus $\mu_i \subset \widetilde{H}_{d,g}$ is an effective divisor whose class in* $\operatorname{Pic}_{\mathbf{Q}} \widetilde{H}_{d,g}$ *is given by*

$$[\mu_i] = A_i\left(6(gd - 6g + d + 6)\cdot\zeta - d(d-12)\cdot\kappa - d^2\cdot\delta\right),$$

where

$$A_i = \binom{d-4}{i-1}^2 \frac{(d-2)(d-3)}{6(i+1)(d-i-1)}.$$

In the theorem, $\widetilde{H}_{d,g}$ is the coarse moduli space of $\alpha\colon C \to \mathbf{P}^1$, where C is an irreducible curve of arithmetic genus g with at worst nodal singularities and α is a finite map of degree d. The classes ζ, κ, and δ are certain tautological divisor classes in $\operatorname{Pic}_{\mathbf{Q}} \widetilde{H}_{d,g}$; they are conjectured to generate $\operatorname{Pic}_{\mathbf{Q}} \widetilde{H}_{d,g}$. See 2 for definitions.

The most surprising feature of the divisor class is that (up to scaling) it is independent of i. However, we do not know if the divisors μ_i themselves are supported on the same set (see 4.4). If they are not, then the picture is reminiscent of the Brill–Noether divisors on $\overline{M}_g$—the classes of all the divisorial Brill–Noether loci are proportional, although the loci themselves are different.

1.1 gives the class of μ_i on a *partial* compactification $\widetilde{H}_{d,g}$ of $H_{d,g}$. It is an interesting (and challenging) problem to compute the class of the closure of μ_i on a full compactification. This was carried out for the Maroni divisor for $d = 3$ in [**7**] and for higher d in [**25**]. It would also be interesting to find replacements for μ_i when d does not divide $g-1$. This would be analogous to the replacement of the Maroni divisor in the case of odd genus trigonal curves found in [**7**].

The paper is organized as follows. In 2, we recall the (largely conjectural) description of the Picard group of $\widetilde{H}_{d,g}$ and describe a number of divisor classes on $\widetilde{H}_{d,g}$. These include the syzygy divisors $\mu_1, \dots, \mu_{d-3}$, whose existence is contingent on the fact that the syzygy bundles N_i are balanced for a generic cover. In 3, we discuss the generic splitting type of the syzygy bundles N_i. In 4, we carry out the main computation of the divisor class of μ_i.

We work over an algebraically closed field of characteristic zero. All schemes and stacks are locally of finite type over this field. A *point* is a closed point, unless mentioned otherwise. The projectivization $\mathbf{P}U$ of a vector bundle U denotes the space of one-dimensional *quotients*. The Hurwitz spaces $\widetilde{H}_{d,g}$ and $H_{d,g}$ have *unparametrized* source and target. That is, $\alpha_1 : C_1 \to \mathbf{P}^1$ and $\alpha_2 : C_2 \to \mathbf{P}^1$ are considered isomorphic if there are isomorphisms $\phi \colon C_1 \to C_2$ and $\psi \colon \mathbf{P}^1 \to \mathbf{P}^1$ with $\alpha_2 \circ \phi = \psi \circ \alpha_1$.

2. Divisors on Hurwitz spaces

The goal of this section is to recall the rational Picard group of the Hurwitz space and some divisor classes in it. Since we are only interested in the rational Picard group, we may work either with the Deligne–Mumford stack $\widetilde{\mathcal{H}}_{d,g}$ or the coarse moduli space $\widetilde{H}_{d,g}$. We will pass from one to the other without comment.

2.1. The rational Picard group.

Denote by

$$\alpha \colon \mathcal{C} \to \mathcal{P}$$

the universal object over $\widetilde{\mathcal{H}}_{d,g}$. Here $\pi \colon \mathcal{C} \to \widetilde{\mathcal{H}}_{d,g}$ is a family of irreducible genus g curves with at worst nodal singularities, $p \colon \mathcal{P} \to \widetilde{\mathcal{H}}_{d,g}$ is a family of smooth genus 0 curves, and α a finite morphism of degree d over $\widetilde{\mathcal{H}}_{d,g}$. The universal family allows us to write the following two 'tautological' divisor classes on $\mathcal{H}_{d,g}$:

$$\pi_* \left(c_1(\omega_\pi^2)\right) \quad \text{and} \quad p_* \left(c_1(\omega_\pi) \cdot c_1(\alpha^* \omega_p)\right).$$

(The third natural product $c_1(\omega_p)^2$ vanishes.) Set

$$\kappa = \pi_* \left(c_1(\omega_\pi^2)\right) \quad \text{and} \quad \zeta = \frac{-1}{2} p_* \left(c_1(\omega_\pi) \cdot c_1(\alpha^* \omega_p)\right).$$

Together with the divisor Δ—the locus of $\alpha \colon C \to \mathbf{P}^1$ where C is singular—we get three divisor classes on $\widetilde{H}_{d,g}$. Conjecturally, these exhaust the Picard group, at least modulo torsion.

CONJECTURE 2.1 (See [**9**]). *The divisor classes κ, ξ, and Δ generate* $\operatorname{Pic}_{\mathbf{Q}}(\widetilde{H}_{d,g})$.

The conjecture has been proved for $d \leq 5$ and for $d > 2g-2$. For $d \leq 5$, the proof uses the unirational parametrization of $\widetilde{H}_{d,g}$ [**8**]. For $d > 2g-2$, the proof uses the fibration $\widetilde{H}_{d,g} \to M_g$ and the deep result of Harer that $\operatorname{Pic}_Q M_g$ has rank 1 [**9, 17**]. The intermediate cases are still open. At any rate, all the divisors we consider in this paper can be written explicitly as linear combinations of κ, ζ, and Δ.

2.2. Divisors from the topology of covers. We have three natural divisors on $\widetilde{H}_{d,g}$ arising from topological considerations. A generic point of $\widetilde{H}_{d,g}$ represents a cover $\alpha\colon C \to \mathbf{P}^1$ that has simple branching. That is, α has $b = 2g + 2d - 2$ distinct branch points and over each branch point, there is a unique ramification point at which the local degree of α is 2.

A simply branched cover specializes in three topologically distinct ways in codimension 1; each possibility gives a divisor on $\widetilde{H}_{d,g}$. The divisor T is the locus of α that have a point of higher ramification—a point $x \in C$ at which the local degree of α is at least 3. The divisor D is the locus of α that have at least two distinct ramification points over the same branch point. The divisor δ is the locus of α whose domain C is singular. It is easy to see that T, D, and δ are irreducible divisors in $\widetilde{H}_{d,g}$.

REMARK 2.2. We can use the topological considerations above to obtain locally closed subsets of $\widetilde{H}_{d,g}$ of higher codimension. Doing so gives a stratification of $\widetilde{H}_{d,g}$ according to the topological type of $\alpha\colon C \to \mathbf{P}^1$. A complete specification of the topological type of α is rather intricate. It includes, for example, the types of ramification profiles for α, the number of singularities of C, and the location of the singularities relative to the ramification profiles.

2.3. Divisors from the algebra of covers. Just as we get special loci in $\widetilde{H}_{d,g}$ from non-generic topological behavior, we get special loci in $\widetilde{H}_{d,g}$ from non-generic algebraic behavior. We make this precise using a structure theorem for finite morphisms due to Casnati and Ekedahl [**4**], which we first recall.

Let X and Y be integral schemes and $\alpha\colon X \to Y$ a finite flat Gorenstein morphism of degree $d \geq 3$. The map α gives an exact sequence

$$0 \to \mathcal{O}_Y \to \alpha_* \mathcal{O}_X \to E_\alpha{}^\vee \to 0, \tag{2.1}$$

where $E = E_\alpha$ is a vector bundle of rank $(d-1)$ on Y, called the *Tschirnhausen bundle* of α. Denote by ω_α the dualizing sheaf of α. Applying $\operatorname{Hom}_Y(-, \mathcal{O}_Y)$ to (2.1), we get

$$0 \to E \to \alpha_* \omega_\alpha \to \mathcal{O}_Y \to 0. \tag{2.2}$$

The map $E \to \alpha_* \omega_\alpha$ induces a map $\alpha^* E \to \omega_\alpha$.

THEOREM 2.3 (See [**4**, Theorem 2.1]). *In the above setup, $\alpha^* E \to \omega_\alpha$ gives an embedding $\iota\colon X \to \mathbf{P}E$ with $\alpha = \pi \circ \iota$, where $\pi\colon \mathbf{P}E \to Y$ is the projection. Moreover, the following hold.*

(1) *The resolution of $\mathcal{O}_X$ as an $\mathcal{O}_{\mathbf{P}E}$ module has the form*

$$\begin{aligned} 0 \to \pi^* N_{d-2}(-d) \to \pi^* N_{d-3}(-d+2) \to \pi^* N_{d-4}(-d+3) \to \dots \\ \dots \to \pi^* N_2(-3) \to \pi^* N_1(-2) \to \mathcal{O}_{\mathbf{P}E} \to \mathcal{O}_X \to 0, \end{aligned} \tag{2.3}$$

where the N_i are vector bundles on Y. Restricted to a point $y \in Y$, this sequence is the minimal free resolution of $X_y \subset \mathbf{P}E_y$.

(2) *The ranks of the N_i are given by*

$$\operatorname{rk} N_i = \frac{i(d-2-i)}{d-1}\binom{d}{i+1},$$

(3) *We have* $N_{d-2} \cong \pi^* \det E$. *Furthermore, the resolution is symmetric, that is, isomorphic to the resolution that is obtained by applying*

$$\operatorname{Hom}_{\mathcal{O}_{\mathbf{P}E}}(-, N_{d-2}(-d)).$$

We call the resolution in (2.3) the *Casnati–Ekedahl* resolution of α.

Let us take $Y = \mathbf{P}^1$. Every vector bundle on $\mathbf{P}^1$ splits as a direct sum of line bundles. The multi-set of degrees of the line bundles appearing in the direct sum decomposition is unique. We refer to this multi-set as the *splitting type* of the bundle. We say that a bundle V is *balanced* if the splitting type is $\{a, \dots, a\}$ for some a.

PROPOSITION 2.4. *Let* $\alpha\colon C \to \mathbf{P}^1$ *be a point of* $\widetilde{H}_{d,g}$. *Denote by* E *the Tschirnhausen bundle and by* N_i *the syzygy bundles in the Casnati–Ekedahl resolution of* α. *Then*

$$\deg E = (g + d - 1), \quad \text{and}$$

$$\deg N_i = (d - 2 - i)(g + d - 1)\binom{d-2}{i-1}.$$

PROOF. The branch divisor of α is cut out by a section of $(\det E)^{\otimes 2}$. Therefore, we get $2 \deg E = 2g + 2d - 2$, from which the first equation follows. We postpone the proof of the second equation to 4 (See 4.3). □

Suppose d divides $g - 1$. Then the rank of N_i divides its degree.

PROPOSITION 2.5. *If* d *divides* $g - 1$, *then for a generic* $\alpha\colon C \to \mathbf{P}^1$ *in* $\widetilde{H}_{d,g}$ *and* $i = 1, \dots, d - 2$, *the bundle* N_i *is balanced.*

We postpone the proof to 3.

DEFINITION 2.6. Suppose d divides $g - 1$. Define the *ith syzygy divisor* $\mu_i \subset \widetilde{H}_{d,g}$ as the locus of $\alpha\colon C \to \mathbf{P}^1$ for which the bundle N_i is unbalanced.

There is a natural scheme structure on $\mu_i \subset \widetilde{\mathcal{H}}_{d,g}$, defined as follows. Let $U \to \widetilde{\mathcal{H}}_{d,g}$ be an étale local chart for the moduli stack over which the conic bundle $\mathcal{P}_U \to U$ admits a relative $\mathcal{O}(1)$. Consider the bundle $\operatorname{End}(N_i) \otimes \mathcal{O}(-1)$ on $\mathcal{P}_U$. Note that $\chi\left(\operatorname{End}(N_i) \otimes \mathcal{O}(-1)\right) = 0$ and $h^1(\operatorname{End}(N_i) \otimes \mathcal{O}(-1) \geq 1$ if and only if N_i is unbalanced. The divisor μ_i is the zero locus of the first Fitting ideal of $R^1 p_* \left(\operatorname{End}(N_i) \otimes \mathcal{O}(-1)\right)$. Henceforth, μ_i is understood to have this scheme structure.

REMARK 2.7. We can use the splitting types of E and N_i to define locally closed subsets of $\widetilde{H}_{d,g}$ of higher codimensions. Doing so gives a stratification of $\widetilde{H}_{d,g}$ according to the isomorphism types of the bundles appearing in the Casnati–Ekedahl resolution. This stratification has a distinctly algebro-geometric favor, and it should be in some sense orthogonal to the topological stratification discussed in 2.2. See [**18**] for more on this stratification.

2.4. Relations between various divisor classes. Assuming 2.1, the divisors defined in 2.2 and 2.3 ought to be expressible as linear combinations of the tautological divisors κ, ζ, and δ. Such an expression for the higher syzygy divisors μ_i is the content of 4. In this section, we give the expressions for all the other divisors.

Denote by E the Tschirnhausen bundle of the universal cover $\alpha\colon \mathcal{C} \to \mathcal{P}$. In addition to the divisors disused so far, it will be useful to also consider the following three auxiliary divisors:

$$p_*c_1(E)^2, \quad p_* \operatorname{ch}_2(E), \quad \pi_*c_1(\omega_\alpha)^2.$$

Lastly, denote by $\lambda = c_1(\pi_*\omega_\pi)$ the class of the Hodge line bundle on $\widetilde{H}_{d,g}$ and by K the canonical divisor class of $\widetilde{H}_{d,g}$. Set

$$b = 2g + 2d - 2.$$

This is the degree of the branch divisor of the covers in $\widetilde{H}_{d,g}$.

PROPOSITION 2.8. *The following identities hold in* $\operatorname{Pic}_{\mathbf{Q}}(\widetilde{H}_{d,g})$*:*

(1) $12\lambda = \kappa + \delta$
(2) $p_*c_1(E)^2 = \frac{b}{2} \cdot \zeta$
(3) $p_* \operatorname{ch}_2(E) = \frac{1}{12} \cdot \kappa + \frac{1}{2} \cdot \zeta + \frac{1}{12} \cdot \delta$
(4) $\pi_*c_1(\omega_\alpha)^2 = \kappa + 4 \cdot \zeta$
(5) $T = 2 \cdot \kappa + 6 \cdot \zeta - \delta$
(6) $D = -3 \cdot \kappa + (b - 10)\zeta + \delta$
(7) $\mu = -\frac{d}{6} \cdot \kappa + \frac{b-2d}{2} \cdot \zeta + \frac{d}{6} \cdot \delta$
(8) $K = \kappa + \zeta - \delta$

PROOF. We compute all the divisor classes on a generic one parameter family $B \to \widetilde{\mathcal{H}}_{d,g}$. Let $\alpha\colon C \to P$ be the pull-back of the universal family to B with the two projections $\pi\colon C \to B$ and $p\colon P \to B$. Set $\sigma = -c_1(\omega_p)/2$.

(1) This is the well-known Mumford relation.

(2) Let $\beta \subset P$ be the branch divisor of α. Since β is cut out by a section of $(\det E)^{\otimes 2}$, we have

$$[\beta] = 2c_1(E).$$

Since $p\colon P \to B$ is a $\mathbf{P}^1$ bundle, we have a relation

$$[\beta] = a\sigma + p^*D$$

for some $a \in \mathbf{Z}$ and $D \in \operatorname{Pic}(B)$. Since $[\beta]$ has degree b on the fibers of p, we get $a = b$. By comparing $\sigma \cdot [\beta]$ and $[\beta]^2$, we get

$$c_1(E)^2 = bc_1(E) \cdot \sigma. \tag{2.4}$$

Since β is the push-forward of the ramification divisor of α, which has class $c_1(\omega_\alpha)$, we have

$$\alpha_*(c_1(\omega_\alpha)) = 2c_1(E).$$

Multiplying the above by σ, noting that $\omega_\alpha \cdot \sigma = \omega_\pi \cdot \sigma$, and using (2.4) yields the second relation.

(3) Applying Rpi_* to both sides of the equation

$$\alpha_*\mathcal{O}_C = \mathcal{O}_P \oplus E^\vee$$

and using Grothendieck–Riemann–Roch for the right hand side yields the third relation.

(4) Using $c_1(\omega_\alpha) = c_1(\omega_\pi) + 2\sigma$ and (2.4) yields the fourth relation.

(5, 6) To get T and D, we sketch the argument from [**18**, Proposition 3.2]. Assuming B is sufficiently generic, the only singularities of β will be nodes and cusps, and the map from the ramification divisor ρ to the branch divisor β will be the normalization. A simple local computation of the branch divisor of a cover specializing to a point of D or T shows that the nodes correspond to intersections of B with D, and the cusps with the intersections of B with T. Therefore, we get

$$p_a(\beta) - p_a(\rho) = T + D.$$

By adjunction on C and P, this leads to

$$(\beta^2 - 2\rho^2)/2 = T + D.$$

The branch points of $\rho \to B$ correspond to the intersections of B with δ or with T. From adjunction on C and Riemann–Hurwitz, we get

$$2\rho^2 + \beta \cdot c_1(\omega_\pi) = T + \delta.$$

Solving for T and D, and using the previous relations yields the fifth and the sixth relations.

(7) The class of μ is given by the Bogomolov expression $\mathrm{c}_1(E)^2 - 2d\,\mathrm{ch}_2(E)$, which yields the seventh relation (See 4.1 for the Bogomolov expression).

(8) We sketch two ways to compute the canonical divisor. Note that the map $\widetilde{\mathcal{H}}_{d,g} \to \widetilde{H}_{d,g}$ is unramified in codimension 1, so the canonical class of the stack is the same as that of the coarse space.

First, consider the morphism $\mathrm{br} : U \to V$, where $V \subset \mathbf{P}(\mathrm{Sym}^b \mathbf{P}^1) /\!\!/ \mathrm{SL}(2)$ is the open locus where at most two of the b marked points coincide, $U \subset \widetilde{H}_{d,g}$ is the locus of covers where at most two branch points coincide, and br is the morphism that assigns to a cover its branch divisor. It is easy to check that the complements of V and U have codimension 2, and hence it suffices to work on V and U for divisor calculations. Let $\Delta \subset U$ be the complement of the locus of b distinct points. A simple local calculation shows that

$$\mathrm{br}^{-1} \Delta = 3T + 2D + \delta.$$

The canonical divisor of U is

$$K_U = -\frac{(b+1)}{2b-2} \cdot \Delta.$$

By Riemann–Hurwitz, we get

$$K_W = \mathrm{br}^* K_U + 2T + D,$$

which combined with the previous relations yield the eighth relation.

Another way is to use the deformation theory of maps developed in [**19**]. We can identify the tangent space to $\widetilde{\mathcal{H}}_{d,g}$ at $\alpha \colon C \to \mathbf{P}^1$ as the the kernel of the induced map

$$\mathrm{Ext}^1(\Omega_C, \mathcal{O}_C) \to \mathrm{Ext}^1(\Omega_{\mathbf{P}^1}, \alpha_* \mathcal{O}_C).$$

We can compute the Chern classes of the bundles on $\widetilde{\mathcal{H}}_{d,g}$ defined by both terms, and their difference yields the Chern class of the tangent bundle of $\widetilde{\mathcal{H}}_{d,g}$.

□

3. The generic splitting type

The goal of this section is to discuss the splitting type of the syzygy bundle N_i for a generic cover, and to prove that it is balanced when d divides $g-1$. Note, however, that the degree of N_i may be divisible by its rank even when d does not divide $g-1$. One may expect N_i to be generically balanced even in this setting. This is not quite true, as the following example shows for the first bundle N_1.

EXAMPLE 3.1. Consider a general degree 6, genus 4 cover $\alpha : C \to \mathbf{P}^1$. We will show that the splitting of N_1 is $\mathcal{O}_{\mathbf{P}^1}(2) \oplus \mathcal{O}_{\mathbf{P}^1}(3)^{\oplus 7} \oplus \mathcal{O}_{\mathbf{P}^1}(4)$. The degree of N_1 is 27, and its rank is 9, so N_1 is balanced if and only if it has a summand of degree ≥ 4.

Let h denote the divisor class of the relative $\mathcal{O}(1)$ on $\mathbf{P}E$, and let f be the class of a fiber of $\mathbf{P}E \to \mathbf{P}^1$. Then the linear system $|h-2f|$ restricts to the complete canonical system on $C \subset \mathbf{P}E$, and furthermore, every element of the linear system $|2h-4f|$ is obtained as a sum of products of elements in $|h-2f|$. Since the canonical model of C lies on a unique quadric Q, we see that there is a unique element of $|2h-4f|$ containing C. This, in turn, translates into an $\mathcal{O}(4)$ summand in N_1.

The example above can be generalized, provided the genus is small compared to the degree. For large g, however, we expect that all bundles in the Casnati–Ekedahl resolution will be balanced. Evidence for this is given by the next theorem.

THEOREM 3.2 (See [**3**]). *The bundle N_1 is balanced for a general branched cover provided g is much larger than d. When d divides $g-1$, all syzygy bundles N_i are balanced for a general branched cover.*

The statement for N_1 is the main result of [**3**]; the statement for d dividing $g-1$ is [**3**, Proposition 2.4].

We now give a brief overview of the proof that the syzygy bundles are generically balanced when this divisibility constraint holds. Since the Hurwitz space is irreducible, and the condition of being balanced is open, it suffices to provide one example of a cover where it holds.

Consider the surface $S = E \times \mathbf{P}^1$, where E is any elliptic curve. Let D be any smooth curve on S with $D \cdot (\{e\} \times \mathbf{P}^1) = k$ and $D \cdot (E \times \{t\}) = d$. We will argue that the projection $D \to \mathbf{P}^1$ has the property that every syzygy bundle N_i is balanced.

The surface S embeds in $\mathbf{P}^{d-1} \times \mathbf{P}^1$ so that the projection to $\mathbf{P}^{d-1}$ is the projection $S \to E$ composed with the embedding of E as an elliptic normal curve of degree d. The curve D is then the intersection of S with a divisor $H \subset \mathbf{P}^{d-1} \times \mathbf{P}^1$ which restricts to a hyperplane in every $\mathbf{P}^{d-1}$.

The main point is that the minimal free resolution of the elliptic normal curve $E \subset \mathbf{P}^{d-1}$ (embedded by any complete linear system of degree d) has the same shape as the Casnati–Ekedahl resolution of a degree d branched cover. This is equivalent to saying that elliptic normal curves are arithmetically Gorenstein. The minimal free resolution of $E \subset \mathbf{P}^{d-1}$ pulls back to a relative minimal free resolution of $\mathcal{O}_S$ as an $\mathcal{O}_{\mathbf{P}^{d-1}\times\mathbf{P}^1}$-module. More precisely, we get a resolution

(3.1)
$$0 \to \mathcal{O}_{\mathbf{P}^{d-1}\times\mathbf{P}^1}(-d) \to V_{d-3} \otimes \mathcal{O}_{\mathbf{P}^{d-1}\times\mathbf{P}^1}(-d+2) \to V_{d-4} \otimes \mathcal{O}_{\mathbf{P}^{d-1}\times\mathbf{P}^1}(-d+3) \to \cdots$$
$$\cdots \to V_2 \otimes \mathcal{O}_{\mathbf{P}^{d-1}\times\mathbf{P}^1}(-3) \to V_1 \otimes \mathcal{O}_{\mathbf{P}^{d-1}\times\mathbf{P}^1}(-2)^{\oplus r_1} \to \mathcal{O}_{\mathbf{P}^{d-1}\times\mathbf{P}^1} \to \mathcal{O}_S \to 0,$$

where the V_i are vector spaces of the same dimension as the rank of the bundles N_i in the Casnati–Ekedahl resolution of a degree d branched cover, and the twists refer to twists by the pullback of $\mathcal{O}_{\mathbf{P}^{d-1}}(1)$. The restriction of this resolution to the relative hyperplane H yields the Casnati–Ekedahl resolution of $D = H \cap S$. Note that the pullback of $\mathcal{O}_{\mathbf{P}^{d-1}}(1)$ to H is $\mathcal{O}_H(1) \otimes \pi^* L$ where $\pi \colon H \to \mathbf{P}^1$ is the projection, and L is a line bundle on $\mathbf{P}^1$. Therefore, the terms in the resolution (3.1) restrict to $\pi^*(V_i \otimes L^{-i-1}) \otimes \mathcal{O}_H(-i-1)$. We thus get $N_i = V_i \otimes L^{-i-1}$, which is balanced.

Since D is a curve of type (d, k) on $E \times \mathbf{P}^1$, its genus g is $d(k-1)+1$. This is where we get the degree-genus restriction $g \equiv 1 \pmod{d}$.

REMARK 3.3. The strategy above required understanding the relative resolution of the (trivial) genus one fibration $S \to \mathbf{P}^1$. In general, if $f \colon X \to \mathbf{P}^1$ is a genus one fibration with simple nodes as singularities, then a relative degree d divisor $D \subset X$ yields a relative embedding

$$X \hookrightarrow \mathbf{P}(f_* \mathcal{O}_X(D)) \to \mathbf{P}^1$$

and X enjoys a relative resolution with exactly the same form as the Casnati–Ekedahl resolution of a degree d branched cover. The bundles appearing in the relative resolution of X and the Casnati–Ekedahl resolution for $D \to \mathbf{P}^1$ are determined by each other, and one is balanced if and only if the other is. In this way, the study of Casnati–Ekedahl resolutions is intimately related to the study of relative resolutions of genus one fibrations.

REMARK 3.4. One might be able to deduce that N_i is as balanced as possible (that is, $h^1(\operatorname{End} E(-1)) = 0$) even when $g \not\equiv 1 \pmod{d}$ as follows. Notice that for a singular $D \subset E \times \mathbf{P}^1$, the argument sketched above still holds without change. If one understands how the syzygy bundles N_i are related for D and its normalization $\widetilde{D}$, one might be able to handle the cases where $g \not\equiv 1 \pmod{d}$.

The strategies outlined in 3.3 and 3.4 have not been fully explored. The authors intend to investigate them in the future. Notice that the idea of using branched covers on elliptic fibrations parallels the idea of using curves on K3 surfaces aprés [**16**].

4. The divisor class of μ_i

The goal of this section is obtain the divisor class of the higher syzygy divisors μ_i.

4.1. The Bogomolov expression. Let B be a smooth curve and $p \colon P \to B$ a $\mathbf{P}^1$ bundle. Let E be a vector bundle of rank r on P which is balanced on the generic fiber of p. Denote by $\mu(E)$ the locus of points in B over which E is unbalanced with the scheme structure given by the first Fitting ideal of $R^1 p_*(\operatorname{End} E \otimes \mathcal{O}(-1))$.

PROPOSITION 4.1. *In the above setup, we have*

$$[\mu(E)] = \mathrm{c}_1^2(E) - 2r \operatorname{ch}_2(E)$$

PROOF. By definition, we have

$$[\mu(E)] = -\mathrm{c}_1 R p_*(\operatorname{End} E \otimes \mathcal{O}(-1))$$

By Grothendieck–Riemann–Roch, we get

$$\begin{aligned}\operatorname{ch} Rp_*(\operatorname{End} E \otimes \mathcal{O}(-1)) &= p_* \left(\operatorname{ch}(E) \otimes \operatorname{ch}(E^\vee) \operatorname{ch} \mathcal{O}(-1) \operatorname{td}(P/B)\right) \\ &= 2r \operatorname{ch}_2(E) - \mathrm{c}_1^2(E).\end{aligned}$$

□

Let us call the expression $\mathrm{c}_1^2(E) - 2r \operatorname{ch}_2(E)$ the *Bogomolov expression* and denote it by $\operatorname{Bog}(N_i)$. Note that $\operatorname{Bog}(N_i) = \operatorname{Bog}(N_i \otimes L)$ for any line bundle L, which should be expected from the geometric interpretation.

4.2. The Koszul resolution. By 4.1, the problem of finding the divisor class of μ_i is reduced to finding $\mathrm{c}_1(N_i)$ and $\operatorname{ch}_2(N_i)$. To calculate the Chern classes of the bundles N_i, we express them as cohomology bundles of a resolution involving more familiar bundles. This is the Koszul resolution, which we now recall.

Let R be a (Noetherian) ring and E a locally free R-module of rank r. Let $S = \operatorname{Sym}^*(E)$ be the symmetric algebra on E and let M be a graded S-module. Suppose we have a graded resolution

$$0 \to F_k \to \cdots \to F_1 \to F_0 \to M \to 0,$$

where

$$F_i = \bigoplus_{j \geq 0} N_{ij} \otimes_R S(-i-j)$$

and the N_{ij} are locally free R-modules. Suppose the resolution is minimal in the sense that all the maps $F_{i+1} \to F_i$ have graded components in positive degree. Then we have the identification

$$N_{ij} = \operatorname{Tor}^i_S(M, R)_{i+j}, \tag{4.1}$$

where the subscript denotes the graded component. The right hand side can be computed in another way. Instead of using an S-resolution of M, we use the S-resolution of R given by the Koszul complex

$$0 \to \wedge^r E \otimes_R S(-r) \to \cdots \to \wedge^p E \otimes_R S(-p) \to \cdots \to E \otimes_R S(-1) \to S \to R \to 0.$$

Tensoring by M and taking the $(i+j)$th graded component yields the complex

$$K_{i+j} : \wedge^r E \otimes_R M_{i+j-r} \to \cdots \xrightarrow{d_{p-1}} \wedge^p E \otimes_R M_{i+j-p} \xrightarrow{d_p} \cdots \to E \otimes_R M_{i+j-1} \to M_{i+j}.$$

Let $H^p(K_{i+j}) = \ker d_p / \operatorname{im} d_{p-1}$ be the cohomology. Then we get the identification

$$\operatorname{Tor}^j_S(M, R)_{i+j} = H^i(K_{i+j}).$$

Combining with (4.1), we get

$$N_{ij} = H^i(K_{i+j}).$$

Let us now turn to the Casnati–Ekedahl resolution of the universal finite cover $\alpha \colon \mathcal{C} \to \mathcal{P}$. Let $E = \ker(\omega_\phi \to \mathcal{O}_{\mathcal{P}})$ be the Tschirnhausen bundle and $\iota \colon \mathcal{C} \to \mathbf{P}E$ the relative canonical embedding. Let $I \subset S = \operatorname{Sym}^* E$ be the homogeneous ideal of $\mathcal{C}$. The Koszul complex K_{i+1} for the S-module S/I is the following

$$K_{i+1} : \wedge^{i+1} E \to \wedge^i E \otimes E \to \wedge^{i-1} E \otimes \alpha_*(\omega_\alpha^2) \to \cdots \to \alpha_*(\omega_\alpha^{i+1}).$$

Denote by $K_{i+1}(j)$ the jth term in the above complex, starting from $j = 0$ and counting from the right to the left.

PROPOSITION 4.2. *Let $1 \leq i \leq d-3$ and let N_i be the ith syzygy bundle of α. Then we have*

$$\operatorname{ch} N_i = \sum_{j=0}^{i+1} (-1)^{j-i-1} \operatorname{ch}(K_{i+1}(j)).$$

PROOF. From the Casnati–Ekedahl resolution of α and the identification of the syzygy bundles with the cohomology of the Koszul complex, we know that

$$H^p(K_{i+1}) = \begin{cases} N_i & \text{if } p = i \\ 0 & \text{otherwise} \end{cases}.$$

Therefore, we have the equality

$$N_i = \sum_{j=0}^{i+1} (-1)^{j-i-1} K_{i+1}(j)$$

in the K-ring, from which the formula for the Chern character follows. □

4.3. The computation. We now compute $\operatorname{ch} N_i$ using the expression in Proposition 4.2. Since we are ultimately only interested in c_1 and ch_2, we ignore all terms of degree higher than 2. We may assume, for example, that the computation is happening over a general curve $B \to \widetilde{\mathcal{H}}_{d,g}$. Denote by $\pi\colon C \to B$ and $p\colon P \to B$ the two projections.

From 4.2, we have

(4.2)

$$\begin{aligned}
\operatorname{ch} N_i &= \sum_{j=0}^{i+1} (-1)^{j-i-1} \operatorname{ch}(K_{i+1}(j)) \\
&= \left(\sum_{j=0}^{i+1} (-1)^{j-1} \operatorname{ch}(\wedge^{i+1-j} E) \operatorname{ch}(\alpha_* \omega_\alpha^j) \right) - \operatorname{ch}(\wedge^i E) + \operatorname{ch}(\wedge^{i+1} E) \operatorname{ch}(E^\vee).
\end{aligned}$$

The two correction terms at the end are needed because the $j = 0$ and $j = 1$ terms in the summation are differ from the corresponding terms of the Koszul resolution in the following way (the computation is in the K-ring):

$$\begin{aligned}
[\wedge^{i+1} E] \otimes [\alpha_* \omega_\alpha^0] &= [\wedge^{i+1} E] \otimes [\mathcal{O} + E^\vee] \\
&= [K_{i+1}(i+1)] + [\wedge^{i+1} E] \otimes [E^\vee], \text{ and} \\
[\wedge^i E] \otimes [\alpha_* \omega_\alpha] &= [\wedge^i E] \otimes [\mathcal{O} + E] \\
&= K_{i+1}(i) + [\wedge^i E].
\end{aligned}$$

Next, by Grothendieck–Riemann–Roch applied to α we get

(4.3)

$$\operatorname{ch} \alpha_* \omega_\alpha^\ell = \alpha_* \left(1 + \ell \cdot c_1(\omega_\alpha) + \frac{\ell^2 c_1(\omega_\alpha)^2}{2} \right) \left(1 - \frac{c_1(\omega_\alpha)^2}{2} + \frac{c_1(\omega_\alpha)^2 + c_2(\Omega_{C/P})}{12} \right).$$

Note that $c_1(\omega_\alpha)$ is the class of the ramification divisor of $\alpha\colon C \to P$. In particular, $\alpha_* c_1(\omega_\alpha)$ is the class of the branch divisor, which is cut out by a section of $(\det E)^{\otimes 2}$. Therefore, we get

$$\alpha_* c_1(\omega_\alpha) = 2 c_1 E. \tag{4.4}$$

Specializing (4.3) to the case $\ell = 0$ and comparing the degree two terms yields

$$\operatorname{ch}_2 E = \alpha_* \left(\frac{c_1(\omega_\alpha)^2 + c_2(\Omega_{C/P})}{12} \right). \tag{4.5}$$

After using (4.4) and (4.5) to simplify (4.3), we get

$$\operatorname{ch} \alpha_* \omega_\alpha^\ell = d + (2\ell - 1)c_1(E) + \left(\operatorname{ch}_2(E) + \frac{\ell^2 + \ell}{2} \pi_* c_1(\omega_\alpha)^2 \right). \tag{4.6}$$

For a vector bundle E of rank $d-1$, we have

$$\begin{aligned}
\operatorname{ch}_0 \wedge^\ell E &= \binom{d-1}{\ell}, \\
c_1(\wedge^\ell E) &= \binom{d-2}{\ell-1} c_1(E), \text{ and} \\
\operatorname{ch}_2(\wedge^\ell E) &= \binom{d-2}{\ell-1} \operatorname{ch}_2(E) + \frac{1}{2}\binom{d-3}{l-2}(c_1(E)^2 - 2\operatorname{ch}_2(E)).
\end{aligned}$$

Using these identities and using (4.6), we expand the terms $\operatorname{ch}(\wedge^{i+1-j}E)\operatorname{ch}(\alpha_*\omega_\alpha^j)$ and carry out the summation. To evaluate the summation in a closed form, we use the following combinatorial identities[1]:

$$\begin{aligned}
\sum_{l=0}^{p} (-1)^l \binom{a}{p-l} &= \binom{a-1}{p} \\
\sum_{l=0}^{p} (-1)^l \binom{a}{p-l} l &= -\binom{a-2}{p-1} \\
\sum_{l=0}^{p} (-1)^l \binom{a}{p-l} l(l-1) &= 2\binom{a-3}{p-2}.
\end{aligned}$$

The result is the following:

$$\begin{aligned}
\operatorname{ch}_0(N_i) &= \frac{i(d-2-i)}{d-1}\binom{d}{i+1} \\
c_1(N_i) &= (d-2-i)\binom{d-2}{i-1} c_1(E) \\
\operatorname{ch}_2(N_i) &= \binom{d-4}{i-1}\left(d\operatorname{ch}_2(E) + \frac{(d-4)i+2}{2(d-i-1)} c_1^2(E) - c_1(\omega_\alpha)^2 \right)
\end{aligned} \tag{4.7}$$

We use this computation to finish a postponed proof from 2.4.

COROLLARY 4.3. $\deg N_i = (d-2-i)(g+d-1)\binom{d-2}{i-1}$.

PROOF. Follows from (4.7) and that $\deg c_1(E) = (g+d-1)$. □

THEOREM 4.4. *The push-forward to* $\widetilde{H}_{d,g}$ *of the Bogomolov expression for* N_i *is the following linear combination the standard divisor classes:*

$$p_* \operatorname{Bog}(N_i) = A_i \left(6(gd - 6g + d + 6)\cdot\zeta - d(d-12)\cdot\kappa - d^2\cdot\delta \right),$$

where the coefficient A_i *is given by*

$$A_i = \binom{d-4}{i-1}^2 \frac{(d-2)(d-3)}{6(i+1)(d-i-1)}.$$

---

[1] along with ample help from the computer algebra system `Maple` with its `sumtools` package

PROOF. This is a direct consequence of the results of the Chern class computation collected in (4.7) and the relations in 2.8. □

Note that $\mathrm{Bog}(N_i)$ is symmetric with respect to the change $i \leftrightarrow d-2-i$, consistent with the fact that N_i and N_{d-2-i} are isomorphic up to twisting and taking duals.

The main theorem (1.1) follows from 2.5, the interpretation of the Bogomolov expression (4.1), and 4.4.

4.4. The supports of μ_i. Given that the divisor classes $[\mu_i]$ are proportional, it is natural to wonder if the divisors μ_i are supported on the same set.

There is some evidence that the μ_i in fact may *not* be supported on the same set, thanks to the work of Christian Bopp. Using his Macaulay2 package [**2**], he has found many examples where the jumping loci of syzygy bundles are not supported on the same set. All such examples so far have involved jumping loci of codimension higher than 1.

References

[1] Dan Abramovich, Alessio Corti, and Angelo Vistoli, *Twisted bundles and admissible covers*, Comm. Algebra **31** (2003), no. 8, 3547–3618, DOI 10.1081/AGB-120022434. Special issue in honor of Steven L. Kleiman. MR2007376

[2] C. Bopp and M. Hoff. `RelativeCanonicalResolution.m2`: Construction of relative canonical resolutions and eagon–northcott type complexes. `Macaulay2` package.

[3] G. Bujokas and A. Patel. Invariants of a general branched cover of $\mathbf{P}^1$. *arXiv:1504.03756 [math.AG]*, Apr. 2015.

[4] G. Casnati and T. Ekedahl, *Covers of algebraic varieties. I. A general structure theorem, covers of degree* 3, 4 *and Enriques surfaces*, J. Algebraic Geom. **5** (1996), no. 3, 439–460. MR1382731

[5] A. Clebsch, *Zur Theorie der Riemann'schen Fläche* (German), Math. Ann. **6** (1873), no. 2, 216–230, DOI 10.1007/BF01443193. MR1509816

[6] Marc Coppens, *Existence of pencils with prescribed scrollar invariants of some general type*, Osaka J. Math. **36** (1999), no. 4, 1049–1057. MR1745644

[7] Anand Deopurkar and Anand Patel, *Sharp slope bounds for sweeping families of trigonal curves*, Math. Res. Lett. **20** (2013), no. 5, 869–884, DOI 10.4310/MRL.2013.v20.n5.a5. MR3207358

[8] Anand Deopurkar and Anand Patel, *The Picard rank conjecture for the Hurwitz spaces of degree up to five*, Algebra Number Theory **9** (2015), no. 2, 459–492, DOI 10.2140/ant.2015.9.459. MR3320849

[9] Steven Diaz and Dan Edidin, *Towards the homology of Hurwitz spaces*, J. Differential Geom. **43** (1996), no. 1, 66–98. MR1424420

[10] M. Fried, *Fields of definition of function fields and Hurwitz families—groups as Galois groups*, Comm. Algebra **5** (1977), no. 1, 17–82, DOI 10.1080/00927877708822158. MR0453746

[11] Michael D. Fried and Helmut Völklein, *The inverse Galois problem and rational points on moduli spaces*, Math. Ann. **290** (1991), no. 4, 771–800, DOI 10.1007/BF01459271. MR1119950

[12] William Fulton, *Hurwitz schemes and irreducibility of moduli of algebraic curves*, Ann. of Math. (2) **90** (1969), 542–575, DOI 10.2307/1970748. MR0260752

[13] Florian Geiss, *The unirationality of Hurwitz spaces of 6-gonal curves of small genus*, Doc. Math. **17** (2012), 627–640. MR3007673

[14] Joe Harris and David Mumford, *On the Kodaira dimension of the moduli space of curves*, Invent. Math. **67** (1982), no. 1, 23–88, DOI 10.1007/BF01393371. With an appendix by William Fulton. MR664324

[15] A. Hurwitz, *Ueber die Anzahl der Riemann'schen Flächen mit gegebenen Verzweigungspunkten* (German), Math. Ann. **55** (1901), no. 1, 53–66, DOI 10.1007/BF01448116. MR1511135

[16] Robert Lazarsfeld, *Brill-Noether-Petri without degenerations*, J. Differential Geom. **23** (1986), no. 3, 299–307. MR852158

[17] Shinichi Mochizuki, *The geometry of the compactification of the Hurwitz scheme*, Publ. Res. Inst. Math. Sci. **31** (1995), no. 3, 355–441, DOI 10.2977/prims/1195164048. MR1355945

[18] A. Patel. Special codimension one loci in Hurwitz spaces. *arXiv:1508.06016 [math.AG]*, Aug. 2015.

[19] Ziv Ran, *Deformations of maps*, Algebraic curves and projective geometry (Trento, 1988), Lecture Notes in Math., vol. 1389, Springer, Berlin, 1989, pp. 246–253, DOI 10.1007/BFb0085936. MR1023402

[20] B. Riemann, *Theorie der Abel'schen Functionen* (German), J. Reine Angew. Math. **54** (1857), 115–155, DOI 10.1515/crll.1857.54.115. MR1579035

[21] Matthieu Romagny and Stefan Wewers, *Hurwitz spaces* (English, with English and French summaries), Groupes de Galois arithmétiques et différentiels, Sémin. Congr., vol. 13, Soc. Math. France, Paris, 2006, pp. 313–341. MR2316356

[22] Francesco Severi, *Vorlesungen über algebraische Geometrie: Geometrie auf einer Kurve, Riemannsche Flächen, Abelsche Integrale* (German), Berechtigte Deutsche Übersetzung von Eugen Löffler. Mit einem Einführungswort von A. Brill. Begleitwort zum Neudruck von Beniamino Segre. Bibliotheca Mathematica Teubneriana, Band 32, Johnson Reprint Corp., New York-London, 1968. MR0245574

[23] Joseph H. Silverman, *The space of rational maps on* $\mathbf{P}^1$, Duke Math. J. **94** (1998), no. 1, 41–77, DOI 10.1215/S0012-7094-98-09404-2. MR1635900

[24] Zvezdelina E. Stankova-Frenkel, *Moduli of trigonal curves*, J. Algebraic Geom. **9** (2000), no. 4, 607–662. MR1775309

[25] Gerard van der Geer and Alexis Kouvidakis, *The cycle classes of divisorial Maroni loci*, Int. Math. Res. Not. IMRN **11** (2017), 3463–3509. MR3693656

Department of Mathematics, University of Georgia, Athens, Georgia 30602

Current address: Mathematical Sciences Institute, The Australian National University, Canberra, Australia

Email address: anand.deopurkar@anu.edu.au

Department of Mathematics, Boston College, Chestnut Hill, Massachusetts 02467

Current address: Department of Mathematics, Oklahoma State University, Stillwater, Oklahoma 74075

Email address: anand.patel@okstate.edu

Selected Published Titles in This Series

703 **Andreas Malmendier and Tony Shaska, Editors,** Higher Genus Curves in Mathematical Physics and Arithmetic Geometry, 2018

701 **Joan-Carles Lario and V. Kumar Murty, Editors,** Number Theory Related to Modular Curves, 2018

700 **Alexandre Girouard, Dmitry Jakobson, Michael Levitin, Nilima Nigam, Iosif Polterovich, and Frédéric Rochon, Editors,** Geometric and Computational Spectral Theory, 2017

699 **Mark L. Agranovsky, Matania Ben-Artzi, Catherine Bénéteau, Lavi Karp, Dmitry Khavinson, Simeon Reich, David Shoikhet, Gilbert Weinstein, and Lawrence Zalcman, Editors,** Complex Analysis and Dynamical Systems VII, 2017

698 **Alexander M. Blokh, Leonid A. Bunimovich, Paul H. Jung, Lex G. Oversteegen, and Yakov G. Sinai, Editors,** Dynamical Systems, Ergodic Theory, and Probability: in Memory of Kolya Chernov, 2017

697 **Fabrizio Broglia, Françoise Delon, Max Dickmann, Danielle Gondard-Cozette, and Victoria Ann Powers, Editors,** Ordered Algebraic Structures and Related Topics, 2017

696 **Ara S. Basmajian, Yair N. Minsky, and Alan W. Reid, Editors,** In the Tradition of Ahlfors–Bers, VII, 2017

695 **Katrina Barron, Elizabeth Jurisich, Antun Milas, and Kailash Misra, Editors,** Lie Algebras, Vertex Operator Algebras, and Related Topics, 2017

694 **Manjul Bhargava, Robert Guralnick, Gerhard Hiss, Klaus Lux, and Pham Huu Tiep, Editors,** Finite Simple Groups: Thirty Years of the Atlas and Beyond, 2017

693 **Michael Cwikel and Mario Milman, Editors,** Functional Analysis, Harmonic Analysis, and Image Processing, 2017

692 **Anatole Katok, Yakov Pesin, and Federico Rodriguez Hertz, Editors,** Modern Theory of Dynamical Systems, 2017

691 **Farrell Brumley, Maria Paula Gómez Aparicio, and Alberto Mínguez, Editors,** Around Langlands Correspondences, 2017

690 **Andrés Eduardo Caicedo, James Cummings, Peter Koellner, and Paul B. Larson, Editors,** Foundations of Mathematics, 2017

689 **Erica Flapan, Allison Henrich, Aaron Kaestner, and Sam Nelson, Editors,** Knots, Links, Spatial Graphs, and Algebraic Invariants, 2017

688 **Jeffrey Bergen, Stefan Catoiu, and William Chin, Editors,** Groups, Rings, Group Rings, and Hopf Algebras, 2017

687 **Fernanda Botelho, Raena King, and T. S. S. R. K. Rao, Editors,** Problems and Recent Methods in Operator Theory, 2017

686 **Alp Bassa, Alain Couvreur, and David Kohel, Editors,** Arithmetic, Geometry, Cryptography and Coding Theory, 2017

685 **Heather A. Harrington, Mohamed Omar, and Matthew Wright, Editors,** Algebraic and Geometric Methods in Discrete Mathematics, 2017

684 **Anna Beliakova and Aaron D. Lauda, Editors,** Categorification in Geometry, Topology, and Physics, 2017

683 **Anna Beliakova and Aaron D. Lauda, Editors,** Categorification and Higher Representation Theory, 2017

682 **Gregory Arone, Brenda Johnson, Pascal Lambrechts, Brian A. Munson, and Ismar Volić, Editors,** Manifolds and K-Theory, 2017

681 **Shiferaw Berhanu, Nordine Mir, and Emil J. Straube, Editors,** Analysis and Geometry in Several Complex Variables, 2017